MITOCHONDRIA AS A KEY INTRACELLULAR TARGET OF THALLIUM TOXICITY

MITOCHONDRIA AS A KEY INTRACELLULAR TARGET OF THALLIUM TOXICITY

SERGEY KOROTKOV

Sechenov Institute of Evolutionary Physiology and Biochemistry, Russian Academy of Sciences, Russian Federation

ELSEVIER

ACADEMIC PRESS

An imprint of Elsevier

Academic Press is an imprint of Elsevier
125 London Wall, London EC2Y 5AS, United Kingdom
525 B Street, Suite 1650, San Diego, CA 92101, United States
50 Hampshire Street, 5th Floor, Cambridge, MA 02139, United States
The Boulevard, Langford Lane, Kidlington, Oxford OX5 1GB, United Kingdom

Notices

Knowledge and best practice in this field are constantly changing. As new research and experience broaden our understanding, changes in research methods, professional practices, or medical treatment may become necessary.

Practitioners and researchers must always rely on their own experience and knowledge in evaluating and using any information, methods, compounds, or experiments described herein. In using such information or methods they should be mindful of their own safety and the safety of others, including parties for whom they have a professional responsibility.

To the fullest extent of the law, neither the Publisher nor the authors, contributors, or editors, assume any liability for any injury and/or damage to persons or property as a matter of products liability, negligence or otherwise, or from any use or operation of any methods, products, instructions, or ideas contained in the material herein.

ISBN: 978-0-323-95531-7

For information on all Academic Press publications visit our website at https://www.elsevier.com/books-and-journals

Publisher: Kathy Padilla
Acquisitions Editor: Megan Ashdown
Editorial Project Manager: Maria Elaine D. Desamero
Production Project Manager: Omer Mukthar
Cover Designer: Miles Hitchen

Typeset by TNQ Technologies

Contents

Introduction *vii*

1. Toxicity of thallium and its compounds **1**

List of abbreviations 1

1.1 Thallium and its compounds: chemical and toxic properties, distribution, and release into the environment 1

1.2 Thallium toxicity to biological organisms (humans, animals, and plants) 7

1.3 Classification, symptomatology, diagnosis, and treatment of thallium poisoning 15

1.4 Methods of thallium analysis in biological samples 32

References 33

2. Toxicology of thallium **53**

List of abbreviations 53

2.1 The toxic effect of thallium on the human body and animals 53

2.2 The toxic effect of thallium compounds on body systems 59

2.3 Toxic effects of thallium on aquatic protozoa and unicellular organisms 66

References 68

3. Toxic thallium effects on cells **79**

List of abbreviations 79

3.1 Molecular mechanisms of thallium transport across cell membranes: use of isotopes and ion channel inhibitors 79

3.2 Thallium toxicity mechanisms on mammalian cells 84

3.3 Interaction of Tl^{+} and Tl^{3+} with molecular thiol groups 91

References 93

4. Effect of thallium on mitochondria **107**

List of abbreviations 107

4.1 Heavy metal effects on isolated mitochondria 108

4.2 Effect of thallium on the inner membrane potential and ion permeability and the mitochondrial respiration 112

4.3 Thallium effects on the concentration of free thiol groups in mitochondrial proteins 138
4.4 Tl^+ induces the opening of the mitochondrial permeability transition pore in the inner membrane of calcium-loaded mitochondria: The effects of the pore inhibitors 144
4.5 The role of calcium sites and potassium channels of the inner membrane in the Tl^+-induced MPTP opening 179
4.6 The dependence of the Tl^+-induced MPTP opening in the inner membrane of rat liver mitochondria on the adenine nucleotide translocase conformation and the phosphate symporter state 199
4.7 Conclusing remarks 236
Acknowledgments 239
References 239

Index *261*

Introduction

List of abbreviations:

5-HD 5-hydroxyodecanoate;
ANT adenine nucleotide translocase;
ATP adenosine triphosphate;
Diam diamide; Dia, diazoxide;
DIDS 4,4′-diisothiocyanostilbene-2,2′-disulfonate;
IMM inner mitochondrial membrane;
mitoBK$_{Ca}$ mitochondrial calcium-activated BK-type potassium channel;
mitoK$_{ATP}$ mitochondrial ATP dependent channel;
MPTP mitochondrial permeability transition pore;
MRC mitochondrial respiratory chain;
PAO phenylarsine oxide;
Pin pinacidil;
RLM rat liver mitochondria;
RR ruthenium red;
$\Delta\Psi_{mito}$ inner membrane potential;

The heavy metal thallium belongs to the rare earth element. It has found wide application in various industries (Schoer, 1984). Thallium isotopes (^{201}Tl, ^{204}Tl) are used in medicine to scan the myocardium and tumor tissues, and thallium sulfate is a part of rodenticides used in rodent control (Chauncey et al., 1977; Blain and Kazantzis, 2015). Thallium releases primarily into the environment during the coal combustion to generate electricity and technological processes in the production of cast iron, steel, nonferrous metals, cement, and sulfuric acid (Schoer, 1984; Wang et al., 2013). The toxicity of thallium to humans and animals is higher than mercury, lead, cadmium, copper, and zinc (Peter and Viraraghavan, 2005; Rodríguez-Mercado and Altamirano-Lozano, 2013). Thallium compounds penetrate the body through the skin, lungs, or food, and damage primarily the nervous and cardiovascular systems, kidneys, liver, gastrointestinal tract, and skin (Goel and Aggarwal, 2007; Blain and Kazantzis, 2015).

Thallium poisoning in humans and animals shows extensive disorders of the intracellular structures in hepatocytes, cardiomyocytes, epithelial cells of renal tubules, glial cells, and neurons (Herman and Bensch, 1967; Woods and Fowler, 1986; Leung and Ooi, 2000; Kiliç and Kutlu, 2010). Mitochondrial swelling, cristae disintegration, intermembrane space expansion, and damage of cytoplasmic membranes were among these disorders. Thallium stimulates apoptosis in Jurkat and PC12 cells and hippocampal and HN9.10e neurons (Bragadin et al., 2003; Hanzel and Verstraeten, 2009; Bramanti et al., 2019; Lin et al., 2020). The apoptosis accompanied by the inner membrane potential ($\Delta\Psi_{mito}$) decline, proteolysis inducing, oxygen radicals' production, lipid peroxidation, and Ca^{2+} and Na^{+} intracellular concentration increase. In addition, thallium reduced the concentration of K^{+} and reduced glutathione in the cytoplasm of these cells (Zierold, 2000; Pourahmad et al., 2010; Bramanti et al., 2019; Lin et al., 2020). Tl^{+}-induced ethanol production did find in hippocampal HN9.10e neurons (Colombaioni et al., 2017). Tl^{+} exposure of isolated synaptosomal/mitochondrial P2 crude fractions from adult rat brains resulted in mitochondrial dysfunction, Na^{+}/K^{+}-ATPase activity inhibition, and lipid peroxidation (Maya-López et al., 2018).

Giving cysteine and methionine to rats decreased toxic thallium effects (Gross et al., 1948; Stavinoha et al., 1959). After suggested that one of the main factors determining thallium toxicity is the interaction of Tl^{+} with thiol groups of cellular and mitochondrial enzymes (Herman and Bensch, 1967; Hanzel and Verstraeten, 2009; Kiliç and Kutlu, 2010), this interaction could lead to a decrease in the cytoplasmic concentration of reduced glutathione and cell damage. Glutathione—SH—dependent system is involved in PC12 cells subjected to Tl^{+}, but both glutathione—SH— and thioredoxin reductase-dependent systems were used to prevent Tl^{3+}-induced oxidative stress in the cells (Puga Molina et al., 2017). However, according to other data, Tl^{+} (in contrast to other heavy metals (Cd^{2+}, Hg^{2+}, Pb^{2+})) did not exhibit a noticeable interaction with the respiratory enzymes of mitochondria and did not inhibit the respiration of these organelles, which contradicts the previously stated assumptions of some authors (Melnick et al., 1976; Woods and Fowler, 1986; Mulkey and Oehme, 1993). These contradictions led to the need to study the interaction of Tl^{+} with the thiol groups of mitochondrial proteins.

Another reason for the high toxicity of Tl^{+} for living organisms is its ability to inhibit Na^{+}/K^{+}-ATPase and pyruvate kinase, which is the essential glycolytic enzyme, as well as to replace K^{+}, which is required to activate protein synthesis in ribosomes (Schoer, 1984; Mulkey and Oehme, 1993; Blain and Kazantzis, 2015). Research in swelling of deenergized mitochondria in nitrate media did reveal weak IMM permeability to K^{+} and H^{+} (Mitchell and Moyle, 1969; Brierley et al., 1977) and high Tl^{+} (Skul'skiĭ et al., 1978; Saris et al., 1981). Tl^{+} is known to be isomorphous to potassium (Williams, 1970; Skul'skiĭ, 1991). In mitochondria, Tl^{+}

electrophoretic transport occurs through the inner membrane K_{ATP}-dependent and K_{Ca}-activated channels (Nikitina and Glazunov, 2003; Wojtovich et al., 2010, 2013). Tl^+, unlike Cd^{2+} (Skul'skiĭ et al., 1988; Korotkov et al., 1998), did not inhibit the respiration of energized mitochondria in sucrose media (Barrera and Gomez-Puyou, 1975; Melnick et al., 1976; Diwan and Lehrer, 1977). In this regard, it was of interest to evaluate the combined effect of Tl^+ in the presence of the main cytoplasmic cations (K^+ or Na^+) or their chemical analogs (NH_4^+ and Li^+) on respiration, swelling, and $\Delta\Psi_{mito}$ in isolated rat liver and rat heart mitochondria in a sucrose-supported 400 mOsm medium containing 25–75 mM $TlNO_3$ and 125 mM nitrates of the above cations (KNO_3, $NaNO_3$, $LiNO_3$, NH_4NO_3). Preswollen succinate-energized mitochondria were earlier found to have contracted in a medium with $TlNO_3$, which made it possible to assume the participation of some Tl^+/H^+ exchange mechanism in this process (Saris et al., 1981). This circumstance made it necessary to investigate the involvement of the mitochondrial K^+/H^+ exchanger in the contraction of mitochondria that had preswollen in the above medium with $TlNO_3$ and nitrates of monovalent cations.

Binding of Ca^{2+} with the matrix-faced calcium sites (localized near the adenine nucleotide translocase [ANT]) followed by $\Delta\Psi_{mito}$ decline has been induced the mitochondrial permeability transition pore (MPTP) opening in the inner mitochondrial membrane (Ichas and Mazat, 1998; Halestrap, 2010; Pozhilova et al., 2014). High-amplitude swelling of mitochondria, a reduction in the matrix Ca^{2+} and ATP concentrations, and a decline in $\Delta\Psi_{mito}$ accompany the pore opening. However, suppose the saturation of Ca^{2+} sites inducing MPTP is not so significant. In that case, the pore opens in a low-conductance state, and the IMM becomes permeable only for inorganic ions (H^+, K^+, Ca^{2+}) and small organic molecules up to 300 Da. As the calcium load increases, the saturation of these sites becomes more, the MPTP opens in a high conductance state, and the IMM becomes permeable to molecules from 300 to 1500 Da. According to modern concepts, the main parts of this pore are the mitochondrial phosphate symporter, cyclophilin D, and some elements of H^+-ATP synthase, and ANT is involved in the regulation of the MPTP (Halestrap, 2010; Bernardi, 2013; Bonora et al., 2017). Divalent heavy metals, particularly Cd^{2+}, stimulated the MPTP opening similar to the action of calcium (Zoratti and Szabo, 1995; He et al., 2000; Belyaeva et al., 2001). Tl^+ was earlier shown to have increased the concentration of Ca^{2+} in the cytoplasm of isolated hepatocytes (Zierold, 2000). As a result, it became necessary to study the effect of Tl^{+} on the MPTP opening in the inner membrane of calcium-loaded mitochondria.

Ruthenium red (RR), Ni^{2+}, Mg^{2+}, and Li^{+} are known to inhibit the transport of Ca^{2+} into mitochondria. On the other hand, Sr^{2+} and Mn^{2+} compete with Ca^{2+} for specific calcium sites facing the matrix. Thus, they contribute to a decrease in free calcium concentration in the matrix that prevents the MPTP opening in the inner membrane (Bernardi et al., 1992; Zoratti and Szabò, 1995; Shalbuyeva et al., 2006, 2007). We have previously shown that Sr^{2+}, Mn^{2+}, and PK prevented Cd^{2+} -induced inhibition of respiration of rat liver mitochondria (RLM) and the MPTP opening (Korotkov and Skul'skiĭ, 1996; Belyaeva et al., 2001). Thus, the involvement of the calcium uniporter and calcium sites of the inner membrane in the opening of thallium-induced MPTP required additional studies.

Inhibition of mitochondrial potassium channels (both ATP-dependent channel (mitoK_{ATP}) by 5-hydroxyodecanoate (5-HD) and Ca^{2+}-activated BK-type potassium channel (mitoBK_{Ca}) with paxillin promoted calcium overload in isolated mitochondria and cardiomyocytes and reasoned by the MPTP opening in the inner mitochondrial membrane (Ardehali and O'Rourke, 2005; Costa et al., 2006; Kang et al., 2007; Kupsch et al., 2007). In addition, mitoK_{ATP} modulators diazoxide (Dia) and pinacidil (Pin) decreased the accumulation of Ca^{2+} by mitochondria. They inhibited the opening of this pore in the mitochondria of the rat heart, rat brain, and rat liver. In this regard, it was of interest to research the effect of the activity of these channels on the opening of thallium-induced pores in the inner membrane of the calcium-loaded rat liver mitochondria.

Stabilization of the ANT "c" conformation in the presence of thiol reagents (phenylarsine oxide [PAO], diamide [Diam], tert-butylhydroperoxide [tBHP], 4,4′-diisothiocyanostilbene-2,2′-disulfonate [DIDS], carboxyatractyloside) promoted the MPTP opening in the inner membrane (Halestrap and Brenner, 2003; Nantes et al., 2011). On the contrary, stabilization of ANT in the "m" conformation decreased the affinity of Ca^{2+} ions to specific calcium sites facing the matrix and the probability of MPTP opening in experiments with ADP, bongkrekic acid, and N-ethylmaleimide.

References

Ardehali, H., O'Rourke, B., 2005. Mitochondrial K_{ATP} channels in cell survival and death. J. Mol. Cell. Cardiol. 39 (1), 7–16. https://doi.org/10.1016/j.yjmcc.2004.12.003.

Barrera, H., Gomez-Puyou, A., 1975. Characteristics of the movement of K^+ across the mitochondrial membrane and the inhibitory action of Tl^+. J. Biol. Chem. 250 (14), 5370–5374.

Belyaeva, E.A., Glazunov, V.V., Nikitina, E.R., Korotkov, S.M., 2001. Bivalent metal ions modulate Cd^{2+} effects on isolated rat liver mitochondria. J. Bioenerg. Biomembr. 33 (4), 303–318. https://doi.org/10.1023/a:1010607307099.

Bernardi, P., 2013. The mitochondrial permeability transition pore: a mystery solved? Front. Physiol. 4 (95), 1–12. https://doi.org/10.1152/physrev.1999.79.4.1127.

Bernardi, P., Vassanelli, S., Veronese, P., Colonna, R., Szabó, I., Zoratti, M., 1992. Modulation of the mitochondrial permeability transition pore. Effect of protons and divalent cations. J. Biol. Chem. 267 (5), 2934–2939.

Blain, R., Kazantzis, G., 2015. Thallium. In: Nordberg, G.F., Fowler, B.A., Nordberg, M. (Eds.), Handbook on the Toxicology of Metals. Volume II: Specific Metals. Elsevier, Amsterdam, pp. 1229–1240.

Bonora, M., Morganti, C., Morciano, G., Pedriali, G., Lebiedzinska-Arciszewska, M., et al., 2017. Mitochondrial permeability transition involves dissociation of F_1F_0 ATP synthase dimers and C-ring conformation. EMBO Rep. 18 (7), 1077–1089. https://doi.org/10.15252/embr.201643602.

Bragadin, M., Toninello, A., Bindoli, A., Rigobello, M.P., Canton, M., 2003. Thallium induces apoptosis in Jurkat cells. Ann. N. Y. Acad. Sci. 1010, 283–291. https://doi.org/10.1196/annals.1299.049.

Bramanti, E., Onor, M., Colombaioni, L., 2019. Neurotoxicity Induced by Low Thallium Doses in Living Hippocampal Neurons: Evidence of Early Onset Mitochondrial Dysfunction and Correlation with Ethanol Production. ACS Chem. Neurosci. 10 (1), 451–459. https://doi.org/10.1021/acschemneuro.8b00343.

Brierley, G.P., Jurkowitz, M., Chávez, E., Jung, D.W., 1977. Energy-dependent contraction of swollen heart mitochondria. J. Biol. Chem. 252 (22), 7932–7939.

Chauncey Jr., D.M., Schelbert, H.R., Halpern, S.E., Delano, F., McKegney, M.L., et al., 1977. Tissue distribution studies with radioactive manganese: a potential agent for myocardial imaging. J. Nucl. Med. 18 (9), 933–936.

Colombaioni, L., Onor, M., Benedetti, E., Bramanti, E., 2017. Thallium stimulates ethanol production in immortalized hippocampal neurons. PLoS One 12 (11), e0188351. https://doi.org/10.1371/journal.pone.0188351. eCollection 2017.

Costa, A.D., Quinlan, C.L., Andrukhiv, A., West, I.C., Jabůrek, M., Garlid, K.D., 2006. The direct physiological effects of mitoK_{ATP} opening on heart mitochondria. Am. J. Physiol. Heart Circ. Physiol. 290 (1), H406–H415. https://doi.org/10.1152/ajpheart.00794.2005.

Diwan, J.J., Lehrer, P.H., 1977. Inhibition of mitochondrial potassium ion flux by thallous ions. Biochem. Soc. Trans. 5 (1), 203–205. https://doi.org/10.1042/bst0050203.

Goel, A., Aggarwal, P., 2007. Pesticide poisoning. Natl. Med. J. India. 20 (4), 182–191.

Gross, P., Runne, E., Wilson, J.W., 1948. Studies on the effect of thallium poisoning of the rat; the influence of cystine and methionine on alopecia and survival periods. J. Invest. Dermatol. 10 (3), 119–134. https://doi.org/10.1038/jid.1948.20.

Halestrap, A.P., 2010. A pore way to die: the role of mitochondria in reperfusion injury and cardioprotection. Biochem. Soc. Trans. 38 (4), 841–860.

Halestrap, A.P., Brenner, C., 2003. The adenine nucleotide translocase: a central component of the mitochondrial permeability transition pore and key player in cell death. Curr. Med. Chem. 10 (16), 1507–1525. https://doi.org/10.2174/0929867033457278.

Hanzel, C.E., Verstraeten, S.V., 2009. Tl(I) and Tl(III) activate both mitochondrial and extrinsic pathways of apoptosis in rat pheochromocytoma (PC12) cells. Toxicol. Appl. Pharmacol. 236 (1), 59–70. https://doi.org/10.1016/j.taap.2008.12.029.

He, L., Poblenz, A.T., Medrano, C.J., Fox, D.A., 2000. Lead and calcium produce rod photoreceptor cell apoptosis by opening the mitochondrial permeability transition pore. J. Biol. Chem. 275 (16), 12175–12184. https://doi.org/10.1074/jbc.275.16.12175.

Herman, M.M., Bensch, K.G., 1967. Light and electron microscopic studies of acute and chronic thallium intoxication in rats. Toxicol. Appl. Pharmacol. 10 (2), 199–222. https://doi.org/10.1016/0041-008x(67)90104-4.

Ichas, F., Mazat, J.P., 1998. From calcium signaling to cell death: two conformations for the mitochondrial permeability transition pore. Switching from low- to high-conductance state. Biochim. Biophys. Acta. 1366 (1-2), 33–50. https://doi.org/10.1016/s0005-2728(98)00119-4.

Kang, S.H., Park, W.S., Kim, N., Youm, J.B., Warda, M., et al., 2007. Mitochondrial Ca^{2+}-activated K^+ channels more efficiently reduce mitochondrial Ca^{2+} overload in rat ventricular myocytes. Am. J. Physiol. Heart Circ. Physiol. 293 (1), H307–H313. https://doi.org/10.1152/ajpheart.00789.2006.

Kiliç, G.A., Kutlu, M., 2010. Effects of exogenous metallothionein against thallium-induced oxidative stress in rat liver. Food. Chem. Toxicol. 48 (3), 980–987. https://doi.org/10.1016/j.fct.2010.01.013.

Korotkov, S.M., Skul'skiĭ, I.A., 1996. Changes in the effect of Cd^{2+} on the respiration of isolated rat liver mitochondria after their preincubation with Ca^{2+}, Sr^{2+}, Ba^{2+}, Mn^{2+} and ruthenium red. Tsitologiia 38 (4-5), 500–509. Russian. https://www.elibrary.ru/item.asp?id=14935301.

Korotkov, S.M., Skul'skiĭ, I.A., Glazunov, V.V., 1998. Cd^{2+} effects on respiration and swelling of rat liver mitochondria were modified by monovalent cations. J. Inorg. Biochem. 70 (1), 17–23. https://doi.org/10.1016/s0162-0134(98)00008-7.

Kupsch, K., Parvez, S., Siemen, D., Wolf, G., 2007. Modulation of the permeability transition pore by inhibition of the mitochondrial K_{ATP} channel in liver vs brain mitochondria. J. Membr. Biol. 215 (2-3), 69–74. https://doi.org/10.1007/s00232-007-9006-3.

Leung, K.M., Ooi, V.E., 2000. Studies on thallium toxicity, its tissue distribution and histopathological effects in rats. Chemosphere 41 (1-2), 155–159. https://doi.org/10.1016/s0045-6535(99)00404-x.

Lin, G., Sun, Y., Long, J., Sui, X., Yang, J., et al., 2020. Involvement of the Nrf2-Keap1 signaling pathway in protection against thallium-induced oxidative stress and mitochondrial dysfunction in primary hippocampal neurons. Toxicol. Lett. 319, 66–73. https://doi.org/10.1016/j.toxlet.2019.11.008.

Maya-López, M., Mireles-García, M.V., Ramírez-Toledo, M., Colín-González, A.L., Galván-Arzate, S., et al., 2018. Thallium-Induced Toxicity in Rat Brain Crude Synaptosomal/Mitochondrial Fractions is Sensitive to anti-excitatory and antioxidants Agents. Neurotox. Res. 33 (3), 634–640. https://doi.org/10.1007/s12640-017-9863-1.

Melnick, R.L., Monti, L.G., Motzkin, S.M., 1976. Uncoupling of mitochondrial oxidative phosphorylation by thallium. Biochem. Biophys. Res. Commun. 69 (1), 68–73. https://doi.org/10.1016/s0006-291x(76)80273-2.

Mitchell, P., Moyle, J., 1969. Translocation of some anions cations and acids in rat liver mitochondria. FEBS J. 9 (2), 149–155. https://doi.org/10.1111/j.1432-1033.1969.tb00588.x.

Mulkey, J.P., Oehme, F.W., 1993. A review of thallium toxicity. Vet. Hum. Toxicol. 35 (5), 445–453.

Nantes, I.L., Rodrigues, T., Caires, A.C., Cunha, R.L., Pessoto, F.S., et al., 2011. Specific effects of reactive thiol drugs on mitochondrial bioenergetics. J. Bioenerg. Biomembr. 43 (1), 11–18. https://doi.org/10.1007/s10863-011-9328-9.

Nikitina, E.R., Glazunov, V.V., 2003. Involvement of K^+-ATP-dependent channel in transport of monovalent thallium (Tl^+) across the inner membrane of rat liver mitochondria. Doklady Biochem. Biophys. 392 (1-6), 244–246. https://doi.org/10.1023/a:1026130527827.

Peter, A.L., Viraraghavan, T., 2005. Thallium: a review of public health and environmental concerns. Environ. Int. 31 (4), 493–501. https://doi.org/10.1016/j.envint.2004.09.003.

Pourahmad, J., Eskandari, M.R., Daraei, B., 2010. A comparison of hepatocyte cytotoxic mechanisms for thallium (I) and thallium (III). Environ. Toxicol. 25 (5), 456–467. https://doi.org/10.1002/tox.20590.

Pozhilova, Ye.V., Levchenkova, O.S., Novikov, V.E., 2014. The regula tory role of mitochondrial pora and the possibility of its pharmac ological modula tion. Rev. Clin. Pharmacol. Drug Therap. 12 (3), 13–19. Russian. https://journals.eco-vector.com/RCF/article/view/803/408.

Puga Molina, L.C., Salvatierra Fréchou, D.M., Verstraeten, S.V., 2017. Early response of glutathione- and thioredoxin-dependent antioxidants defense systems to Tl(I)- and Tl(III)-mediated oxidative stress in adherent pheochromocytoma (PC12adh) cells. Arch. Toxicol. 92 (1), 195–211. https://doi.org/10.1007/s00204-017-2056-0.

Rodríguez-Mercado, J.J., Altamirano-Lozano, M.A., 2013. Genetic toxicology of thallium: a review. Drug Chem. Toxicol. 36 (3), 369–383. https://doi.org/10.3109/01480545.2012.710633.

Saris, N.E., Skul'skiĭ, I.A., Savina, M.V., Glasunov, V.V., 1981. Mechanism of mitochondrial transport of thallous ions. J. Bioenerg. Biomembr. 13 (1-2), 51–59. https://doi.org/10.1007/BF00744746.

Schoer, J., 1984. Thallium. In: Hutzinger, O. (Ed.), Handbook of Environmental Chemistry. Anthropogenic Compounds. Springer, New York, pp. 143–214.

Shalbuyeva, N., Brustovetsky, T., Bolshakov, A., Brustovetsky, N., 2006. Calcium-dependent spontaneously reversible remodeling of brain mitochondria. J. Biol. Chem. 281 (49), 37547–37558. https://doi.org/10.1074/jbc.M607263200.

Shalbuyeva, N., Brustovetsky, T., Brustovetsky, N., 2007. Lithium desensitizes brain mitochondria to calcium, antagonizes permeability transition, and diminishes cytochrome C release. J. Biol. Chem. 282 (25), 18057–18068. https://doi.org/10.1074/jbc.M702134200.

Skul'skiĭ, I.A., 1991. The isomorphism of univalent thallium and potassium in membrane transport processes. Tsitologiia 33 (11), 118–129. Russian. PMID: 1668045.

Skul'skiĭ, I.A., Korotkov, S.M., Glazunov, V.V., Ivanova, T.I., Savina, M.V., 1988. Effect of cadmium ions on the respiration, cation transport and morphology of the liver mitochondria in the rat and the lamprey. Tsitologiia 30 (8), 956–962. Russian. PMID: 2849819.

Skulskii, I.A., Savina, M.V., Glasunov, V.V., Saris, N.E., 1978. Electrophoretic transport of Tl^+ in mitochondria. J. Membr. Biol. 44 (2), 187–194. https://doi.org/10.1007/BF01976038.

Stavinoha, W.B., Emerson, G.A., Nash, J.B., 1959. The effects of some sulfur compounds on thallotoxicosis in mice. Toxicol. Appl. Pharmacol. 1, 638–646. https://doi.org/10.1016/0041-008x(59)90068-7.

Wang, C., Chen, Y., Liu, J., Wang, J., Li, X., et al., 2013. Health risks of thallium in contaminated arable soils and food crops irrigated with wastewater from a sulfuric acid plant in western Guangdong province, China. Ecotoxicol. Environ. Saf. 90, 76–81. https://doi.org/10.1016/j.ecoenv.2012.12.014.

Williams, R.J.V., 1970. The biochemistry of sodium, potassium, magnesium, and calcium. Q. Rev. Chem. Soc. 3, 331–365. https://doi.org/10.1039/qr9702400331.
Wojtovich, A.P., Nadtochiy, S.M., Urciuoli, W.R., Smith, C.O., Grunnet, M., et al., 2013. A non-cardiomyocyte autonomous mechanism of cardioprotection involving the SLO1 BK channel. Peer J. 1 (e48), 1–23. https://doi.org/10.7717/peerj.48.
Wojtovich, A.P., Williams, D.M., Karcz, M.K., Lopes, C.M., Gray, D.A., et al., 2010. A novel mitochondrial K_{ATP} channel assay. Circ. Res. 106 (7), 1190–1196. https://doi.org/10.1161/CIRCRESAHA.109.215400.
Woods, J.S., Fowler, B.A., 1986. Alteration of hepatocellular structure and function by thallium chloride: ultrastructural, morphometric, and biochemical studies. Toxicol. Appl. Pharmacol. 83 (2), 218–229. https://doi.org/10.1016/0041-008x(86)90299-1.
Zierold, K., 2000. Heavy metal cytotoxicity studied by electron probe X-ray microanalysis of cultured rat hepatocytes. Toxicol. in Vitro. 14 (6), 557–563. https://doi.org/10.1016/s0887-2333(00)00049-7.
Zoratti, M., Szabò, I., 1995. The mitochondrial permeability transition. Biochim. Biophys. Acta. 1241 (2), 139–176. https://doi.org/10.1016/0304-4157(95)00003-a.

CHAPTER 1

Toxicity of thallium and its compounds

List of abbreviations

ADP	adenosine diphosphate
ATP	adenosine triphosphate
CNS	central nervous system
EDTA	ethylenediaminetetraacetic acid
EGTA	ethylene guanine tetraacetic acid
LD100	lethal dose in all exposed subjects
LD50	median lethal dose
MPC	maximum permissible concentration

1.1 Thallium and its compounds: chemical and toxic properties, distribution, and release into the environment

1.1.1 Chemical and toxic properties of thallium and its compounds

This section is devoted to a brief description of various aspects of the toxic effects of thallium and its compounds on living organisms. Thallium was discovered by the English chemist William Crookes by the spectral method in 1861. The atomic number of thallium in the periodic table is 81. Tl atom electron configuration is [Xe] $4f^{14}$ $5d^{10}$ $6s^2$ $6p^1$. Thallium has two oxidation states, +1 and +3. The inert-pair effect determines thallium chemical properties. This effect is the outermost s-electron pair of the thallium atom remain unsharing in the formation of chemical bonds of this metal compounds. These electrons show stronger binding to the atom nucleus, especially kainosymmetry atoms like Tl and Hg, which are more difficult to ionize or share. The inert-pair effect results in naturally occurring thallium compounds having an oxidation state of +1 but not +3 as would follow from this metal position in the periodic table. The chemical behavior of thallium, on the one hand, resembles the behavior of heavy metals (Pb, Ag, Cu, Bi), and on the other hand, of alkali metals (K, Rb, Cs). Thallium has

Mitochondria as a Key Intracellular Target of Thallium Toxicity
ISBN 978-0-323-95531-7
https://doi.org/10.1016/B978-0-323-95531-7.00004-5

an increased affinity for sulfur and selenium, and for this reason, this metal classifies as a chalcophile chemical element. It forms stable chelates with L-cysteine and N-acetyl-L-cysteine (Bugarin et al., 1989). The stability constants of the complex compounds of Tl^{+} with ADP, ATP, thiourea, mercaptosuccinic, and oxalic acids turned out to be two orders of magnitude lower than the analogous constants for Ag^{+} and bivalent heavy metals (Hg^{2+}, Pb^{2+}, Cd^{2+}, Sn^{2+}, Cu^{2+}, Zn^{2+}) and, respectively, by six orders of magnitude lower with ethylenediamine and diethyltriamine, and also 10–12 orders of magnitude less with EGTA and EDTA (Perrin, 1979). So, it should be expected that the affinity of Tl^{+} for free SH-groups of proteins will be less than the similar affinity found for the aforementioned divalent metals.

Thallium belongs to the group of trace elements. It occurs naturally only as an admixture to minerals (feldspar, mica) and ores of heavy and nonferrous metals (Zn, Cu, Fe, Pb, Cd; Schoer, 1984; Mulkey and Oehme, 1993; Kazantzis, 2000). On the other hand, Tl^{3+} containing compounds, due to their high oxidizing ability, can be obtained exclusively in laboratory conditions and do not exist in free form in nature (Harris and Messori, 2002). The thallium concentration in the Earth's crust ranges from 0.1 to 1.7 mg/kg, and Tl is more common in nature than mercury, silver, and gold (Schoer, 1984; Mulkey and Oehme, 1993; Kazantzis, 2000). Unlike toxic divalent metals, thallium does not bind to metallothioneins, and it belongs to the first class of hazard to humans due to the high toxicity of its compounds (Mulkey and Oehme, 1993; Ramsden, 2007; Gad and Pham, 2014). Thallium gets into water and air from various mineral rocks during the production of cast iron, steel, nonferrous metals, cement, and sulfuric acid, but the primary way thallium gets into the air is the combustion of coal in electricity production (Schoer, 1984; Wang et al., 2013). The industry uses metal thallium and its compounds. Thallium isotopes (^{201}Tl and ^{204}Tl) have been applied in instrumentation and medicine to scan the myocardium and tumor tissues (Chauncey et al., 1977; McCall et al., 1986; Mulkey and Oehme, 1993; Abdel-Dayem et al., 1994; Fukumoto et al., 1998). Rodenticides (rodent poisons) produced using highly toxic thallium sulfate have some delayed action on rodents (Munch, 1931; Peter and Viraraghavan, 2005; Ramsden, 2007; Kiliç and Kutlu, 2010; Blain and Kazantzis, 2015; Li et al., 2015). However, the negative consequences of this practice were the death of many wild and domestic animals who ate rodents poisoned by thallium.

Being an analog of K^+ due to the proximity of ionic radii, Tl^+ can easily penetrate the human and animal body through the skin, respiratory or digestive systems, damaging the cardiovascular, central nervous, and renal systems, as well as the gastrointestinal tract, and skin, causing hair loss (Schoer, 1984; Clarkson, 1987; Leung and Ooi, 2000; Mulkey and Oehme, 2000; Goel and Aggarwal, 2007; Rodríguez-Mercado and Altamirano-Lozano, 2013; Blain and Kazantzis, 2015). Thallium compounds are colorless, tasteless, and odorless chemicals. Thallium compounds were used earlier as pharmaceutical agents to treat various diseases such as syphilis, gonorrhea, gout, dysentery, night sweats, malaria, tuberculosis, mycoses of the head, and ringworm. In addition, these compounds were used as a cosmetic for depilation (thallium acetate), in the commission of crimes, in suicidal and abortive attempts. Such various applications made it possible to find toxicity of these compounds to humans (Remy, 1956; Reed et al., 1963; Chandler and Scott, 1986; Goel and Aggarwal, 2007; Huang et al., 2012; Rodríguez-Mercado and Altamirano-Lozano, 2013; Riyaz et al., 2013; Blain and Kazantzis, 2015).

Most fatal poisoning occurs when 10–15 mg/kg thallium penetrates the human body (Schoer, 1984; Worthley, 2002a,b). The most toxic to humans are water-soluble thallium compounds—sulfate, acetate, nitrate, and carbonate (Downs et al., 1960; Moeschlin, 1980; Moore et al., 1993). The toxicity of thallium compounds for humans and mammals turned out to be higher than the toxicity of mercury, cadmium, lead, and zinc, and it increased with the joint penetration of this metal into the body along with lead, copper, indium, and arsenic (Zitko, 1975; Delvalls et al., 1999; Peter and Viraraghavan, 2005; Rodríguez-Mercado and Altamirano-Lozano, 2013). The first studies carried out on mammals and birds indicated already the high toxicity of thallium compared to mercury, lead, and arsenic (Lamy, 1863; Paulet, 1863; Stadion, 1865).

Thallium belongs to cytotoxic poisons, and this metal is excreted from the body much more slowly than potassium (Gehring and Hammond, 1967; Mulkey and Oehme, 1993). Thallium ingested in an organism can react with albumin, globulins, and other blood proteins, inactivating SH-groups of these proteins. That is prone to the formation of metalloprotein toxins (Sheyman et al., 2014). According to various sources, the half-life of thallium from the human body is approximately three to five days (Mulkey and Oehme, 1993; Blain and Kazantzis, 2015). Thallium is excreting from the body of humans and animals with urine and feces (Thyresson, 1951a; Truhaut, 1960; Ziskoven et al., 1983; Pelclová et al., 2009).

1.1.2 Distribution of thallium compounds in the environment

Thallium occurs naturally as an admixture to various minerals (granite, shale, volcanic rock, blendes, pyrites, potassium minerals, micas) and other metal ores (manganese nodules). Thallium minerals themselves (crookesite $(Cu,Tl,Ag)_2Se$, lorandite ($TlAsS_2$), vrbaite $[Tl(As,Sb)_3S_5]$, gutchinsonite $((Tl,Ag)_2S.PbS.2As_2S_3)$, etc.) are extremely rare in nature. The concentration of thallium in the Earth's crust is in the range of 0.1−1.7 mg/kg, and this metal is more common in nature than mercury, silver, and gold (Remy, 1956; Spiridonova and Shabalina, 1977a; Schoer, 1984; Mulkey and Oehme, 1993; Kazantzis, 2000; Lopez-Arce et al., 2017). This metal belongs to the group of trace elements (0.3−0.7 ppm in the Earth's crust) and, unlike toxic divalent metals, does not biodegrade (Mulkey and Oehme, 1993; Peter and Viraraghavan, 2005; Ramsden, 2007; Gad and Pham, 2014). The thallium content was somewhat higher (up to 4 mg/kg) in some silty sediments and ferromanganese nodules of the Gulf of Riga. Natural thallium content increases from tenfold to hundredfold, and it reaches 229 mg/kg in glandular manganese nodules and ocean phosphorites (Baturin, 2004; Baturin and Dubinchuk, 2009). The thallium content increase (up to 0.9 mg/kg) was found in silty sediments, lacustrine, and terrestrial plants near the Rybnitsa Cement Plant (Munteanu and Munteanu, 2006). Thallium (20−130 mg/kg) did find in the soot of pipes of brick factories in Western Ukraine (Vodyanitskii, 2012). Pakistani woman (39-year-old) recently received thallium poisoning while consuming dry herbs used in traditional medicine (Curto-Barredo et al., 2015).

Thallium is the first hazard class to humans (Polyakova et al., 1977; Spiridonova and Shabalina, 1977a; Chukhlovina, 1999). Along with other metals (Be, Se, Sb, Cd), thallium is the first highly hazardous environmental pollutants group. These metals' soil maximum permissible concentration (MPC) is at a level below 1 mg/kg versus 1−10 mg/kg for Hg, Cu, As, and Ba (group II), as well as over 10 mg/kg for Zn, Co, Sn, Pb and Mo (group III) (Vodyanitskii, 2012). The usual thallium content in the sea and freshwater is 0.1 and 0.01 mg/L, respectively, and in soil, it is 0.1−1 mg/kg or 1 g per m^3 (Spiridonova and Shabalina, 1977a; Heim et al., 2002; Blain and Kazantzis, 2015). MPC for thallium in fresh water is 0.1 μg/L

(Krasovsky et al., 1980; Munteanu and Munteanu, 2006). The thallium content in the soils of the Kurai-Yustyd mercury-polymetallic zone of Gorny Altai was within 3.0—4.8 mg/kg, and in the area of zinc-copper-lead manifestations in the basins of the Peschanaya, Charysh, and Shargayta rivers, it was 25.0—30.0 mg/kg (Ushakova, 2009). However, in some areas of China, industrial activity has resulted in a marked increase in soil thallium levels to over 100 mg/kg (Xiao et al., 2004a,b). In Yongfu (China), where there is an outlet of pyrites to the surface, an increase in the concentration of thallium in river water was observed to 9.2 μg/L, and the maximum concentration of thallium in the mine runoff reached 194 μg/L (Li et al., 2011; Liu et al., 2016a). The contribution of thallium and arsenic to the Hazard Index of the waters of the Pardo River (Brazil) was the highest among metals such as, Be, Cd, Cr, Cu, Pb, Mn, Hg, Ni, Tl, Sn, V и, and Zn (Alves et al., 2014).

1.1.3 Thallium release routes into the environment

The thallium release into the environment results in the following reasons: weathering of rocks, when the metal washed out of clay and calcium deposits and basalts, the oxidation of sulfide deposits, the production of cement, cast iron, steel, sulfuric acid, the production of certain metals (Cu, Zn, Cd, Pb) from their sulfide ores and during industrial processing and combustion of coal (Grigorovich, 1960; Tikhova, 1967; Schoer, 1984; Mulkey and Oehme, 1993; Nielsen et al., 2006; Ushakova, 2009; Wang et al., 2013; Liu et al., 2016b, 2020a, 2020b). *Shewanella putrefaciens* bacteria metabolized the artificially synthesized jarosite, resulting in mineral biodegradation with Fe^{2+} concentration increase in a medium. At the same time, the Tl^{+} concentration did not change. This phenomenon indicates that the liberation of Tl^{+} impurities from minerals into the environment may not necessarily occur due to the minerals' biodegradation (Smeaton et al., 2012). At the same time, the thallium content was exceptionally high in calcium-containing rocks (2.8 mg/L) or anthracite with an admixture of sulfides (135 vs. 4 mg/kg). Soil thallium is associated with the mineralization of hydrothermal and pyrite minerals and coal deposits (Shvartsev and Gomonova, 1975; Taheri et al., 2015). Unlike silver, thallium adsorbs in the environment from the following sources: hematite, peaty and soil samples, and mineral soils contaminated with mining or metallurgical wastes (Jacobson et al., 2005; Lopez-Arce et al., 2017; Cruz-Hernández et al., 2018; Zhou et al., 2021).

Pure thallium annual production is 380 tons, but from 2 to 5 tons of this metal in total is used in industry (Mulkey and Oehme, 1993; Kazantzis, 2000; Guberman, 2015). When examining three enterprises for the production of metallic thallium and its salts, including the Clerici solution, increased concentrations of this metal were found on various surfaces of industrial premises (Spiridonova and Shabalina, 1977b). Thallium compounds apply in the next: the electronics industry, in the production of glass, scintillation counters, gas discharge lamps, alloys, selenium rectifiers, catalysts, and thermometers. The isotopes of ^{201}Tl and ^{204}Tl are used in the following medicine areas: instrumentation, clinical cardiology in cardiac tomography, myocardial perfusion and cell dosimetry, in the treatment of diseases peripheral nervous system, as well as in the study of human tumors, treatment of otitis media, infectious diseases of the eyes, and diseases of the peripheral nervous system of infectious-toxic etiology (Bibikova and Darvoyd, 1959; Grigorovich, 1960; Denisyuk, 1964; Darvoyd et al., 1968; Zaychenkov, 1970, 1971; Gridnev et al., 1971; McCall et al., 1986; Matveev, 1989; Sehweil et al., 1989; Mulkey and Oehme, 1993; Nadel, 1993; Abdel-Dayem et al., 1994; Fukumoto et al., 1998; Yudin et al., 2010; Kawakami et al., 2011; Hsieh et al., 2012; Shiraishi et al., 2015; Osytek et al., 2021). Tl^+-filled peptide-modified bacteriophage MS2 preparations resulted in the death of cultured human breast cancer cells and necrosis of these tumor xenografts in mice (Kolesanova et al., 2019). Perhaps in the future, such drugs will use in the treatment of human tumors. Thallium compounds have found wide application in various fields of industry: the optical, jewelry (bijouterie), photographic industry, and the cement industry in the production of specially resistant grades of cement, as catalysts for organic synthesis, and their high toxicity and "delayed" nature of action on the body made it possible to use these compounds to combat rodents and insects (Lee, 1972; Nachman and Hartley, 1975; Spiridonova and Shabalina, 1978; Schaller et al., 1980; Dipalma, 1981; Dorman, 1990; Peter and Viraraghavan, 2005; Ramsden, 2007; Kiliç and Kutlu, 2010; Blain and Kazantzis, 2015; Li et al., 2015). Unfortunately, rodenticides in some countries are available without a prescription, and they, for this reason, may even be open to children (Dipalma, 1981). The quenching by Tl^+ of fluorophore's fluorescence (1,3,6,8-pyrene-tetra sulfonic acid) was used to study the cholinergic receptor to understand better the nature of the autoimmune disease myasthenia gravis (Donnelly et al., 1984). Recently, the US and EU have replaced the use of thallium sulfate as a rodenticide with warfarin. Still, the emergence of resistance in rats to this compound has forced some countries to revert to the use of thallium sulfate as a rodenticide (Brewer and Haggerty, 1958; Peter and Viraraghavan, 2005; Blain and Kazantzis, 2015). Thallium is used in green fireworks, low-temperature thermometers, semiconductors, tissue, and paper thickness gauges, industrial ceramics, scintillation counters, and switchgear (Abramova and Cherny, 1977; Moore et al., 1993; Mulkey and Oehme, 1998).

Noticeable harm to health was caused by applying thallium salts using outdated therapeutic methods (Blain and Kazantzis, 2015). Being K^+ surrogate (due to the proximity of ion radii), Tl^+ can easily penetrate the body of animals and humans, being absorbed on their skin, lungs, and digestive system (Sessions and Goren, 1947; Rodríguez-Mercado and Altamirano-Lozano, 2013; Alves et al., 2014). The ingestion of thallium compounds into the human body can occur due to contact with thallium-contaminated air, water, and food. This ingestion can also be in copper, lead, cadmium, zinc smelting, power plants' coal burning, and thallium compounds' use as catalysts in organic synthesis (Krasovsky et al., 1978; Blain and Kazantzis, 2015). The layered disodium zirconium phosphate successfully carried out thallium-contaminated wastewater treatment (Cheng et al., 2019). To purify the drinking water from thallium trace impurities, new methods recently proposed. They are next: the ion exchange processes, adsorption, macrocyclic compounds, oxidation-reduction precipitation, the $MnFe_2O_4$-biochar composite, and solvent extraction (Xu et al., 2019; Zhao et al., 2020; Liu et al., 2021).

1.2 Thallium toxicity to biological organisms (humans, animals, and plants)

1.2.1 Thallium accumulation in humans, animals, and plants

Workers in conditions of high thallium content in the air (0.07 vs. 0.01 mg/m^3 in standards) showed an increased range of this metal in the urine (3–560 μg/L vs. 0.13–1.69 μg/L in standards). A medical study revealed symptoms of chronic poisoning thallium, which were hyperhidrosis, the electroretinogram changes, diffuse red dermographism, revitalization of tendon reflexes, and tachycardia (Polyakova et al., 1977; Spiridonova and Shabalina, 1977b, 1978; Shamshinova et al., 1990; Mulkey and Oehme, 1993). The thallium concentration in the urine of people who lived near cement plants in Germany and Russia was 10 times higher than the same concentration for people from areas not contaminated with thallium, and it was determined respectively 0.2–0.4 and 2.6–76.5 μg/L

(Brockhaus et al., 1981; Schoer, 1984; Kamiljanov, 1993). The use of atomic absorption spectroscopy revealed an increase in the levels of creatinine and cystatin C in the urine of workers in cement, zinc, and lead plants due to the accumulation of thallium in their bodies (Schaller et al., 1980; Shelley et al., 2012; Garcia-Vargas et al., 2014). According to the results of these studies, the chronic poisoning threshold value for factory workers was proposed to establish within thallium levels from 300 μg/L (Schoer, 1984).

The thallium concentration in the urine of people living near a lead-zinc plant in northern Mexico was in the range of 0.4–1.2 μg/L (Garcia-Vargas et al., 2014). Medical studies have shown that long-term use of opioids can lead to the accumulation of thallium in the body. This metal begins to be determined in the urine against the background of clinical manifestations of thallotoxicosis (Ghaderi et al., 2015). Up to 300–480 μg of thallium found in the skin washes of workers' hands involved in this metal production (Tikhova, 1967). Thallium content in tobacco smoke is 24–100 μg/kg (Spiridonova and Shabalina, 1977a). The thallium concentration was significantly higher in the smoker group than in the nonsmoker group (Shakeri et al., 2021). The thallium content in the kidneys, liver, and hair of people living near the incinerator in Constant (Tarragona, Spain) was shallow compared to other heavy metals (As, Cd, and Hg) found in fish and seafood (Bocio et al., 2005; Nadal et al., 2005; Martí-Cid et al., 2009; Mari et al., 2014). Thallium content near the incineration plant increased markedly between 1998 and 2011 (Vilavert et al., 2012). In addition to manganese, thallium and arsenic were the most prominent pollutants of the Pardo River (Brazil) from industrial discharges associated with sugar cane cultivation (Alves et al., 2014). The maximum permissible concentration of thallium in different countries varies in the air from 0.02 (Sweden) to 0.1 (USA) $\mu g/m^3$, in drinking water up to 0.8 μg/L and in soil up to 1 mg/kg dry weight (Sessions and Goren, 1947; Peter and Viraraghavan, 2005).

The widespread use of thallium sulfate as a rodenticide has resulted in the deaths of large numbers of cats, dogs, martens, badgers, and wild foxes, and thallium-rich vegetation has been the cause of widespread poisoning in farm animals (Dipalma, 1981; Peter and Viraraghavan, 2005). The thallium content in the kidneys and liver of mice caught near a zinc plant in Poland was 34.3 and 14.5 mg/kg, respectively, versus 0.07 mg/kg in standards (Dmowski et al., 1998). Thallium oxide toxicity (LD100) for dogs and small mammals (rats, guinea pigs, and rabbits) was 20–30 mg/kg (Downs et al., 1960). Experiments on white rats showed that the thallium sulfate bioavailability was noticeably higher than a ceramic semiconductor material (Mulkey and Oehme, 1998). The amount of thallium found in the kidneys of river ducks and diving ducks caught in different prefectures of Japan, respectively, determined in the range of 4.0–20.4 and 2.1–3.0 μg/kg of dry weight, and in the liver of these birds, this concentration of thallium was respectively 4.3–6.5 and 2.1–5.3 μg/kg of dry weight (Mochizuki et al., 2013). At the same time, fluctuations in thallium concentration in these organs correlated with similar cadmium, vanadium, and lead concentrations. The concentration of thallium in the liver of dolphins (Brazil), in the liver of ocean moonfish (USA and Ontario, Canada), and liver of thick-headed minnows (Saskatchewan, Canada) was shallow, and it was at the limit of the sensitivity of the devices (Kunito et al., 2004; Ouellet et al., 2013; Perrault et al., 2014; Ponton et al., 2016). The thallium content in the kidneys and especially in the liver of Pacific loons found dead on the coast of the Japanese island of Tsushima significantly exceeded those of the surviving birds (Mochizuki et al., 2005). The thallium toxicity for the earthworm *Eisenia fetida* and the snail *Arianta arbustorum*, eating watercress grown on artificial soils containing 1–500 mg/kg thallium versus 0.1–0.13 mg/kg in standards, was 100 times higher than the cadmium one (Heim et al., 2002). The number of living worms *Caenorhabditis elegans* and their longevity visibly reduced in 100–1000 μM Tl-acetate, which bioaccumulated in the L1 larval stage (Hurtado-Díaz et al., 2020; Varão et al., 2021). The maximum thallium concentrations in toad tissues and cells were about 18.6 mg/kg, corresponding to 100 μM Tl in body water (Dmowski et al., 2015).

The use of *Hyalella azteca* crustaceans as an in vivo model for assessing thallium toxicity found that the survival of these organisms after four weeks, respectively, decreased by 25% at a Tl^{+} concentration in water and their body of 48 nM and 32 μM (Borgmann et al., 1998). The toxicity of thallium for crustaceans calculated on dry weight was comparable to the toxicity of Cd, Hg, and Pb. At the same time, K^{+} water concentration increase significantly reduced Tl^{+} toxicity for these crustaceans, while Ca^{2+}, Mg^{2+}, and Na^{+} did not affect it. However, the toxic thallium effects increased much more when Tl^{+} was used with other heavy metals (Cd^{2+}, Hg^{2+}).

Thallium accumulates well in some plant species, mollusks, mussels, and fish (Filipov, 2015). The bioaccumulation of thallium by fish and other aquatic organisms increased in water with a low potassium and calcium concentration. It depended on the age and length of the fish (Gantner et al., 2009; Rickwood et al., 2015). The significant increase of metallothionein

(MT)2 synthesis, superoxide dismutase enzymatic activities, and Na^+/K^+-ATPase activity found in the chronic experiments with Zebrafish *Danio rerio* kept in water with 100–500 ng/L thallium (Hou et al., 2017). Thallium uptake by plants depends on the type of plant, the type of thallium-containing mineral, and the soil acidity, and it is higher in acidic and mineralized soils with low cation exchange capacity (Pavlíčková et al., 2006; Kazantzis, 2007; Liu et al., 2011; Vodyanitskii, 2012; Grösslová et al., 2015). Thallium bioaccumulation by green cabbage leaves or soil fungi was directly proportional to the content of this metal in the soil (Ning et al., 2015; Sun et al., 2015). Thallium accumulated more strongly in roots than in broad bean shoots, and this metal resulted in oxidative damage to lipids and proteins and induced DNA strand breaks (Radić et al., 2009). Thallium slowed down some plants' growth that eliminated with the subsequent introduction of potassium salts into the soil (Siegel and Siegel, 1975).

1.2.2 Comparative toxicity of thallium compounds to humans

Thallium compounds are colorless, tasteless, and odorless chemicals. The toxicity of thallium to humans was found when using its compounds as a pharmaceutical (treatment of syphilis, gonorrhea, gout, dysentery, refractory diarrhea, night sweats, malaria, tuberculosis, mycoses of the head, and ringworm), as a cosmetic for depilation (thallium acetate), as a rodenticide and insecticide to control rodents and insects (thallium sulfate), in producing cement, ceramic semiconductors and optical glasses, Clerici liquid, committing crimes, as well as in suicidal and abortive attempts (Uspenskaya, 1940; Remy, 1956; Link and Mohr, 1953; Brewer and Haggerty, 1958; Erbslöh, 1960; Johnson, 1960; Petersohn, 1960; Reed et al., 1963; Patterson, 1975; Spiridonova and Shabalina, 1977a; Shabalina et al., 1980; Sabbioni and Manzo, 1980; Schaller et al., 1980; Dipalma, 1981; Chandler and Scott, 1986; Mulkey and Oehme, 1998; Hoffman, 2003; Goel and Aggarwal, 2007; Livanov et al., 2011, 2015; Huang et al., 2012; Rodríguez-Mercado and Altamirano-Lozano, 2013; Riyaz et al., 2013; Blain and Kazantzis, 2015; Filipov, 2015). Moreover, most fatal poisoning occurs when 10–15 mg/kg of thallium enters the body (Budrin and Meshcherskaya, 1975; Schoer, 1984). Later, there were attempts to use thallium compounds in relatively small quantities in the so-called Boucher doses (Prick, 1979). This practice was discontinued later due to the cumulative accumulation of thallium in the body. Water-soluble thallium compounds (sulfate, acetate, nitrate, and carbonate) are the most toxic to humans (Downs et al., 1960; Moeschlin, 1980; Moore et al., 1993). The toxicity of thallium compounds except methylmercury for humans and mammals was higher than mercury, cadmium, lead, and zinc (Zitko, 1975; Peter and Viraraghavan, 2005; Rodríguez-Mercado and Altamirano-Lozano, 2013). In this regard, it is interesting to note that the earliest studies conducted on mammals and birds first reported the higher toxicity of thallium compared to mercury, lead, and arsenic (Paulet, 1863; Lamy, 1863; Stadion, 1865). For a comparative assessment of thallium and other heavy metals' toxicity, the following series were proposed: gonadotoxicity one Tl > Cd > Hg > As; neurotoxicity 1 Tl > Hg > Pb > Ag; nephrotoxicity one Cd > Hg > Tl > Zn (Ushakova, 2009). According to some data, the toxicity of trivalent thallium compounds for mammals is above, and according to other data, it does not exceed the toxicity of Tl(I) compounds (Schoer, 1984).

1.2.3 Ways the human body route of thallium

Thallium compounds are absorbed quickly enough after entering the human body with food, from the air, or through the skin (Polyakova et al., 1977; Alves et al., 2014; Blain and Kazantzis, 2015). Occupational thallium poisoning (production of rodenticides and other thallium-containing products) has a lower specific weight than accidental poisoning during suicide attempts and criminal cases (Spiridonova and Shabalina, 1978). The mortality rate from thallotoxicosis after medical use (depilation) and in case of poisoning with thallium-containing rodenticides reached 6%, incredibly high among children (Munch, 1931, 1934; Brewer and Haggerty, 1958). The extreme thallium toxicity was established even in the first Russian work devoted to studying the thallium chloride effect on dogs. In this regard, Stadion proposed to refrain from using thallium compounds in the children's treatment (Stadion, 1865). The daily thallium intake into the human body from the environment through food is about 2–5 μg, and 5 ng with air which does not pose any danger to human health (Schoer, 1984; Peter and Viraraghavan, 2005). The usual thallium content in various human organs does not exceed 2–3 ng/kg (Weinig and Zink, 1967). These concentrations increased to 10–20 mg/kg with fatal thallium poisoning (Cavanagh et al., 1974). During the production and purification of thallium compounds, the thallium adsorption through the skin reached 2.5 μg/cm^2/min when a paste containing 2% Tl_2O_3 was used (Sanotskiy, 1961). The urine thallium concentration in standards was within 0.1—1.7 μg/L, in blood 2.0—3.0 μg/L, and the internal organs 0.1—29.5 μg/kg (Spiridonova and Shabalina, 1978; Mulkey and Oehme, 1993; Malbrain et al., 1997; Chukhlovina, 1999; Lyubchenko et al., 2004; Blain and Kazantzis, 2015). At the same time, the thallium concentration in the urine of smokers and vegetarians (0.9—1.7 and 0.6—1.4 μg/L) was higher than the concentration of thallium in nonsmokers who eat regular food (0.1—0.5 μg/L) (Spiridonova and Shabalina, 1977a).

Regular food (fruits, vegetables, and meat) contains less than 0.5 mg/kg Tl and its daily intake with food is less than 5 μg TL and with clean air up to 0.005 μg per day (WHO, 1996; Delvalls et al., 1999; Das et al., 2006). The thallium content in some products is even less (μg/kg): milk: 10—30, pork: 8.5, 20—300; vegetables and cereals, red, and white wine: 40—480; bream: 800—8600; trout: more than 3000; and sea shellfish and fish: 350—2900 (Rodríguez-Mercado and Altamirano-Lozano, 2013). Health risks exist from the dietary intake of common aquatic species such as tilapia and carp besides crustaceans due to high arsenic, cadmium, lead, and thallium levels (Chan et al., 2021). The thallium did detect in shrimp at levels that pose significant human health risks (Dietrich and Ayers, 2021). The thallium concentration in urine increased to 5.2 μg/L in areas close to sources of thallium in the atmosphere (Spiridonova and Shabalina, 1977a). Accidental human poisoning with thallium is associated with the consumption of containing thallium foods or green vegetables and fruits grown in soils, which show a high content of thallium compounds (1 mg/kg vs. 0.03—0.1 mg/kg in uncontaminated areas). These soils are usually located near cement, brick, and smelting plants or coal-fired power plants (Schoer, 1984; Galván-Arzate, Santamaría, 1998; Peter and Viraraghavan, 2005; Gad and Pham, 2014; Liu et al., 2019). The thallium content ranged from 0.4 to 1.7 mg/kg in the meat of animals (lamb) fed with such plants and grown on soils with a high thallium content (Schoer, 1984). Symptoms of polyneuritis, sleep disturbances, headache, fatigue, and some psychasthenic symptoms, as well as the increased urine thallium concentration, were found in residents who grew vegetables and fruits in plots located near the cement plant (Brockhaus et al., 1981; Kamiljanov, 1993; Peter and Viraraghavan, 2005). A significant positive correlation was found between thallium content in two moss species (*Pleurozium schreberi* and *Polytrichum formosum*) and the prevalence of circulatory system diseases (Heim et al., 2002). Thallium ingress into the body through the lungs occurred during the processing of blast furnace dust and with dust generated during the burning of pyrite, during the production of thallium-containing rodenticides and lenses, during the industrial separation of diamonds (Peter and Viraraghavan, 2005).

1.2.4 Toxicological characteristics of thallium compounds

The ingestion of about 1 g of thallium is the human lethal dose (Worthley, 2002a,b). At the same time, the minimum lethal dose of thallium for large mammals and humans is about 5–20 mg/kg, and for small mammals (rats, mice, guinea pigs, rabbits) and birds, about 5–35 mg/kg (Downs et al., 1960; Skelley and Gabriel, 1964; Moeschlin, 1980; Schoer, 1984; Mulkey and Oehme, 1993; OT, 2014; Gresham et al., 2014; Gad and Pham, 2014). The thallium concentration in human blood in standards does not exceed 1.1 μg/L. It rises to 100–13,000 μg/L in case of poisoning with Tl salts, and the concentration reaches a maximum of 2 h after this metal enters the body (Innis and Moses, 1978; Huang et al., 2012; Rodríguez-Mercado and Altamirano-Lozano, 2013; Sheiman et al., 2014). The lethal thallium dose in the human blood is approximately 0.28–5.37 mg/mL, and this one is respectively 1.12 ± 0.43 and 0.72 ± 0.11 mM Tl_2SO_4 for isolated humans and rats hepatocytes (Ponsoda et al., 1995; Riyaz et al., 2013). The tissue thallium concentrations (mg/kg) were equal 2.3 in urine, 5.2 in the brain, 7 in bile, and 7.8 in kidneys in the case of fatal poisoning (Smith and Doherty, 1964). Thallium concentrations in blood above 100 μg/L and urine above 200 μg/L can be considered toxic (Malbrain et al., 1997). Immediately after the poisoning, the thallium concentration in the urine reached 11,400–11,900 μg/L. Still, the blood and urine concentration quickly decreased after a month, and this one became close to standards only after three months (Huang et al., 2012). Thallium concentration in blood and urine reached 219–1414 and 957–11,285 μg/L by two weeks of later poisoning with thallium salts (Malbrain et al., 1997; Sun et al., 2012). The thallium concentration in the urine of a 67-year-old woman was 1800 μg/L five weeks after poisoning (Pau, 2000).

Thallium accumulation in humans and mammals (rats and mice) is cumulative (Thyresson, 1951a; Tikhova, 1967). No noticeable differences were found in thallium toxicity and metabolism in humans and the white rat (Krasovsky and Kenesariev, 1984). The toxicity of thallium salts for all mammals (white rats, rabbits, guinea pigs, dogs) is practically the same and the LD100 after a single dose is 7–19 mg/kg for Tl-acetate and 5–92 mg/kg for Tl_2O_3 (Abramova and Cherny, 1977). In this regard, the extrapolation is justified in comparative analyzing toxicological experiments with white rats exposed to thallium salts and human thallium poisoning consequences (Krasovsky and Kenesariev, 1984). Thallium oxide at a concentration of 100–350 μM was highly toxic to rats, rabbits, and dogs (Schoer, 1984). For the first time, the death of rabbits four to five days after applying a paste containing 145–207 mg of thallium chloride to their skin showed by Russian researcher Stadion (Stadion, 1865). Due to the scarcity of knowledge about the toxicity of thallium compounds in the last century, there were attempts to justify a method of obtaining wool from sheep subjected to 16 mg/kg of thallium acetate (Uspenskaya, 1940). After that, the animal's meat had fed to cats, which showed an enlargement of the spleen 1.5–2 times and hyperplastic changes in this organ.

According to the results of experiments with white rats, the MPC for thallium carbonate was established, which turned out to be equal to 0.0004 mg/m^3 (Kamiljanov, 1993). In experiments on rats, LD50 for thallium carbonate applied to the skin was about 117 mg/kg. This Clerici solution containing equal proportions of thallium formate and thallium malonate was 15–18.8 mg/kg when administered orally and 57.7 mg/kg on the skin (Shabalina and Spiridonova, 1976; Krasovsky et al., 1980; Shabalina et al., 1980). Similar LD50 values (21.0–23.7 mg/kg) for thallium carbonate, sulfate, and chloride were found in chronic experiments in rats and mice (Tikhova, 1967). According to other data (Krasovsky et al., 1978; Zasukhina et al., 1980), the lethal dose of thallium carbonate for rats was 15 mg/kg. LD50 for thallium carbonate, bromide, and iodide was respectively 18, 29, and 28 mg/kg (Sanotskiy, 1961). The toxicity was lower in experiments with rats and mice after intraabdominal or intragastric administration of 30% thallium inorganic complex compound or high-temperature superconductor thallium-barium-calcium cuprite ($Tl_2Ca_2Ba_2Cu_3O_{10}$) than the toxicity of thallium oxide (Tl_2O) (Neizvestnova et al., 1992). LD50 concentrations (mg/kg) were correspondingly equal to 141, 95, and 36 for the complex, cuprite, and thallium oxide in the experiments with mice through intragastric and intraperitoneal administration. Based on these and some other experiments for cuprite-thallium-barium-calcium, the MPC was set equal to 0.04 mg/m^3 in the air of industrial premises (Neizvestnova et al., 1992). For comparison, LD50 for $AgNO_3$ in silver was also 15.3 mg/kg in experiments on rats and mice (Lyublina and Dvorkin, 1983). The lethal concentration (mg/L) for thallium nitrate was 10–15 for rainbow trout, 60 for perch, and 40–60 for roach (Nehring, 1962). The lethal thallium concentration (mg/L) was 10–60 for aquatic fish, two to four for marine insects and invertebrates, and only 0.4 for tadpoles (Zitko et al., 1975a; Peter and Viraraghavan, 2005). On the other hand, mollusks and mussels turned out to be correspondingly less sensitive to thallium, which did eliminate mainly from the body of these animals after one or three weeks (Zitko et al., 1975b).

1.3 Classification, symptomatology, diagnosis, and treatment of thallium poisoning

1.3.1 Thallium poisoning classification

The thallium poisoning classification includes two main groups: professional and nonprofessional; the latter group is the largest (Spiridonova and Shabalina, 1977a). In turn, nonprofessional thallium poisoning can divide into accidental, intentional, and criminal. At the same time, the largest group is suicides. Then there is a group of patients treated with thallium for syphilis, night sweats for tuberculosis, and people who suffered from thallium compounds by mistake, and in addition, from murder and abortion (Spiridonova and Shabalina, 1977a).

After a short time, the ingestion of thallium compounds into the human body causes nausea and vomiting, but the first poisoning symptoms appear after 3–4 days. A rapidly progressive painful peripheral neuropathy develops, which begins in the legs. After a while, the hands involved in this process, brown or black formations appear in the bulb and the root part of the hair shaft (Paulson et al., 1972; Cavanagh, 1991; Ibrahim et al., 2006; Rusyniak et al., 2002, 2010; Curto-Barredo et al., 2015; Liu and Liao, 2021). Acute alopecia may be further evidence of acute thallium poisoning (Senthilkumaran et al., 2017). During this period, patients most often paid attention to abdominal pain and diarrhea, which almost always followed by constipation due to the possible involvement of the vagus nerve in this process (Reed et al., 1963; Mulkey and Oehme, 1993; Hoffman, 2003). This poisoning stage occasionally observed glossitis, pharyngoesophagitis, gastritis, enteritis, and colitis (Galván-Arzate and Santamaría, 1998). Quite often, subfebrile or high body temperature occurs from the very onset of the disease, caused by the thallium ingress into the body (Spiridonova and Shabalina, 1977a).

Discomfort neuralgic pains in both legs, behind the breastbone, and abdomen, great thirst, and insomnia were emerged on the second week after poisoning (Moeschlin, 1980; Worthley, 2002a,b). Hypertension, ophthalmoplegia, circulatory disorders, and tachycardia develop due to toxic damage to the myocardium; a deficiency of vitamin B1 and niacin is found; motor and respiratory nerves are affected; damage to nerve fibers and chromatolysis of neurons occurs; dryness of the skin (anhidrosis) appears due to the disappearance of sweat secretion owing to the destruction of the sweat glands; pleural pain, diffuse myalgias, nonspecific dizziness, decreased activity, memory impairment, confusion, incoherent speech, chronic neuropsychiatric manifestations, tremors, convulsions, and ataxia occur; hair loss (alopecia) begins; the heart rate increases; there is a sluggish paraparesis; acute toxic polyneuritis appears and in some cases hand reflexes are affected, since some cranial nerves and, in particular, pairs II, III, and VI are affected (Brewer and Haggerty, 1958; Reed et al., 1963; Budrin and Meshcherskaya, 1975; Spiridonova and Shabalina, 1977a; Cavanagh, 1979, 1991; Prick, 1979; Moeschlin, 1980; McMillan et al., 1997; Worthley, 2002a,b; Lyubchenko et al., 2004; Ibrahim et al., 2006; Tsai et al., 2006; Cvjetko et al., 2010; Rusyniak et al., 2010; Pavlova et al., 2012; Chukhlovina, 1999; Huang et al., 2012; Sheiman et al., 2014; Blain and Kazantzis, 2015). Perioral numbness, the face erythematous eruptions, hair roots' damage, polyneuropathy, parakeratosis, the basal layer vacuolar degeneration, and epidermal nerves' loss initially developed in two patients after ingestion of thallium-containing water (Lu et al., 2007). We can note that impaired coordination, gait, and extensive alopecia appeared on the third day after thallium ingestion into a three-year-old child's body (Taber, 1964).

After three to four weeks of thallium poisoning often observed weight loss (Schoer, 1984). Symptoms of polyneuropathy and, in particular, distal axonal polyneuropathy, convulsive syndrome, agitation, delirium, coma also appear here (Innis and Moses, 1978; Cavanagh, 1979; Chukhlovina, 1999; Vorobyev, 2003; Sun et al., 2012). The resembling neuralgia pain, colic-like pain in the abdomen and behind the sternum is the main symptoms at this stage (Schoer, 1984). These symptoms of thallotoxicosis were much stronger in two people simultaneously poisoned with thallium and arsenic (Rusyniak et al., 2002). Increased anxiety, excitement, and confusion with a simultaneous two to threefold increase in thallium clearance occurred just after it entered the body of a 64-year-old man (Papp et al., 1969). The aforementioned symptoms prolonged over time in chronic thallium poisoning.

The following symptoms accompanied this poisoning: eye diseases, including catarrhal conjunctivitis and loss of vision, distorted color perception, varying degrees of pain in the legs without noticeable signs of polyneuritis, and the very subtle onset of hair loss, and decreased activity of Mg^{2+}-ATPase and Na^{+}/K^{+}-ATPase (Bulavintseva, 1975; Schoer, 1984; Tabandeh and Thompson, 1993). Chronic thallium poisoning primarily damages the extrapyramidal system, which choreic and athetotic movements often characterized (Prick, 1979). Along with this, indigestion, movement disorders, endocrine disruptions, lack of resistance to infection, lack of appetite, weight loss, brain and nervous disorders with intellectual disabilities, and behavioral abnormalities appear (Schoer, 1984). Recent studies of a large number of patients have not found an association between elevated thallium concentrations in humans and a subsequent increase in the risk of developing diabetes (Menke et al., 2016).

The thallium concentration (μg/L) in the blood may indicate the degree of poisoning of a person with thallium with the related symptoms. These symptoms can subdivide: mild degree of damage is less than 30 (asthenia, hyperesthesia, electrolyte disorders in the heart muscle, gastroduodenitis); the average degree of damage is 30–100 (asthenia, diffuse brain damage, impaired sensitivity, pain in the distal extremities, an increase in dysmetabolic disorders in the myocardium); severe lesion is 100–500 (depression of consciousness, an increase in pain syndrome, the appearance of cardiac arrhythmias and conduction disorders, the formation of zones of ischemia of the left ventricle); extremely severe degree is more than 500 (coma of the first degree, damage to the cranial nerves; Kuzmich, 2006). The thallium concentration in blood plasma reaches a maximum of 2 h after this metal ingestion into the body. Then thallium is quickly distributed to other organs, and its excretion in the urine is observed (Oehme, 1978). Thallium ingested into the body accumulates in the muscles, and its minimum amount is in adipose and connective tissues (Spiridonova and Shabalina, 1977a). The liver dysfunction symptoms (the urobilinogen increase, albuminuria, the increase of iron and copper serum levels) can observe in severe thallium poisoning. Poisoning in later stages noted severe liver obesity and degeneration of the heart and kidneys (Spiridonova and Shabalina, 1977a).

Albuminuria and hematuria were after toxic kidney damage following thallium salt poisoning, but renal function was not significantly impaired (Reed et al., 1963). Histological findings from thallium-poisoned patients showed centrilobular necrosis with fatty changes in most liver lobes and abnormal amounts of porphyrins in the urine (Cavanagh et al., 1974).

Two to three months after thallium poisoning, typical Meese stripes appear on the nails of the hands and feet, and bluish lines may appear on the gums (Reed et al., 1963; Budrin and Meshcherskaya, 1975; Huang et al., 2012; Liu et al., 2015). Acute confusion and severe painful neuropathy, along with skin lesions, developed in two thallium-poisoned patients from Taiwan (Huang et al., 2012). All thallium-affected people showed changes in blood vessels and blood flow as well as minor hemorrhages, stasis, parenchymal dystrophy (myocardium, kidneys, liver), widespread bronchopneumonia with foci of pulmonary destruction, and edema with the spinal cord tension (Treshchinsky et al., 1989; Vorobyev, 2003; Kuzmich, 2006). Thallium's toxic effects are its ability to disrupt kidney and liver function and its ability to react directly with proteins (Tikhova, 1967).

Thallium belongs to cytotoxic poisons. It is much slower than potassium and is released from the body and secreted by the epithelium of the stomach, salivary glands, and renal tubules (Mulkey and Oehme, 1993; Sheiman et al., 2014; Gad and Pham, 2014). Thallium trapped into the cell is not released as quickly as potassium (Gehring and Hammond, 1967). The concentration of thallium in the saliva of a 28-year-old woman poisoned by it was 15 times higher than the same concentration in urine (Richelmi et al., 1980). As soon as thallium enters the body, the formation of metalloprotein toxins occurs due to the thallium reaction with albumins, globulins, and other blood proteins in subsequent inactivating these proteins' SH-groups (Sheiman et al., 2014). In this case, toxins with a molecule size from 10 to 200 nm had the maximum damaging effect. According to various sources, the half-life of thallium from the human body and experimental animals is approximately 3–5 days (Lie et al., 1960; Schaller et al., 1980; Mulkey and Oehme, 1993; Kazantzis, 2007; Cvjetko et al., 2010; Livanov et al., 2013, 2015). Urine thallium level can judge the metal amount is taken (Spiridonova and Shabalina, 1977a). A man and a woman (China) got poisoned by an unknown amount of thallium (Huang et al., 2012; Liu et al., 2013). As a result, these people found impaired consciousness, speed of perception, and memory, but they gradually restored during the first year. There have been cases when thallium was detected in the urine even five months after poisoning with a rodenticide containing Tl_2SO_4 (Blain and Kazantzis, 2015).

1.3.2 Symptoms and characteristics of the toxic thallium effects on the animal body

The first studies of this newly discovered metal, thallium, found the same toxicity of thallium carbonate, sulfate, and chloride for mammals and birds (Paulet, 1863; Lamy, 1863; Stadion, 1865). Poisoning with thallium compounds in dogs was accompanied by a decrease in appetite, nervous disorders, seizures, ataxia, weakness, depression, impaired proprioception, muscle tremor, acute tubular necrosis of the kidneys, nausea, vomiting, respiratory and liver failure, hypoproteinemia, anxiety, conjunctivitis, intestinal colic, and diarrhea, often hemorrhagic. Tachycardia, pneumonia, hypertension, paresis or paralysis of the extremities, blood clots, necrosis of erythemal areas of the skin, and alopecia there were in the later stages of poisoning (Lamy, 1863; Paulet, 1863; Stadion, 1865; Skelley and Gabriel, 1964; Zook and Gilmore, 1967; Hakala, 1984; Ruhr and Andries, 1985; Waters et al., 1992; Volmer et al., 2006; Blain and Kazantzis, 2015). Mucopurulent discharge from the eyes was common in dogs, but febrile symptoms were absent in 94%, and rectal temperature often became subnormal as the disease progressed (Stadion, 1865; Zook and Gilmore, 1967; Waters et al., 1992). A one-year-old dog swallowed in agar thallium acetate tablets (5 mg/kg body weight) (Puschner et al., 2012). The following clinical signs (diarrhea, vomiting, weight loss, alopecia, dysphagia, ataxia, paresthesia, tremor, the esophagus enlargement followed by aspiration pneumonia, and several seizures episodes) developed within two to three weeks (Farley, 1965). Autopsy of a dog that died from thallium poisoning reveals necrotic changes throughout the gastrointestinal tract from the stomach to the large intestine (Hakala, 1984). The thallium concentration in the poisoned dogs' urine and liver samples reached 7.68 mg/L and 7.6 mg/kg accordingly. Here observed proteinuria and bilirubinuria, pulmonary edema, the spleen and necrotic damage increase in the myocardium, especially Purkinje fibers, and the damage to skeletal muscles, liver, and kidneys (Farley, 1965; Zook and Gilmore, 1967). In addition, these animals observed severe focal swelling of myelin sheaths containing granules with axon debris. In clinical syndromes of febrile and multisystem disease caused by poisoning with thallium compounds, signs of gastrointestinal tract disorders, respiratory organs, skin, and nerves prevail, which are typical for both humans and dogs (Zook and Gilmore, 1967).

Subcutaneous injections of thallium acetate in cats for seven to eight weeks (2–4 mg/kg per week) led to the onset of symptoms of hypotension and ataxia in these animals. Various neurological damage in the structure of sensory neurons and the death of central and peripheral axons results in these symptoms (Kennedy and Cavanagh, 1977). Thallium poisoning in cats was diagnosed based on the detection of severe and microscopic damage to internal organs and the presence of thallium in the urine or tissues of these animals (Zook et al., 1968). Cats poisoned with thallium compounds showed apathy, weakness, loss of strength, thirst, appetite loss, pulmonary edema, vomiting, hemorrhagic gastroenteritis, liver and kidney damage, and skin changes consistently accompanied by redness crusting, flaking, and alopecia (Zook et al., 1968). Cat liver found necrotic changes, and deposits of the iron-containing pigment hemosiderin occurred. The later poisoning symptoms were the next: following nephrosis, polyneuritis, hypersensitivity, brain tissues' focal edema and chromatolysis, degenerative changes in peripheral nerves (especially in the area of Ranvier interceptions), axonal swelling, necrosis of skeletal and the myocardium muscle fibers, tremor, ataxia, paresis, and contortion with signs of reflexive peripheral and central neuropathy. At the same time, approximately 20–30 mg/kg thallium led to the death of all experimental cats (Zook et al., 1968). Dogs and cats, poisoned accidently with rodenticides or ate thallium-poisoned animals, showed hemorrhagic gastroenteritis with liver and kidney damage, skin lesions with inflammation, bleeding, crusting and shedding of fur, and later lack of coordination, trembling, and paralysis (Zook and Gilmore, 1967; Zook et al., 1968; Schoer, 1984; Waters et al., 1992). At the same time, the following dysfunctions were observed at the organ level: kidneys (tubular necrosis with glomerular hyalinization); brain (focal edema and chromatolysis); peripheral nerves (degenerative changes); skeletal and cardiac muscle fibers (necrosis); lungs (edema and bronchopneumonia); and the intestines (hemorrhage and ulceration).

The thallium concentration in the liver of thallium-poisoned martens and badgers was from 4.7 to 57 mg/kg, corresponding to a concentration of 23–280 μM (Schoer, 1984). The thallium concentration on the first and third days after a single injection of Tl-acetate (50 mg/kg) in hamsters was 1330 and 470 μM in the kidneys and 180 and 140 μM in the liver (Aoyama et al., 1988). The dose of thallium salts in the aforementioned cases for martens and badgers and the body of hamsters and rats exposed to thallium salts was approximately 120–900 μmol/L, calculated on the total water content in the body (Aoyama et al., 1988; Aoyama, 1989; Leung and Ooi, 2000). An increase in the daily dose of thallium in sheep from 0.03 to 0.1 mg/kg of live weight led to a significant (10—20-fold) increase in the concentration of this metal in the spleen, muscles, heart, liver, and kidneys, and the last two organs, it was maximum (Schoer, 1984).

Chronic experiments on rabbits treated for six months with thallium carbonate or sulfate (0.25 mg/kg) resulted in the appearance of CNS damage in the first two months. There expressed the increased aggressiveness of animals, paralysis of the hind limbs (in some individuals), and the decreased concentration of SH-groups and albumins in the blood serum (Tikhova, 1967). These animals (already after six months) showed liver dystrophy and lymphoid infiltration of the gastric mucosa. Granular dystrophy and a sharp plethora of glomeruli are revealed there in the kidneys. Even a single injection of thallium provoked alopecia in sheep and angora rabbits (Uspenskaya, 1940; Thyresson, 1950). The epilating effect of thallium in animals increased in the presence of p-bromophenyl isothiocyanate (Gurevich, 1965). Daily exposure to thallium carbonate or sulfate (0.25 mg/kg, orally or subcutaneously for six months) in domestic rabbits or thallium acetate (4—8 mg/kg orally for seven days) in rats has resulted in behavioral disturbances (Brown et al., 1985; Schoer, 1984). Also, the rabbits observed aggressiveness, deformation of the hind limbs, diarrhea, hair loss, decreased activity of alkaline phosphatase and blood serum SH-groups, as well as histological changes in the membranes of the liver, kidneys, and stomach.

Prenatal and postnatal studies of the thallium sulfate effects on rats found that this metal disrupted the development of the rat vascular autonomic nervous system in decreasing alpha- and beta-adrenergic and muscarinic vasomotor reactivity (Rossi et al., 1987). The research on the thallium sulfate effects on rats found dystrophic changes in the brain, liver, and kidneys (Shabalina et al., 1980). The exposure to thallium in rats and mice led to severe protein dysfunction of the liver and epithelium of the convoluted renal tubules (Tikhova, 1967). The administration of thallium acetate to rats and rabbits caused baldness of skin areas to which paraffin heated to 55°C was applied (Gerasimov, 1959). Exposure of rats to $TlCl_3$ (50—60 mg/kg) provoked the development of anemia in them, reduced the concentration of iron in the blood, and caused hair loss (Fatemi et al., 2007).

White rats inhaled the combustion products of a magnesium alloy containing 1% thallium or copper-cadmium concentrate dust containing thallium. The weight lag, pneumonia, bronchitis, formed-fibrous nodules in the lungs, the decreased number of red blood cells, dystrophic changes in the lungs, liver, kidneys, spleen, heart, and brain, and the serum SH-groups' content decrease, the decreased concentration of albumins, and the increased concentration of α- and γ-globulins were there as a result (Roshchina, 1966, 1970; Burkhanov, 1980). Rats received in experiments in vivo 0.4–4.0 mg of thallium carbonate per kg of live weight. In the blood serum of these animals, the content of SH-groups decreased, the activity of alkaline phosphatase and catalase decreased, the concentration of albumin, hemoglobin, and iron decreased. The concentration of globulins, on the contrary, increased (Truhaut, 1960; Tikhova, 1967; Inturrisi, 1969b; Kamiljanov, 1993). Experiments with rats subjected to chronic thallium intoxication found damage to intracellular membranes in the kidneys, liver, brain, and other organs, including microvilli in the kidneys and the endoplasmic reticulum, the retina dystrophic alterations, as well as mitochondrial membranes due to massive swelling of mitochondria (Herman and Bensch, 1967; Danilewicz et al., 1979; Woods and Fowler, 1986; Shamshinova et al., 1990). At the same time, the blood urea content and the urine protein of these animals increased. Rat thallotoxicosis chronic form has the following symptoms: pelage alopecia, cataract, slow growth, gastric mucosa proliferation, impaired sexual function, paresis, rickets due to ossification disorders, and blood changes in clinical and chemical composition (Thyresson, 1950, 1952). It was shown in experiments with white mice males that the various thallium salts' toxicity is directly proportional to the amount of thallium itself, expressed in micromoles (LD100 20–30 mg/kg). This effect does not depend on the type of anion (inorganic or organic) in the thallium salt used in the experiments (Korsunskaya, 1970; Spiridonova and Shabalina, 1977a; Mulkey and Oehme, 1998). At the same time, white mice and rats were more susceptible to the thallium action than rabbits and guinea pigs (Korsunskaya, 1970).

The thallium content in the internal organs of rats subjected to a single injection of thallium sulfate increased in the order blood < brain < lungs, skeletal muscles < liver, spleen, testes, intestines < heart < kidneys (Barclay et al., 1953; Sabbioni et al., 1980; Cavailloles et al., 1983; Chandler and Scott, 1986; Ríos et al., 1989; Mulkey and Oehme, 1998, 2000). ^{204}Tl accumulated in all organs of rats, including the skin and especially in the hair follicles (Thyresson, 1951a, 1952). The ^{201}Tl distribution was weakly dependent on the age of male rats. It decreased half an hour after in the row of kidneys, heart > adrenal glands > lungs, aorta, liver > spleen, thymus, muscle > testes, brain, adipose tissue > erythrocytes, wool (Barclay et al., 1953; Llaurado et al., 1978, 1981; Madden et al., 1982). A similar series were found in experiments with rats treated with 20 mg/kg $TlNO_3$ (Leloux et al., 1987). The ^{201}Tl content in the organs of these rats exceeded that for ^{43}K in all cases (Llaurado et al., 1981). In this case, a redistribution of the isotope occurred after 2.5 h. Its concentration decreased in the kidneys, heart, and adrenal glands, but at the same time, it increased markedly in the muscles, testes, brain, and adipose tissue. The concentration of ^{201}Tl and ^{43}K in the internal organs of these rats increased during the exercise of these animals (Llaurado et al., 1978, 1981). However, the statistical processing of many studies did not reveal a complete correlation of the distribution of thallium isotopes in the myocardium with a similar distribution of potassium (Rinck and Beckmann, 1982). Similar experiments did carry out with white mice female subjected to a single injection of $^{204}Tl_2SO_4$ (5 mg/kg). The same $^{204}Tl^+$ distribution did find 1 h after in the female organs, which increased in the series brain < blood, lungs < heart, skeletal muscle < liver < stomach < femur < appendages < large intestine < kidney with a maximum in the cortex (Andre et al., 1960). Like above results were obtained in experiments with mice injected with Tl_2SO_4 (80–160 mg/kg) in the stomach (Achenbach et al., 1980) and pregnant rats exposed to 2 μg thallium sulfate and 50 μCi ^{201}Tl (Rade et al., 1982). However, the metal after eight days was redistributed toward a noticeable increase in the $^{204}Tl^+$ content in the brain, appendages, wool hairs, and teeth, while maintaining a high concentration in the kidney cortex and the large intestine. Bones, eye lenses, and the tongue epithelium showed noticeable thallium accumulation (Andre et al., 1960). At the same time, thallium showed poorly excretion from the myocardium and skeletal muscles. However, according to other data obtained in the study of the ^{204}Tl distribution, the general part of the ingested thallium is excreted from the body of rats with feces and, to a lesser extent, through the kidneys (Thyresson, 1951a; Barclay et al., 1953). The excretion (%) of 70 silver, 46 copper, 45 thallium, 38 lead, 32 mercury, 16 cadmium, and 16 arsenic was detected four days later after administering heavy metal salts to rats (Gregus and Klaassen, 1986). However, other experiments showed that the principal amount of thallium (34%) did excrete with feces and possibly since pure soluble salt (TlCl) did apply in this experiment (Gregus and Klaassen, 1986). The thallium content (%) increased 2 h after administration in the organs of experimental rats in the row kidney

(5.4) < liver (7.6) < intestine (8.7). Thallium was found to be excreted from the body of humans, rats, and mice through kidneys and a lesser extent with feces (Andre et al., 1960; Truhaut, 1960; Ziskoven et al., 1983; Leloux et al., 1987; Galván-Arzate, Santamaría, 1998; Mulkey and Oehme, 1998; Pelclová et al., 2009). The excretion of thallium from the body of rats and dogs in experiments in vivo occurred mainly through the kidneys. It accelerated with potassium concentration increase in food or stimulation by diuretics of the excretion of water and ions (Na^+, K^+, $C1^-$) (Stadion, 1865; Gehring and Hammond, 1967; Lameijer and van Zwieten, 1977). Removing thallium from the body of rats and dogs showed first-order kinetics (Gehring and Hammond, 1967).

1.3.3 Ways to prevent thallium poisoning

The strategy of preventing thallotoxicosis for a long time was to try to convert thallium that entered the body into less toxic compounds, as well as to stimulate the natural excretion of this metal through the kidneys and gastrointestinal tract (Barclay et al., 1953; Reed et al., 1963; Schoer, 1984; Appenroth et al., 2001). Gastric and intestinal lavage (lavage), ingestion of activated charcoal (3—5 mg/kg for 5—10 days) and potassium chloride, brewer's yeast, and vitamin B1 used to achieve this purpose (Sessions and Goren, 1947; Thyresson, 1950, 1951a; Lund, 1956; Chamberlain et al., 1958; Smith and Doherty, 1964; Taber, 1964; Zook and Gilmore, 1967; Zook et al., 1968; Rauws et al., 1969; Paulson et al., 1972; Patterson, 1975; Spiridonova and Shabalina, 1977a; Thompson, 1981; Hakala, 1984; Meggs et al., 1997; Hoffman et al., 1999; Hoffman, 2003; Ramsden, 2007; Rusyniak et al., 2002, 2010; Livanov et al., 2011). On the other hand, some reagents did apply to decrease thallium toxicity. These reagents include: sulfur-containing chelators (BAL - 2,3-dimercaptopropanol, N-acetylcysteine, methionine), ion exchange resin (Dowex 50), thiomerine diuretic, EDTA, dimercaptopropane sulfonic acid, dithiocarb, sodium thiosulfate, used to reduce toxic heavy metals, dithizone, increased the thallium clearance. However, the above chemicals were ineffective for treating humans and animals (dogs, cats, rats, guinea pigs, rabbits, mice) or even increased the thallium toxicity in experiments with animals due to the redistribution of the metal in the central nervous system. Simultaneous intravenous administration of $TlNO_3$ and dithiocarb to rats increased thallium concentration in the brain by 5—10 times compared with a similar injection of thallium self (Rauws et al., 1969). Dimercaptol and mercaptosuccinic acid did not affect the distortion of taste and odor perception in rats treated with 5 mg/kg thallium sulfate, in contrast to similar experiments with lead acetate where these reagents were effective (Peele et al., 1986, 1987).

Attempts to use sodium polystyrene sulfonate for binding Tl^+ were ineffective. In experiments in vitro, the binding of Tl^+ to this agent in media with KCl significantly decreased compared to the binding of activated carbon and Prussian blue to this metal (Hoffman et al., 1999). On the other hand, the toxicity of thallium salts noticeably weakened when animals, treated by cystine, methionine, betaine, S-aminoethylisothiouronium, and S-aminopropylisothiuronium bromides, as well as the racemic γ-tocopherol derivative (Gross et al., 1948; Meyer and Tal, 1957; Stavinoha et al., 1959; Appenroth et al., 2001). Cystine excess partially reduced thallium toxicity in chronic poisoning (Abramova and Cherny, 1977). Previously, a similar cystine effect showed for white rats receiving chronically 2–3 mg/kg thallium acetate (Gross et al., 1948). Ehallium-induced hair loss virtually disappeared in the cystine-treated animals. However, according to other data (Lund, 1956), cystine and cysteine did not reduce thallium toxicity for rats.

On the other hand, the thallium secretion from the human body and experimental animals increased in experiments using KCl, dimercaptosuccinic acid, deferiprone, desferrioxamine, penicillamine (cuprenyl), and diethyldithiocarbamate. Still, the price for this was thallium concentration increase in the blood cytoplasm and redistribution of this metal in the body vital areas, and especially the nervous system and, as a consequence, exacerbation of neurological symptoms such as chorea-like movements (Lund, 1956; Schwetz et al., 1967; Kamerbeek et al., 1971a; Patterson, 1975; Thompson, 1981; Schoer, 1984; Chandler and Scott, 1986; Leloux et al., 1990; Rìos and Monroy-Noyola, 1992; Hoffman, 2003; Amiri et al., 2007; Fedotov, 2008; Adams et al., 2009; Goldschmidt et al., 2010; Wanger et al., 2012). Levodopa reduced the neurological thallium poisoning sequelae with KCl therapy in the case of a 37-year-old man (Patterson, 1975). Desferrioxamine reduced the thallium concentration in the body of rats treated with 50–60 mg/kg $TlCl_3$ (Fatemi et al., 2007). Thallium excretion from the body of rats receiving thallium sulfate (12 mg/kg) was enhanced on a diet enriched with potassium, but not sodium, and was higher than in similar experiments with furosemide (Lameijer and van Zwieten, 1978). Furosemide accelerated the thallium excretion in experiments with rats treated with thallium sulfate (12 mg/kg) in contrast to potassium-sparing diuretics (triamterene and canrenoate) (Lameijer and van Zwieten, 1979; Lameijer, 1979).

Daily injection of 40—60 mg/kg of thioacetamide into the ear vein prevented the death of rabbits who received 5 mg/kg of thallium sulfate orally (Edelman and Lipiec, 1955). In this regard, attempts to apply thioacetamide in the thallotoxicosis human treatment made (Edelman, 1955; Joszpa and Batko, 1957). The therapeutic use of potassium chloride or its other preparations successfully treated people poisoned with thallium compounds, the excretion of which in the urine increased two to three times (Papp et al., 1969; Hoffman, 2003; Livanov et al., 2013). However, the KCl use in this therapy is limited due to the probable onset or aggravation of neurological symptoms, consisting in an increase in dysesthesia, weakness of the distal leg muscles up to their paralysis with the disappearance of knee reflexes, as well as the possible the thallium redistribution to vital organs (Innis and Moses, 1978; Hoffman, 2003). In addition, attempts with thallium poisoning treatment consisted of the use of 3%—5% solutions (taken orally) of NaI or KI salts, which form poorly soluble thallium iodide to reduce the metal absorption from the intestine (Livanov et al., 2011, 2013).

The currently used treatment in poisoning with thallium salts is, first of all, its removal from the body: ingestion of activated carbon, washing the gastrointestinal tract and carrying out forced diuresis, hemofiltration, hemosorption, and hemodialysis, as well as the use of enterosorption and the drug (unitol), that promotes the excretion of thallium from the body (Richelmi et al., 1980; Thompson, 1981; Treshchinsky et al., 1989; Lyubchenko et al., 2004; Kuzmich, 2006; Livanov et al., 2013; Sheiman et al., 2014). The maximum effect did observe with hemodialysis treatment (Treshchinsky et al., 1989). The plasmapheresis used for the treatment of patients poisoned with thallium compounds gave a positive result. The thallium concentration in the blood of these patients began to decrease rapidly, and their condition improved markedly within a few days (Livanov et al., 2013). Painful neuropathy in thallium poisoning could be correct with using the pharmaceutical agents: antidepressants (tricyclic antidepressants), antiepileptic drugs (carbamazepine, gabapentin, diphenylhydantoin), and a serotonin/norepinephrine reuptake inhibitor (duloxetine) (Huang et al., 2012).

1.3.4 Prussian blue using to reduce the thallotoxicosis effects

A large number of scientific studies are devoted to using Prussian blue to prevent thallium poisoning. This compound is a continuous series of ferriferrocyanides from $KFe^{III}[Fe^{II}(CN)_6]$ (soluble form) through intermediate colloidal forms to $Fe^{III}_4[Fe^{II}(CN)_6]_3$ (insoluble form) (Kamerbeek et al., 1971b; Schoer, 1984; Hoffman et al., 1999; Hoffman, 2003; Liu and Liao, 2021). Colloidal forms proved to be the most effective in the thallium poisoning treatment in humans and domestic animals, as well as to do the thallotoxicosis consequences in experiments with rats and mice (Dvorak, 1969, 1970; Günther, 1971; Kamerbeek, 1972; Hakala, 1984; Treshchinsky et al., 1989; Díaz et al., 1990; Kravzov et al., 1993; Meggs et al., 1997). The soluble form affinity of this complex for the monovalent cation increases with its radius, and for this reason, this one for Tl^+ (0.147 nm) turned out to be higher than the affinity for K^+ (0.133 nm). The binding of Tl^+ with $Fe^{III}_4[Fe^{II}(CN)_6]_3$ in vitro was maximum at physiological pH (7.5) in maximally watered forms of this substance with 17 water molecules and for particles with a size of 32–90 μm (Yang et al., 2008). A similar binding did find in reacting Tl^+ with sodium hexacyanoferrate ($Na_4[Fe^{II}(CN)_6]_3$) (Dvorak, 1970). The absorption parameters found in experiments in vitro for Prussian blue and activated carbon are in good agreement with the efficacy of these substances in the thallotoxicosis treatment in vivo (Lehmann and Favari, 1984). The instability constant of the $[TlFe(CN)_6]_3^-$ complex is $6 \cdot 10^{-4}$ (Korenman, 1960). The Prussian blue use significantly reduced the consequences of poisoning with thallium compounds in humans and animals. This reagent prevented the thallium accumulation through the gastrointestinal tract, reduced the metal concentration in the body, accelerated the thallium excretion in feces, and reduced mortality from poisoning with this metal (Kamerbeek et al., 1971b; Thompson, 1981; Lehmann and Favari, 1985; Díaz et al., 1990; Mulkey and Oehme, 2000; Hoffman, 2003; Locatelli at al., 2003; Rusyniak et al., 2003, 2010; Ibrahim et al., 2006; Pelclová et al., 2009; Al Hammouri et al., 2011; Montes et al., 2011). The thallium content in various organs (kidneys, heart, muscles, spleen, liver, lungs, and brain) in rats receiving 39 mg/kg of thallium sulfate decreased visibly on the third or fourth day in experiments with Prussian blue (Kamerbeek et al., 1971b; Manninen et al., 1976). Insofar as KCl prevents Prussian blue with Tl^+ complexation, combined use of these two reagents in thallotoxicosis treatment is not recommended (Kamerbeek, 1972). Prussian blue nanoparticles showed good biocompatibility, and they are suitable adsorbents for thallium poisoning antidotes (Busquets and Estelrich, 2020). Recent experiments carried out with Wistar rats intraperitoneally injected to Tl (16 mg/kg) with simultaneous administrating metallothionein and Prussian blue (Anaya-Ramos et al., 2021). Prussian blue effectively reduced thallium concentration in all rat organs, including the brain, with the maximum effect in the presence of metallothionein.

The half-life of thallium from the body without treatment was found to be about eight days, and with the use of Prussian blue (3 g orally three times a day or 150—250 mg/kg daily) in combination with the introduction of a 15% mannitol solution is reduced to three days (Kamerbeek, 1972; Hoffman, 2003; Livanov et al., 2013). The Prussian blue use in combination with the diuretic furosemide or penicillamine (cuprenil) increased the thallium excretion from the rat's body through the kidneys or the gastrointestinal tract with activated carbon (Lehmann and Favari, 1985; de Groot and van Heijst, 1988; Leloux et al., 1990; Rìos and Monroy-Noyola, 1992; Vrij et al., 1995; Hoffman, 2003; Fedotov, 2008; Riyaz et al., 2013). On the other hand, the Prussian blue protective effect did not increase after intraperitoneal administration of N-acetylcysteine in experiments with mice injected subcutaneously with 70 mg/kg (LD_{90}) of thallium acetate (Meggs et al., 1997). Thallium poisoning did treat successfully using Prussian blue combined with both plasma exchange (Qi et al., 2019) and multi-dose activated charcoal with airway protection (Yumoto et al., 2017). It was reported recently about a successful treatment example of thallium poisoning (Zavaliy et al., 2021). Patients got daily oral ferrocin ($K_4Fe(CN)_6$) combined with laxative preparations and sodium dimercaptopropane sulfonate intramuscular injections. Microcirculation preparations did use to correct neurological disorders (Zavaliy et al., 2021). Another successful example was using Prussian blue treatment in combination with blood purification treatments (Lin et al., 2019). The thallium poisoning treatment decreased blood and urine thallium concentrations and reduced neurological symptoms (Lin et al., 2019; Zavaliy et al., 2021).

Optimal therapy in the thallium intoxication treatment was achieved with the simultaneous use of Prussian blue with the diuretics furosemide and mannitol, followed by hemodialysis (Spiridonova and Shabalina, 1977a; Galván-Arzate, Santamaría, 1998; Worthley, 2002a,b; Blain and Kazantzis, 2015; Campbell et al., 2016). Immediate Prussian administration produced a positive therapeutic effect combined with intravenous administration of KCl solution, diuresis, and hemodialysis to a 38-year-old thallium-poisoned woman (Malbrain et al., 1997). The impact of rapid thallium withdrawal from the digestive tract did not achieve when taking Prussian blue alone. In this case, it is necessary to decrease the thallium blood concentration by using combined hemoperfusion, continuous venovenous hemofiltration, and a neuroprotective agent 2,3-dimercaptopropane-1-sulfonate (Mulkey and Oehme, 2000; Sun et al., 2012; Huang et al., 2014). By itself, 2,3-dimercaptopropane-1-sulfonate reduced the thallium concentration in the blood of rats but did not have any effect on the concentration in other organs of these animals (Mulkey and Oehme, 2000). Prussian blue has proven to be a safe and effective method to significantly reduce ^{201}Tl radiation exposure during thallium isotope myocardial scintigraphy (Bhardwaj et al., 2006). The simultaneous use of Prussian blue with enterosorbents, potassium preparations, and oral water load produced a positive effect (Vorobyev, 2003). However, the Prussian blue use in combination with diuretics on the third day after the student's case, poisoned with Clerici's solution, did not give a positive result (Hologgitas et al., 1980). Prussian blue and activated charcoal therapy immediately after ingesting an unknown amount of thallium sulfate rodenticide by a 36-year-old man was ineffective and could not prevent his death (Riyaz et al., 2013).

Numerous animal studies exist in using Prussian blue and ferrocin to decrease thallium toxicity. Their result was the following positive benefits: the animal mortality reducing, the increased excretion of thallium and its isotopes (^{201}Tl, ^{202}Tl) from their body with urine and feces, and the thallium concentration significant decreasing in internal organs (kidneys, liver, spleen, brain) (Heydlauf, 1969; Kamerbeek et al., 1971b; Rauws, 1974; Sabbioni et al., 1982; Borisov et al., 1984; Rusyniak et al., 2003; Ibrahim et al., 2006; Montes et al., 2007). The Prussian blue use in experiments with pregnant rats receiving thallium sulfate (10 mg/kg) and 50 μCi ^{201}Tl reduced the mortality of animals. It prevented the thallium accumulation in the organs of the mother and fetus, and it promoted the removal of accumulated metal from them (Sabbioni et al., 1982). The application with L-methionine or L-cysteine, regardless of the presence of Prussian blue, increased the mortality of rats exposed to thallium acetate (Montes et al., 2007, 2011). The combined use of 2,3-dimercaptosuccinic acid or D-penicillamine (cuprenil) in combination with Prussian blue or preparations based on it (ferrocin) significantly reduced the thallotoxicosis manifestations in rats and increased the survival rate of animals by 40%–45% (Barroso-Moguel et al., 1994; Rusyniak et al., 2003; Fedotov, 2008). At the same time, the thallium neurotoxic effect on the organism of these animals decreased with a significant decrease in the number of damaged Purkinje cells. The combined use of ferrocin and zincacine (zinc-trisodium salt of diethylenetriaminepentaacetic acid) against thallotoxicosis was more effective than similar experiments with cuprenil and ferrocin. In this case, compared with the control (without chelating agents), the deposition of thallium in the rat brain significantly decreased, which, respectively, was equally excreted through the kidneys and the gastrointestinal tract, and not only through the kidneys alone, as in experiments with cuprenil and ferrocin (Fedotov, 2008). At the same time, the use of these complexones (cuprenil and zincacine) reduced thallium deposition in the rat skeleton and spleen. Using a nanoemulsion containing p-tert-butylthiacalix[4]arene significantly reduced the content of ^{99m}Tc, ^{131}I, and ^{201}Tl isotopes on artificial skin that mimics human skin much as possible and on the skin of experimental rats (Rana et al., 2014; Sharma et al., 2017). The tridentate chelator deferasirox resulted in the thallium elimination from the body of rats receiving 20–140 μM $TlCl_3$ daily for a month. It significantly reduced the manifestations of thallotoxicosis in these animals, expressed in red coloration around the eyes, greenish macular liver damage, weakness, hair, and weight loss (Saljooghi et al., 2011).

1.3.5 The thallium poisoning diagnostics

Early diagnosis of thallium poisoning is often difficult due to the simultaneous involvement of various organs (heart, kidneys) in this process and the nervous and digestive systems. Therefore, it is necessary to diagnose nerve conduction and microscopic analysis of hair to detect thallium poisoning. In this regard, daily urine analysis is instrumental, in which increased concentrations of thallium begin to be determined within an hour after the metal ingestion into the body (Lyubchenko et al., 2004; Ibrahim et al., 2006; Luzanova et al., 2008; Rusyniak et al., 2010; Sheiman et al., 2014; Livanov et al., 2015). Other thallium poisoning diagnostic methods can be electroretinography, delta-aminolevulinic acid, and capillaroscopy (Shabalina et al., 1996). We could talk about significant thallium poisoning the body if the metal urine concentrations were about 15 times higher than the usual content of this metal 0.3–0.4 μg/L (Blain and Kazantzis, 2015). It should note that the second peak exists in the thallium concentration increase in the blood serum on days six to eight after a gradual decrease in the thallium urine concentration, in many cases. The repeated increase in the thallium urine excretion rate is associated with the reentry of deposited thallium into the blood from the internal organs (Sheiman et al., 2014). The determination of thallium in the blood may not provide the necessary information since this metal quickly distributes from the blood to various organs (Luzanova et al., 2008).

In the differential diagnosis of the early stages of thallium poisoning, it is necessary to take into account the presence of such signs as the presence of paresthesia of the fingers, the green coloration of urine in the first hours after thallium ingestion in the body, a blood test for thallium, a violation of the bioelectrical activity of the retina (Gridnev et al., 1971; Shabalina and Spiridonova, 1976; Livanov et al., 2011, 2015). In case of poisoning, the thallium urine concentration can reach 167–4600 μg/L (Lyubchenko et al., 2004; Luzanova et al., 2008). Unfortunately, thallium poisoning in many cases is diagnosed with a significant delay, in the late stages of thallotoxicosis and often two to three weeks after this metal ingestion into the human body, when treatment becomes less effective (Livanov et al., 2013; Cvjetko et al., 2010; Blain and Kazantzis, 2015). Initially, dermatological signs such as the plaque formation on the palms and soles and acne on the face may appear, but they are not specific for human poisoning with thallium compounds (Tromme et al., 1998; Peter and Viraraghavan, 2005). On the contrary, painful paresthesias, brown or black-purple formations in the bulb and root part of the hair shaft, excretion of thallium in the urine (up to 300 μg/L) after three to four days later, poisoning can be used as an early diagnosis of such poisoning with this metal (Moeschlin, 1980; Pavlova et al., 2012; Blain and Kazantzis, 2015).

The alopecia associated with thallium poisoning did often detect after two weeks, and the main symptoms of thallium poisoning appear already four days after thallium ingestion into the body. It becomes essential in the differential diagnosis of this poisoning to pay attention to a combination of many symptoms that may cause suspicion for poisoning with thallium compounds (Worthley, 2002a,b; Livanov et al., 2013; Campbell et al., 2016). For example, after getting its salts with water or food in the first hours, thallium can be detected in the urine, which acquires a greenish coloration, and this can serve as one of the early diagnostic methods for poisoning with this metal (Malbrain et al., 1997; Livanov et al., 2013). Thallium poisoning was detected by analyzing individual hairs using the combined mass spectrometry method and inductively coupled plasma with laser ablation (Ash and He, 2018). A new method for diagnosing thallium poisoning was proposed, which consists of the simultaneous detection of K^+ and Tl^+ in aqueous solutions and serum using gold nanoparticles modified with single-stranded DNA (Huang et al., 2020).

1.4 Methods of thallium analysis in biological samples

One of the ways to diagnose thallium in urine is to use a reagent containing sodium bismuthate, nitric acid, and sodium iodide (Taber, 1964). A fractional method did propose to analyze the thallium content in various organs. This method detects colored compounds with different organic compounds, dithizone and brilliant green (Krylova, 1971). Determination of thallium content in biological samples using various analytical methods has acquired great interest due to its high toxicity, especially in the case of monovalent thallium compounds (Das et al., 2006). Different analytical techniques are under use to determine Tl^+. They are the next: atomic absorption spectroscopy, emission spectral analysis, inductively coupled plasma mass spectrometry, photometric determination of Tl^{3+} using a diamond green indicator after oxidation of Tl^+ with bromine, analysis using Reinecke salt, neutron activation analysis, X-ray fluorescence analysis, mass spectroscopy during desorption in an electric field, and gas chromatography (Bykhovskaya and Babina, 1956; Muravyova, 1960; Shvartsev and Gomonova, 1975; Schoer, 1984; Luzanova et al., 2008; Gileva et al., 2016). A susceptible voltammetric method did offer to determine thallium in samples (up to 0.8 ng/mL) using a specially synthesized polymer and carbon nanotubes (Nasiri-Majd et al., 2015). The determination of thallium in blood samples can do by a modified graphite furnace atomic absorption spectrometry (Solovyev et al., 2011). Currently, a very accurate diagnostic technique (1.5 μg/L Tl) created to assess the thallium content in the body, which requires only 50 μL of blood (Lehner et al., 2013), which makes it possible to study small animals (birds, reptiles, amphibians, and small mammals). Another modern method for determining the Tl^+ content in samples is using a combined fluorescence method and nanoparticles containing gold atoms with a Tl^+ determination threshold of up to 4.6 μM (Hoang et al., 2016). An assessment of the toxic effects of thallium on aquatic organisms created using the algae *Pseudokirchneriella subcapitata, Ceriodaphnia dubia,* and the black fathead *Pimephales promelas* (Rickwood et al., 2015). The inductively coupled plasma-mass spectrometry detection method invented to assess the intracellular thallium level (0.097 μg/L) in the ultrasonic lysed hepatoma HepG2 cells (Krivohlavek et al., 2021).

References

Abdel-Dayem, H.M., Scott, A.M., Macapinlac, H.A., El-Gazzar, A.H., 1994. Role of 201Tl chloride and 99mTc sestamibi in tumor imaging. In: Freeman, L.M. (Ed.), Nuclear Medicine Annual. Raven Press, New York, pp. 181–234.

Abramova, Z.I., Cherny, Z.K., 1977. [Thallium and its compounds]. In: Lazarev, N.V., Gadaskina, I.D. (Eds.), Harmful Substances in Industry. Leningrad, vol. 3, pp. 423–427 (Russian). https://bookree.org/reader?file=542086&pg=4.

Achenbach, C., Hauswirth, O., Heindrichs, C., Ziskoven, R., Köhler, F., et al., 1980. Quantitative measurement of time-dependent thallium distribution in organs of mice by field desorption mass spectrometry. J. Toxicol. Environ. Health A 6 (3), 519–528. https://doi.org/10.1080/15287398009529870.

Adams, J.B., Baral, M., Geis, E., Mitchell, J., Ingram, J., et al., 2009. Safety and efficacy of oral DMSA therapy for children with autism spectrum disorders: Part A - medical results. BMC Clin. Pharmacol. 9 (16), 1–22. https://doi.org/10.1186/1472-6904-9-16.

Al Hammouri, F., Darwazeh, G., Said, A., Ghosh, R.A., 2011. Acute thallium poisoning: series of ten cases. J. Med. Toxicol. 7 (4), 306–311. https://doi.org/10.1007/s13181-011-0165-3.

Alves, R.I., Sampaio, C.F., Nadal, M., Schuhmacher, M., Domingo, J.L., Segura-Muñoz, S.I., 2014. Metal concentrations in surface water and sediments from Pardo River, Brazil: human health risks. Environ. Res. 133, 149–155. https://doi.org/10.1016/j.envres.2014.05.012.

Amiri, A., Fatemi, S.J., Fatemi, S.N., 2007. Removal of thallium by combining desferrioxamine and deferiprone chelators in rats. Biometals 20 (2), 159–163. https://doi.org/10.1007/s10534-006-9023-1.

Anaya-Ramos, L., Díaz-Ruíz, A., Ríos, C., Mendez-Armenta, M., Montes, S., et al., 2021. The acute systemic toxicity of thallium in rats produces oxidative stress: attenuation by metallothionein and Prussian blue. Biometals. https://doi.org/10.1007/s10534-021-00343-8.

Andre, T., Ullberg, S., Winqvist, G., 1960. The accumulation and retention of thallium in tissues of the mouse. Acta Pharmacol. Toxicol. 16, 229–234. https://doi.org/10.1111/j.1600-0773.1960.tb01207.x.

Aoyama, H., 1989. Distribution and excretion of thallium after oral and intraperitoneal administration of thallous malonate and thallous sulfate in hamsters. Bull. Environ. Contam. Toxicol. 42 (3), 456–463. https://doi.org/10.1007/BF01699976.

Aoyama, H., Yoshida, M., Yamamura, Y., 1988. Induction of lipid peroxidation in tissues of thallous malonate-treated hamster. Toxicology 53 (1), 11–18. https://doi.org/10.1016/0300-483x(88)90232-6.

Appenroth, D., Karge, E., Kiessling, G., Wechter, W.J., Winnefeld, K., Fleck, C., 2001. LLU-alpha, an endogenous metabolite of gamma-tocopherol, is more effective against metal nephrotoxicity in rats than gamma-tocopherol. Toxicol. Lett. 122 (3), 255–265. https://doi.org/10.1016/s0378-4274(01)00368-x.

Ash, R.D., He, M., 2018. Details of a thallium poisoning case revealed by single hair analysis using laser ablation inductively coupled plasma mass spectrometry. Forensic Sci. Int. 292, 224–231. https://doi.org/10.1016/j.forsciint.2018.10.002.

Barclay, R.K., Peacock, W.C., Karnofsky, D.A., 1953. Distribution and excretion of radioactive thallium in the chick embryo, rat, and man. J. Pharmacol. Exp. Ther. 107 (2), 178–187.

Barroso-Moguel, R., Villeda-Hernández, J., Méndez-Armenta, M., Ríos, C., Monroy-Noyola, A., 1994. Combined D-penicillamine and Prussian blue as antidotal treatment against thallotoxicosis in rats: evaluation of cerebellar lesions. Toxicology 89 (1), 15–24. https://doi.org/10.1016/0300-483x(94)90129-5. PMID: 8178320.

Baturin, G.N., Dubinchuk, V.T., 2009. [Composition of ferromanganese nodules from Riga bay (Baltic Sea)]. Oceanology. 49(1): 111-120. Russian.

Baturin, G.N., 2004. [Thallium in oceanic phosphorites]. Doklady Earth Sciences. 394(1): 68-72. Russian.

Bhardwaj, N., Bhatnagar, A., Pathak, D.P., Singh, A.K., 2006. Dynamic, equilibrium and human studies of adsorption of 201Tl by Prussian blue. Health Psychol. 90 (3), 250–257. https://doi.org/10.1097/01.HP.0000180771.66884.d3.

Bibikova, V.I., Darvoyd, T.I., 1959. Thallium. In: Bibikova, V.I., Gornova, Z.A., Darvoyd, T.I., Kasatkina, N.A., Nikitina, Z.N., et al. (Eds.), Scattered Metals (Indium, Gallium, Thallium, Rhenium). Areas of Mastered and Possible Application. Moscow, pp. 27–40 (Russian). https://rusneb.ru/catalog/000200_000018_rc_4960747/.

Blain, R., Kazantzis, G., 2015. Thallium. In: Nordberg, G.F., Fowler, B.A., Nordberg, M. (Eds.), Handbook on the Toxicology of Metals. Volume II: Specific Metals. Elsevier, Amsterdam, pp. 1229–1240.

Bocio, A., Nadal, M., Garcia, F., Domingo, J.L., 2005. Monitoring metals in the population living in the vicinity of a hazardous waste incinerator: concentrations in autopsy tissues. Biol. Trace Elem. Res. 106 (1), 41–50. https://doi.org/10.1385/BTER:106:1:041.

Borgmann, U., Cheam, V., Norwood, W.P., Lechner, J., 1998. Toxicity and bioaccumulation of thallium in Hyalella azteca, with comparison to other metals and prediction of environmental impact. Environ. Pollut. 99 (1), 105–114. https://doi.org/10.1016/s0269-7491(97)00181-4.

Borisov, V.P., Seletskaya, L.I., Skomorokhova, T.N., Popov, V.A., 1984. [The effectiveness of ferrocin in reducing the resorption of radioactive thallium]. Med. Radiol. 29 (9), 15–18 (Russian. PMID: 6482701).

Brewer, E., Haggerty, R.J., 1958. Toxic hazards. Rat poisons. III. Thallium, strychnine & ANTU. N. Engl. J. Med. 259 (21), 1038–1040. https://doi.org/10.1056/NEJM195811202592115.

Brockhaus, A., Dolgner, R., Ewers, U., Krämer, U., Soddemann, H., Wiegand, H., 1981. Intake and health effects of thallium among a population living in the vicinity of a cement plant emitting thallium containing dust. Int. Arch. Occup. Environ. Health 48 (4), 375–389. https://doi.org/10.1007/BF00378686.

Brown, D.R., Callahan, B.G., Cleaves, M.A., Schatz, R.A., 1985. Thallium induced changes in behavioral patterns: correlation with altered lipid peroxidation and lysosomal enzyme activity in brain regions of male rats. Toxicol. Ind. Health 1 (1), 81–98. https://doi.org/10.1177/074823378500100109.

Budrin, Y.P., Meshcherskaya, K.A., 1975. [Poisoning with thallium compounds]. Sud. Med. Ekspert. 4, 37–38 (Russian). https://www.forens-med.ru/book.php?id=963.

Bugarin, M.G., Casas, J.S., Sordo, J., Filella, M., 1989. Thallium (I) interactions in biological fluids: a potentiometric investigation of thallium(I) complex equilibria with some sulphur-containing amino acids. J. Inorg. Biochem. 35 (2), 95–105. https://doi.org/10.1016/0162-0134(89)80002-9.

Bulavintseva, O.A., 1975. [The effect of thallium on the activity of human blood enzymes]. Features of resuscitation in acute poisoning. Nauchn. Tr. Irkutsk. Med. Inst. 127, 159–161 (Russian).

Burkhanov, A.I., 1980. [To the assessment of the pathogenic effect of mixed dust of cadmium production]. J. Occup. Health Ind. Ecol. (12), 39–41 (Russian). https://www.elibrary.ru/item.asp?id=21474436.
Busquets, M.A., Estelrich, J., 2020. Prussian blue nanoparticles: synthesis, surface modification, and biomedical applications. Drug Discov. Today 25 (8), 1431–1443. https://doi.org/10.1016/j.drudis.2020.05.014.
Bykhovskaya, M.S., Babina, M.D., 1956. [Determination of the content of lead, zinc and thallium in the air by the spectrographic method]. Hyg. Sanit. (7), 26–30 (Russian). https://cyberleninka.ru/article/n/opredelenie-soderzhaniya-svintsa-tsinka-i-talliya-v-vozduhe-spektrografifeskim-metodom.
Campbell, C., Bahrami, S., Owen, C., 2016. Anagen effluvium caused by thallium poisoning. JAMA Dermatol. 152 (6), 724–726. https://doi.org/10.1001/jamadermatol.2016.0194.
Cavailloles, F., Vitaux, F., Petiet, A., Vitaux, J., 1983. Thallium uptake in fasted and nonfasted rats. Eur. J. Nucl. Med. 8 (2), 87–88. https://doi.org/10.1007/BF00252563.
Cavanagh, J.B., 1979. The 'dying back' process. A common denominator in many naturally occurring and toxic neuropathies. Arch. Pathol. Lab. Med. 103 (13), 659–664.
Cavanagh, J.B., 1991. What have we learnt from Graham Frederick Young?: reflections on the mechanism of thallium neurotoxicity. Neuropathol. Appl. Neurobiol. 7 (1), 3–9. https://doi.org/10.1111/j.1365-2990.1991.tb00687.x.
Cavanagh, J.B., Fuller, N.H., Johnson, H.R., Rudge, P., 1974. The effects of thallium salts, with particular reference to the nervous system changes. A report of three cases. Q. J. Med. 43 (170), 293–319.
Chamberlain, P.H., Stavinoha, W.B., Davis, H., Kniker, W.T., Panos, T.C., 1958. Thallium poisoning. Pediatrics 22 (6), 1170–1182.
Chan, W.S., Routh, J., Luo, C., Dario, M., Miao, Y., et al., 2021. Metal accumulations in aquatic organisms and health risks in an acid mine-affected site in South China. Environ. Geochem. Health 43 (11), 4415–4440. https://doi.org/10.1007/s10653-021-00923-0.
Chandler, H.A., Scott, M., 1986. A review of thallium toxicology. J. Roy. Nav. Med. Serv. 72 (2), 75–79.
Chauncey Jr., D.M., Schelbert, H.R., Halpern, S.E., Delano, F., McKegney, M.L., et al., 1977. Tissue distribution studies with radioactive manganese: a potential agent for myocardial imaging. J. Nucl. Med. 18 (9), 933–936.
Cheng, Y., Chui, S.S., Wang, X.D.T., Jaenicke, S., Chuah, G.K., 2019. One-pot synthesis of layered disodium zirconium phosphate: crystal structure and application in the remediation of heavy-metal-contaminated wastewater. Inorg. Chem. 58 (19), 13020–13029. https://doi.org/10.1021/acs.inorgchem.9b01938.
Chukhlovina, M.L., 1999. [Medical and hygienic aspects of thallium neurotoxicity]. Hyg. Sanit. 4, 38–40 (Russian). https://www.elibrary.ru/item.asp?id=22787601.
Clarkson, T.W., 1987. Metal toxicity in the central nervous system. Environ. Health Perspect. 75 (1), 59–64. https://doi.org/10.1289/ehp.877559.
Cruz-Hernández, Y., Ruiz-García, M., Villalobos, M., Romero, F.M., Meza-Figueroa, D., et al., 2018. Fractionation and mobility of thallium in areas impacted by mining-metallurgical activities: identification of a water-soluble Tl(I) fraction. Environ. Pollut. 237, 154–165. https://doi.org/10.1016/j.envpol.2018.02.031.
Curto-Barredo, L., Segura, S., Martín-Ezquerra, G., Lloveras, B., Gallardo, F., Pujol, R.M., 2015. Anagen effluvium due to thallium poisoning derived from the intake of Chinese herbal medicine and rodenticide containing thallium salts. J. Dermatol. 42 (10), 1027–1029. https://doi.org/10.1111/1346-8138.12997.

Cvjetko, P., Cvjetko, I., Pavlica, M., 2010. Thallium toxicity in humans. Arh. Hig. Rada Toksikol. Croatia. 61 (1), 111–119. https://doi.org/10.2478/10004-1254-61-2010-1976.

Danilewicz, M., Danilewicz, M., Kurnatowski, A., 1979. Histopathological changes in the kidney of experimental animals caused by lethal doses of thallium (in Polish). Med. Pr. 30 (4), 257–265.

Darvoyd, T.I., Dorofeeva, A.K., Popova, M.A., Kolomiets, B.G., Petrovsky, G.T., Polukhin, S.N., 1968. [Thallium and its Application in Modern Technology]. Central Research Institute of Information and Technical and Economic Research of Non-ferrous Metallurgy, Moscow, pp. 1–76 (Russian).

Das, A.K., Chakraborty, R., Cervera, M.L., de la Guardia, M., 2006. Determination of thallium in biological samples. Anal. Bioanal. Chem. 385 (4), 665–670. https://doi.org/10.1007/s00216-006-0411-8.

de Groot, G., van Heijst, A.N., 1988. Toxicokinetic aspects of thallium poisoning. Methods of treatment by toxin elimination. Sci. Total Environ. 71 (3), 411–418. https://doi.org/10.1016/0048-9697(88)90213-6.

Delvalls, T.A., Saenz, V., Arias, A.M., Blasco, J., 1999. Thallium in the marine environment: first ecotoxicological assessments in the Guadalquivir 500 estuary and its potential adverse effect on the Donana European natural reserve after the Aznalcollar mining spill. Cienc. Mar. 25 (2), 161–175.

Denisyuk, S.D., 1964. [Treatment with Radioactive Thallium of Some Inflammatory Diseases of the Sclera and Cornea]. The Dissertation Abstract of the Candidate of Medical Sciences. Odessa, Ukraine. Russian.

Díaz, V., Tapia, R., Brinck, G., Gutiérrez, M., Hurtado, C., del Peso, G., 1990. Thallium poisoning: Prussian blue treatment in 4 cases. Rev. Med. Chile 118 (2), 183–185 (Spain).

Dietrich, M., Ayers, J., 2021. Geochemical partitioning and possible heavy metal(loid) bioaccumulation within aquaculture shrimp ponds. Sci. Total Environ. 788, 147777. https://doi.org/10.1016/j.scitotenv.2021.147777.

Dipalma, J.R., 1981. Human toxicity from rat poisons. Am. Fam. Physician 24 (4), 186–189.

Dmowski, K., Kozakiewicz, A., Kozakiewicz, M., 1998. Small mammal populations and community under conditions of extremely high thallium contamination in the environment. Ecotoxicol. Environ. Saf. 41 (1), 2–7. https://doi.org/10.1006/eesa.1998.1660.

Dmowski, K., Rossa, M., Kowalska, J., Krasnodebska-Ostrega, B., 2015. Thallium in spawn, juveniles, and adult common toads (*Bufo bufo*) living in the vicinity of a zinc-mining complex, Poland. Environ. Monit. Assess. 187 (1), 1–8. https://doi.org/10.1007/s10661-014-4141-7.

Donnelly, D., Mihovilovic, M., Gonzalez-Ros, J.M., Ferragut, J.A., Richman, D., Martinez-Carrion, M., 1984. A noncholinergic site-directed monoclonal antibody can impair agonist-induced ion flux in *Torpedo californica* acetylcholine receptor. Proc. Natl. Acad. Sci. U.S.A. 81 (24), 7999–8003. https://doi.org/10.1073/pnas.81.24.7999.

Dorman, D.C., 1990. Toxicology of selected pesticides, drugs, and chemicals. Anticoagulant, cholecalciferol, and bromethalin-based rodenticides. Vet. Clin. North Am. Small Anim. Pract. 20 (2), 339–352. https://doi.org/10.1016/s0195-5616(90)50028-5.

Downs, W.L., Scott, J.K., Steadman, L.T., Maynard, E.A., 1960. Acute and sub-acute toxicity studies of thallium compounds. Am. Ind. Hyg. Assoc. J. 21 (5), 399–406. https://doi.org/10.1080/00028896009344093.

Dvorak, P., 1969. Colloidal hexacyanoferrates (II) as antidotes in thallium poisoning. Z. Gesamte Exp. Med. 151 (1), 89–92.

Dvorák, P., 1970. Hexacyanoferrates (II) as thallium antidotes. Preparation and properties. Arzneim. Forsch. 20 (12), 1886–1888.
Edelman, M., 1955. Two cases of thallium poisoning cured by thioacetamide. Pol. Tyg. Lek. 10 (15), 465 (Polish).
Edelman, M., Lipiec, T., 1955. Therapeutic application of thioacetic acid amide in heavy metal poisoning. Pol. Tyg. Lek. 10 (27), 883–884 (Polish).
Erbsloeh, J., 1960. Thallium poisoning in the second half of pregnancy (in Germany). Arch. Toxicol. 18, 156–159.
Farley, M., 1965. A case report: thallium toxicity in the dog. Iowa State Univ. Vet. 27 (3), 117–118.
Fatemi, S.J., Amiri, A., Bazargan, M.H., Tubafard, S., Fatemi, S.N., 2007. Clinical evaluation of desferrioxamine (DFO) for removal of thallium ions in rat. Int. J. Artif. Organs 30 (10), 902–905. https://doi.org/10.1177/039139880703001007.
Fedotov, I.A., 2008. [Influence of Complexones on the Course of Severe Poisoning with Thallium Salts]. The Candidate Dissertation Abstract of PhD in Medicine. St. Petersburg, Russia. Russian.
Filipov, N.M., 2015. Thallium. In: Harbison, R.D., Bourgeois, M.M., Johnson, G.T. (Eds.), Hamilton & Hardy's Industrial Toxicology, sixth ed. Wiley, pp. 239–246.
Fukumoto, M., Kurohara, A., Yoshimura, N., Yoshida, D., Akagi, N., Yoshida, S., 1998. Relationship between ATP synthesis and ^{201}Tl uptake in transformed and non-transformed cell lines. Nucl. Med. Commun. 19 (12), 1169–1175. https://doi.org/10.1097/00006231-199812000-00009.
Gad, S.C., Pham, T., 2014. Thallium. In: Wexler, P. (Ed.), Encyclopedia of Toxicology, third ed. Academic Press, pp. 527–529.
Galván-Arzate, S., Santamaría, A., 1998. Thallium toxicity. Toxicol. Lett. 99 (1), 1–13. https://doi.org/10.1016/s0378-4274(98)00126-x.
Gantner, N., Power, M., Babaluk, J.A., Reist, J.D., Köck, G., et al., 2009. Temporal trends of mercury, cesium, potassium, selenium, and thallium in Arctic char (*Salvelinus alpinus*) from Lake Hazen, Nunavut, Canada: effects of trophic position, size, and age. Environ. Toxicol. Chem. 28 (2), 254–263. https://doi.org/10.1897/08-054.1.
Garcia-Vargas, G.G., Rothenberg, S.J., Silbergeld, E.K., Weaver, V., Zamoiski, R., et al., 2014. Spatial clustering of toxic trace elements in adolescents around the Torreón, Mexico lead-zinc smelter. J. Expo. Sci. Environ. Epidemiol. 24 (6), 634–642. https://doi.org/10.1038/jes.2014.11.
Gehring, P.J., Hammond, P.B., 1967. The interrelationship between thallium and potassium in animals. J. Pharmacol. Exp. Therapeut. 155 (1), 187–201.
Gerasimov, M.I., 1959. [The effect of physical and chemical agents on the epilation effect of thallium acetate]. Vestn. Dermatol. Venerol. 33 (1), 27–30. Russian. PMID: 13635983.
Ghaderi, A., Vahdati-Mashhadian, N., Oghabian, Z., Moradi, V., Afshari, R., Mehrpour, O., 2015. Thallium exists in opioid poisoned patients. Daru 23 (39), 1–4. https://doi.org/10.1186/s40199-015-0121-x.
Gileva, O.V., Ulanova, T.S., Veikhman, G.A., Nedoshitova, A.V., Stenno, E.V., 2016. [Methodological support for the determination of toxic and essential elements in human biological environments for the tasks of social and hygienic monitoring and biomedical research]. Hyg. Sanit. 95 (1), 116–121. https://doi.org/10.18821/00169900-2016-95-1-116-121 (Russian).
Goel, A., Aggarwal, P., 2007. Pesticide poisoning. Natl. Med. J. India. 20 (4), 182–191.
Goldschmidt, J., Wanger, T., Engelhorn, A., Friedrich, H., Happel, M., et al., 2010. High-resolution mapping of neuronal activity using the lipophilic thallium chelate complex TlDDC: protocol and validation of the method. Neuroimage 49 (1), 303–315. https://doi.org/10.1016/j.neuroimage.2009.08.012.

Gregus, Z., Klaassen, C.D., 1986. Disposition of metals in rats: a comparative study of fecal, urinary, and biliary excretion and tissue distribution of eighteen metals. Toxicol. Appl. Pharmacol. 85 (1), 24–38. https://doi.org/10.1016/0041-008x(86)90384-4.

Gresham, C., Lawrey, E.A., Paeds, D., 2014. Thallium Toxicity. Medscape. http://emedicine.medscape.com/article/821465-overview#showall.

Gridnev, V.A., Shurupov, B.C., Sergienko, L.V., Telnov, Y.A., 1971. [Radioactive thallium treatment of patients with chronic suppurative otitis media]. Zh. Ushn. Nos. Gorl. Bolezn. 31 (6), 21–25. Russian. PMID: 5162069.

Grigorovich, A.N., 1960. [Thallium and its Industrial Production]. Alma-Ata (Republic of Kazakhstan), pp. 1–173. Russian. https://search.rsl.ru/ru/record/01006094803.

Gross, P., Runne, E., Wilson, J.W., 1948. Studies on the effect of thallium poisoning of the rat; the influence of cystine and methionine on alopecia and survival periods. J. Invest. Dermatol. 10 (3), 119–134. https://doi.org/10.1038/jid.1948.20.

Grösslová, Z., Vaněk, A., Mihaljevič, M., Ettler, V., Hojdová, M., et al., 2015. Bioaccumulation of thallium in a neutral soil as affected by solid-phase association. J. Geochem. Explor. 159, 208–212. https://doi.org/10.1016/j.gexplo.2015.09.009.

Guberman, D.E., 2015. Thallium. In: U.S. Geological Survey, Mineral Commodity Summaries, January, pp. 164–165. http://minerals.usgs.gov/minerals/pubs/commodity/thallium/mcs-2015-thall.pdf.

Günther, M., 1971. Influence of colloidal ferrihexacyanoferrate(II) on distribution and toxicity of thallium. Arch. Toxicol. 28 (1), 39–45.

Gurevich, E.E., 1965. [The first Russian work on the study of the toxic-pharmacodynamic action of thallium]. In: Proceedings IIIst Scientific. Conference Dermatovenereologists, Mariupol (Formerly Zhdanov), Ukraine, November. Russian, pp. 103–104.

Hakala, J.E., 1984. Thallium poisoning in a dog. Mod. Vet. Pract. 65 (10), 783, 786.

Harris, W.R., Messori, L., 2002. A comparative study of aluminum(III), gallium(III), indium(III), and thallium(III) binding to human serum transferring. Coord. Chem. Rev. 228, 237–262.

Heim, M., Wappelhorst, O., Markert, B., 2002. Thallium in terrestrial environments - occurrence and effect. Ecotoxicology 11 (5), 369–377. https://doi.org/10.1023/a:1020513523462.

Herman, M.M., Bensch, K.G., 1967. Light and electron microscopic studies of acute and chronic thallium intoxication in rats. Toxicol. Appl. Pharmacol. 10 (2), 199–222. https://doi.org/10.1016/0041-008x(67)90104-4.

Heydlauf, H., 1969. Ferric-cyanoferrate (II): an effective antidote in thallium poisoning. Eur. J. Pharmacol. 6 (3), 340–344. https://doi.org/10.1016/0014-2999(69)90194-0.

Hoang, M., Huang, P., J., Liu, J., 2016. G-quadruplex DNA for fluorescent and colorimetric detection of thallium(I). ACS Sens. 1 (2), 137–143. https://doi.org/10.1021/acssensors.5b00147.

Hoffman, R.S., 2003. Thallium toxicity and the role of Prussian blue in therapy. Toxicol. Rev. 22 (1), 29–40. https://doi.org/10.2165/00139709-200322010-00004.

Hoffman, R.S., Stringer, J.A., Feinberg, R.S., Goldfrank, L.R., 1999. Comparative efficacy of thallium adsorption by activated charcoal, prussian blue, and sodium polystyrene sulfonate. J. Toxicol. Clin. Toxicol. 37 (7), 833–837. https://doi.org/10.1081/clt-100102462.

Hologgitas, J., Ullucci, P., Driscoll, J., Grauerholz, J., Martin, H., 1980. Thallium elimination kinetics in acute thallotoxicosis. J. Anal. Toxicol. 4 (2), 68–75. https://doi.org/10.1093/jat/4.2.68.

Hou, L.P., Yang, Y., Shu, H., Ying, G.G., Zhao, J.L., et al., 2017. Changes in histopathology, enzyme activities, and the expression of relevant genes in zebrafish (*Danio rerio*) following long-term exposure to environmental levels of thallium. Bull. Environ. Contam. Toxicol. 99 (5), 574–581. https://doi.org/10.1007/s00128-017-2176-5.

Hsieh, P.J., Su, H.Y., Lo, H.S., Chen, M.L., 2012. Dipyridamole ^{201}Tl myocardial SPECT in the assessment of a patient with myocardial bridging and concomitant atherosclerotic coronary artery disease. Clin. Nucl. Med. 37 (10), e257−e262. https://doi.org/10.1097/RLU.0b013e318244429f.

Huang, C., Zhang, X., Li, G., Jiang, Y., Wang, Q., Tian, R., 2014. A case of severe thallium poisoning successfully treated with hemoperfusion and continuous veno-venous hemofiltration. Hum. Exp. Toxicol. 33 (5), 554−558. https://doi.org/10.1177/0960327113499039.

Huang, C.C., Shih, T.S., Liu, C.H., 2012. Acute thallium intoxication: an experience in Taiwan. Curr. Top. Toxicol. 8, 33−41.

Huang, L., Xiang, L., Zhang, Y., Wang, Y., Nie, Z., 2020. Simultaneous quantitative analysis of K^+ and Tl^+ in serum and drinking water based on UV-Vis spectra and chemometrics. Spectrochim. Acta A Mol. Biomol. Spectrosc. 238, 118392. https://doi.org/10.1016/j.saa.2020.118392.

Hurtado-Díaz, M.E., Estrada-Valencia, R., Rangel-López, E., Maya-López, M., Colonnello, A., et al., 2020. Thallium toxicity in *Caenorhabditis elegans*: involvement of the SKN-1 pathway and protection by S-allylcysteine. Neurotox. Res. 38 (2), 287−298. https://doi.org/10.1007/s12640-020-00220-1.

Ibrahim, D., Froberg, B., Wolf, A., Rusyniak, D.E., 2006. Heavy metal poisoning: clinical presentations and pathophysiology. Clin. Lab. Med. 26 (1), 67−97. https://doi.org/10.1016/j.cll.2006.02.003.

Innis, R., Moses, H., 1978. Clinical conferences at the Johns Hopkins hospital. Thallium poisoning. Johns Hopkins Med. J. 142 (1), 27−31.

Inturrisi, C.E., 1969. Thallium activation of K^+-activated phosphatases from beef brain. Biochim. Biophys. Acta 173 (3), 567−569. https://doi.org/10.1016/0005-2736(69)90022-4.

Jacobson, A.R., McBride, M.B., Baveye, P., Steenhuis, T.S., 2005. Environmental factors determining the trace-level sorption of silver and thallium to soils. Sci. Total Environ. Amsterdam. 345 (1−3), 191−205. https://doi.org/10.1016/j.scitotenv.2004.10.027.

Johnson, W., 1960. A case of thallium poisoning during pregnancy. Med. J. Aust. 47 (2), 540−542. https://doi.org/10.5694/j.1326-5377.1960.tb68680.x.

Joszpa, O., Batko, B., 1957. Two cases of severe thallium poisoning treated with thioacetamide. Pol. Tyg. Lek. 12 (5), 179−181 (Polish).

Kamerbeek, H.H., 1972. Special communication: thallium poisoning. Clin. Toxicol. 5 (l), 89−93. https://doi.org/10.3109/15563657208990509.

Kamerbeek, H.H., Rauws, A.G., ten Ham, M., van Heijst, A.N., 1971a. Dangerous redistribution of thallium by treatment with sodium diethyldithiocarbamate. Acta Med. Scand. 189 (3), 149−154. https://doi.org/10.1111/j.0954-6820.1971.tb04356.x.

Kamerbeek, H.H., Rauws, A.G., ten Ham, M., van Heijst, A.N., 1971b. Prussian blue in therapy of thallotoxicosis. An experimental and clinical investigation. Acta Med. Scand. 189 (4), 321−324. https://doi.org/10.1111/j.0954-6820.1971.tb04383.x.

Kamiljanov, A.K., 1993. [Experimental substantiation of the maximum permissible concentration of thallium carbonate in the atmospheric air]. Hyg. Sanit. (5), 8−10 (Russian).

Kawakami, N., Kunisada, T., Sato, S., Morimoto, Y., Tanaka, M., et al., 2011. Thallium-201 scintigraphy is an effective diagnostic modality to distinguish malignant from benign soft-tissue tumors. Clin. Nucl. Med. 36 (11), 982−986. https://doi.org/10.1097/RLU.0b013e3182177407.

Kazantzis, G., 2000. Thallium in the environment and health effects. Environ. Geochem. Health 22 (4), 275−280. https://doi.org/10.1023/A:1006791514080.

Kazantzis, G., 2007. Thallium. In: Nordberg, G.F., Fowler, B.A., Nordberg, M., Friberg, L. (Eds.), Handbook on the Toxicology of Metals. Elsevier, New York, pp. 827−837.

Kennedy, P., Cavanagh, J.B., 1977. Sensory neuropathy produced in the cat with thallous acetate. Acta Neuropathol. 39 (1), 81–88. https://doi.org/10.1007/BF00690389.

Kiliç, G.A., Kutlu, M., 2010. Effects of exogenous metallothionein against thallium-induced oxidative stress in rat liver. Food. Chem. Toxicol. 48 (3), 980–987. https://doi.org/10.1016/j.fct.2010.01.013.

Kolesanova, E.F., Melnikova, M.V., Bolshakova, T.N., Rybalkina, E.Y., Sivov, I.G., 2019. Bacteriophage MS2 as a tool for targeted delivery in solid tumor chemotherapy. Acta Naturae 11 (2), 98–101. https://doi.org/10.32607/20758251-2019-11-2-98-101.

Korenman, I.M., 1960. [Analytical Chemistry of Thallium]. Moscow. Russian. https://studizba.com/files/show/djvu/2342-1-i-m-korenman–analiticheskaya-himiya.html.

Korsunskaya, V.V., 1970. [To the question of the comparative toxicity of thallium compounds]. In: Preobrazhensky, O.N., Yurikov, S.A. (Eds.), Proceedings of Scientific Papers of Postgraduate Students and Residents of 1st Moscow Medical Institute, Moscow, pp. 25–26 (Russian).

Krasovsky, G.N., Kenesariev, U.I., Vasyukovich, L.Y., Bonashevskaya, T.I., Solv'eva, T.G., Nosacheva, L.F., 1978. [On the toxicity and hazard of thallium in water bodies]. Hygienic aspects of environmental protection. Moscow Times 6, 113–114 (Russian). https://www.elibrary.ru/item.asp?id=26119703.

Krasovsky, G.N., Kenesariev, U.I., 1984. [Methodological scheme of experimental substantiation of the system of indicators of the unfavorable effect of metals on the health of the population (for example, thallium)]. Hyg. Sanit. 2, 22–25 (Russian).

Krasovsky, G.N., Kenesariev, U.I., Vasyukovich, L.Y., Bonashevskaya, T.I., Solv'eva, T.G., Nosacheva, L.F., 1980. [Experimental data on the hygienic regulation of thallium in the water of reservoirs]. Hyg. Sanit. 9, 11–13 (Russian). https://www.elibrary.ru/item.asp?id=26119703.

Kravzov, J., Rios, C., Altagracia, M., Monroy-Noyola, A., López, F., 1993. Relationship between physicochemical properties of prussian blue and its efficacy as antidote against thallium poisoning. J. Appl. Toxicol. 13 (3), 213–216. https://doi.org/10.1002/jat.2550130313.

Krivohlavek, A., Kuharić, Ž., Marjanović Čermak, A.M., Šikić, S., Pavičić, I., Domijan, A.M., 2021. Assessment of intracellular accumulation of cadmium and thallium. J. Pharmacol. Toxicol. Methods 110, 107087. https://doi.org/10.1016/j.vascn.2021.107087.

Krylova, A.I., 1971. [Forensic chemical detection and determination of thallium in biological material by fractional method]. Sud. Med. Ekspert. 14 (3), 38–42 (Russian). https://www.forens-med.ru/auth.php?auth=%CA%F0%FB%EB%EE%E2%E0+%C0.%CD.

Kunito, T., Nakamura, S., Ikemoto, T., Anan, Y., Kubota, R., et al., 2004. Concentration and subcellular distribution of trace elements in liver of small cetaceans incidentally caught along the Brazilian coast. Mar. Pollut. Bull. 49 (7–8), 574–587. https://doi.org/10.1016/j.marpolbul.2004.03.009.

Kuzmich, V.G., 2006. [Comparative Effectiveness of Extracorporeal Detoxification Methods for Poisoning with Metal Salts and Arsenic]. The Dissertation Abstract of the Candidate of Medical Sciences. St. Petersburg, Russian.

Lameijer, W., 1979. Accelerated renal elimination of thallium in the rat due to treatment with furosemide or potassium ions. Arch. Toxicol. 2, 365–366.

Lameijer, W., van Zwieten, P.A., 1977. Kinetic behavior of thallium in the rat. Accelerated elimination of thallium owing to treatment with potent diuretic agents. Arch. Toxicol. 37 (4), 265–273. https://doi.org/10.1007/BF00330818.

Lameijer, W., van Zwieten, P.A., 1978. Accelerated elimination of thallium in the rat due to subchronic treatment with furosemide. Arch. Toxicol. 40 (1), 7–16. https://doi.org/10.1007/BF00353275.

Lamy, M., 1863. Sur les effets toxiques du thallium. C. R. Acad. Sci. 57, 442–445.
Lee, A.G., 1972. The coordination chemistry of thallium(I). Coord. Chem. Rev. 8 (4), 289–349. https://doi.org/10.1016/s0010-8545(00)80001-6.
Lehmann, P.A., Favari, L., 1984. Parameters for the adsorption of thallium ions by activated charcoal and Prussian blue. J. Toxicol. Clin. Toxicol. 22 (4), 331–339. https://doi.org/10.3109/15563658408992564.
Lehmann, P.A., Favari, L., 1985. Acute thallium intoxication: kinetic study of the relative efficacy of several antidotal treatments in rats. Arch. Toxicol. 57 (1), 56–60. https://doi.org/10.1007/BF00286576.
Lehner, A.F., Rumbeiha, W., Shlosberg, A., Stuart, K., Johnson, M., et al., 2013. Diagnostic analysis of veterinary dried blood spots for toxic heavy metals exposure. J. Anal. Toxicol. 37 (7), 406–422. https://doi.org/10.1093/jat/bkt048.
Leloux, M.S., Nguyen, P.L., Claude, J.R., 1987. Experimental studies on thallium toxicity in rats. I-Localization and elimination of thallium after oral acute and sub-acute intoxication. J. Toxicol. Clin. Exp. 7 (4), 247–257.
Leloux, M.S., Nguyen, P.L., Claude, J.R., 1990. Experimental studies on thallium toxicity in rats. II–The influence of several antidotal treatments on the tissue distribution and elimination of thallium, after subacute intoxication. J. Toxicol. Clin. Exp. 10 (3), 147–156.
Leung, K.M., Ooi, V.E., 2000. Studies on thallium toxicity, its tissue distribution and histopathological effects in rats. Chemosphere 41 (1–2), 155–159. https://doi.org/10.1016/s0045-6535(99)00404-x.
Li, S., Huang, W., Duan, Y., Xing, J., Zhou, Y., 2015. Human fatality due to thallium poisoning: autopsy, microscopy, and mass spectrometry assays. J. Forensic Sci. 60 (1), 247–251. https://doi.org/10.1111/1556-4029.12623.
Li, X.P., Qi, J.Y., Wang, C.L., Chen, Y.H., 2011. Distribution of heavy metals in waters and pollution assessment in thallium contaminated area of Yunfu, Guangdong (in Chinese). Huanjing Kexue 32 (5), 1321–1328.
Lie, R., Thomas, R.G., Scott, J.K., 1960. The distribution and excretion of thallium-204 in the rat, with suggested MPC's and a bio-assay procedure. Health Psychol. 2, 334–340. https://doi.org/10.1097/00004032-195910000-00004.
Lin, G., Yuan, L., Peng, X., Long, J., Wang, C., et al., 2019. Clinical characteristics and treatment of thallium poisoning in patients with delayed admission in China. Medicine (Baltim.) 98 (29), e16471. https://doi.org/10.1097/MD.0000000000016471.
Link, V.B., Mohr, C.O., 1953. Rodenticides in bubonic-plague control. Bull. World Health Organ. 9 (5), 585–596.
Liu, C.H., Lin, K.J., Wang, H.M., Kuo, H.C., Chuang, W.L., et al., 2013. Brain fluorodeoxyglucose positron emission tomography (^{18}FDG PET) in patients with acute thallium intoxication. Clin. Toxicol. 51 (3), 167–173. https://doi.org/10.3109/15563650.2013.773008.
Liu, E.M., Rajagopal, R., Grand, M.G., 2015. Optic nerve atrophy and hair loss in a young man. JAMA Ophthalmol. 133 (12), 1469–1470. https://doi.org/10.1001/jamaophthalmol.2015.1957.
Liu, H., Liao, G., 2021. Long-term misdiagnosis and neurologic outcomes of thallium poisoning: a case report and literature review. Brain Behav. 11 (3), e02032. https://doi.org/10.1002/brb3.2032.
Liu, J., Li, N., Zhang, W., Wei, X., Tsang, D.C.W., et al., 2019. Thallium contamination in farmlands and common vegetables in a pyrite mining city and potential health risks. Environ. Pollut. 248, 906–915. https://doi.org/10.1016/j.envpol.2019.02.092.
Liu, J., Lippold, H., Wang, J., Lippmann-Pipke, J., Chen, Y., 2011. Sorption of thallium(I) onto geological materials: influence of pH and humic matter. Chemosphere 82 (6), 866–871.

Liu, J., Ren, S., Cao, J., Tsang, D.C.W., Beiyuan, J., et al., 2021. Highly efficient removal of thallium in wastewater by $MnFe_2O_4$-biochar composite. J. Hazard Mater. 401, 123311. https://doi.org/10.1016/j.jhazmat.2020.123311.

Liu, J., Wang, J., Chen, Y., Lippold, H., Xiaoc, T., et al., 2016a. Geochemical transfer and preliminary health risk assessment of thallium in a riverine system in the Pearl River Basin, South China. J. Geochem. Explor. 176, 64–75. https://doi.org/10.1016/j.gexplo.2016.01.011.

Liu, J., Wang, J., Chen, Y., Xie, X., Qi, J., et al., 2016b. Thallium transformation and partitioning during Pb-Zn smelting and environmental implications. Environ. Pollut. 212, 77–89. https://doi.org/10.1016/j.envpol.2016.01.046.

Liu, J., Yin, M., Xiao, T., Zhang, C., Tsang, D.C.W., et al., 2020b. Thallium isotopic fractionation in industrial process of pyrite smelting and environmental implications. J. Hazard Mater. 384, 121378. https://doi.org/10.1016/j.jhazmat.2019.121378.

Liu, J., Zhou, Y., She, J., Tsang, D.C.W., Lippold, H., et al., 2020a. Quantitative isotopic fingerprinting of thallium associated with potentially toxic elements (PTEs) in fluvial sediment cores with multiple anthropogenic sources. Environ. Pollut. 266 (Pt 3), 115252. https://doi.org/10.1016/j.envpol.2020.115252.

Livanov, G.A., Batotsyrenov, B.V., Ostapenko, Y.N., Shestova, G.V., Rutkovsky, G.V., Malygin, A.Y., 2013. [Features of early diagnosis and treatment of acute poisoning with thallium compounds]. Gen. Reanimatol. 9 (3), 35–40 (Russian). https://www.elibrary.ru/item.asp?id=19138423.

Livanov, G.A., Ostapenko, Y.N., Shestova, G.V., Goldfarb, Y.S., Rutkovsky, G.V., et al., 2011. [The importance of early diagnosis of acute severe poisoning with thallium compounds at the initial stages of intoxication]. Toxicol. Rev. Moscow 5, 2–7 (Russian). https://www.elibrary.ru/item.asp?id=23307286.

Livanov, G.A., Shestova, G.D., Sukhodolova, G.N., Batotsyrenov, B.V., Lodyagin, A.N., Sizova, K.V., 2015. [Clinical picture, diagnosis and treatment of acute severe poisoning with thallium compounds]. Emerg. Med. Care. Moscow. 16 (3), 55–61 (Russian). https://www.elibrary.ru/item.asp?id=25446438.

Llaurado, J.G., Madden, J.A., Meade, R.C., Smith, G.A., 1978. Distribution of thallium-201 injected into rats following stress: imaging, organ to plasma uptake ratios, and myocardial kinetics. J. Nucl. Med. 19 (2), 172–177.

Llaurado, J.G., Madden, J.A., Smith, G.A., 1981. Differences between ^{201}Tl and ^{43}K tissue distribution in rested and exercises rats. Nuklearmedizin 20 (1), 40–45.

Locatelli, C., Petrolini, V., Butera, R., Bernareggi, G., Arrigoni, S., et al., 2003. Long-lasting polyneuropathy and psychiatric disorders in thallium poisoning. Study of six cases. Toxicol. Lett. 144 (1), s72. https://doi.org/10.1016/s0378-4274(03)90259-1.

Lopez-Arce, P., Garcia-Guinea, J., Garrido, F., 2017. Chemistry and phase evolution during roasting of toxic thallium-bearing pyrite. Chemosphere 181, 447–460. https://doi.org/10.1016/j.chemosphere.2017.04.109.

Lu, C.I., Huang, C.C., Chang, Y.C., Tsai, Y.T., Kuo, H.C., et al., 2007. Short-term thallium intoxication: dermatological findings correlated with thallium concentration. Arch. Dermatol. 143 (1), 93–98. https://doi.org/10.1001/archderm.143.1.93.

Lund, A., 1956. The effect of various substances on the excretion and the toxicity of thallium in the rat. Acta Pharmacol. Toxicol. 12 (3), 260–268. https://doi.org/10.1111/j.1600-0773.1956.tb01386.x.

Luzanova, I.S., Pleteneva, T.V., Maksimova, T.V., Salomatin, E.M., Lutskiy, Y.M., 2008. [Determination of thallium content in human urine in acute poisoning by atomic absorption spectrometry]. Sud. Med. Ekspert. 51 (5), 41–43 (Russian). https://www.elibrary.ru/item.asp?id=11754017.

Lyubchenko, P.N., Fomin, A.M., Polyakova, E.A., 2004. [A case of familial thallium poisoning from an unknown source]. J. Occup. Health Ind. Ecol. 1, 29–31 (Russian). https://www.elibrary.ru/item.asp?id=17633514.

Lyublina, E.I., Dvorkin, E.A., 1983. [On the toxicity of metal salts]. In: Dudareva, A.Y. (Ed.), Hygienic Toxicology of Metals: Collection of Scientific Papers, F.F. Erisman Moscow Research Institute of Hygiene, Moscow. Russian, pp. 25–29.

Madden, J.A., Smith, G.A., Llaurado, J.G., 1982. Myocardial kinetics and organ distribution of thallium-201 in two different age groups of rats. J. Gerontol. 37 (1), 16–23. https://doi.org/10.1093/geronj/37.1.16.

Malbrain, M.L., Lambrecht, G.L., Zandijk, E., Demedts, P.A., Neels, H.M., et al., 1997. Treatment of severe thallium intoxication. J. Toxicol. Clin. Toxicol. 35 (1), 97–100. https://doi.org/10.3109/15563659709001173.

Manninen, V., Mälkönen, M., Skul'skiĭ, I.A., 1976. Elimination of thallium in rats as influenced by Prussian Blue and sodium chloride. Acta Pharmacol. Toxicol. 39 (2), 256–261. https://doi.org/10.1111/j.1600-0773.1976.tb03177.x.

Mari, M., Nadal, M., Schuhmacher, M., Barbería, E., García, F., Domingo, J.L., 2014. Human exposure to metals: levels in autopsy tissues of individuals living near a hazardous waste incinerator. Biol. Trace Elem. Res. 159 (1–3), 15–21. https://doi.org/10.1007/s12011-014-9957-z.

Martí-Cid, R., Perelló, G., Domingo, J.L., 2009. Dietary exposure to metals by individuals living near a hazardous waste incinerator in Catalonia, Spain: temporal trend. Biol. Trace Elem. Res. 131 (3), 245–254. https://doi.org/10.1007/s12011-009-8368-z.

Matveev, V.A., 1989. [Biological behavior of thallium-201. Methods of obtaining radioisotopes of thallium and their application in nuclear medicine]. In: Tarasova, N.F., Malinin, A.B. (Eds.), Conference Proceedings. Moscow, pp. 86–95 (Russian).

McCall, D., Zimmer, L.J., Katz, A.M., 1986. Effect of ischemia-related metabolic factors on thallium exchange in cultured rat myocardial cells. Can. J. Cardiol. 2 (3), 176–183.

McMillan, T.M., Jacobson, R.R., Gross, M., 1997. Neuropsychology of thallium poisoning. J. Neurol. Neurosurg. Psychiatry 63, 247–250. https://doi.org/10.1136/jnnp.63.2.247.

Meggs, W.J., Cahill-Morasco, R., Shih, R.D., Goldfrank, L.R., Hoffman, R.S., 1997. Effects of Prussian blue and N-acetylcysteine on thallium toxicity in mice. J. Toxicol. Clin. Toxicol. 35 (2), 163–166. https://doi.org/10.3109/15563659709001187.

Menke, A., Guallar, E., Cowie, C.C., 2016. Metals in urine and diabetes in U.S. Adults. Diabetes 65 (1), 164–171. https://doi.org/10.2337/db15-0316.

Meyer, H., Tal, M., 1957. Studies on thallium poisoning. I. The detoxifying action of some compounds containing sulphur and labile methyl groups. J. Sci. Food Agric. 8, 516–519.

Mochizuki, M., Mori, M., Akinaga, M., Yugami, K., Oya, C., et al., 2005. Thallium contamination in wild ducks in Japan. J. Wildl. Dis. 41 (3), 664–668. https://doi.org/10.7589/0090-3558-41.3.664.

Mochizuki, M., Yamamoto, H., Yamamura, R., Suzuki, T., Ochiai, Y., et al., 2013. Contents of various elements in the organs of seabirds killed by an oil spill around Tsushima Island, Japan. J. Vet. Med. Sci. 75 (5), 667–670. https://doi.org/10.1292/jvms.12-0386.

Moeschlin, S., 1980. Thallium poisoning. Clin. Toxicol. 17 (1), 133–146. https://doi.org/10.3109/15563658008985073.

Montes, S., Pérez-Barrón, G., Rubio-Osornio, M., Ríos, C., Diaz-Ruíz, A., et al., 2011. Additive effect of DL-penicillamine plus Prussian blue for the antidotal treatment of thallotoxicosis in rats. Environ. Toxicol. Pharmacol. 32 (3), 349–355. https://doi.org/10.1016/j.etap.2011.07.002.
Montes, S., Soriano, L., Ríos, C., Monroy-Noyola, A., 2007. Endogenous thiols enhance thallium toxicity. Arch. Toxicol. 81 (10), 683–687. https://doi.org/10.1007/s00204-007-0203-8.
Moore, D., House, I., Dixon, A., 1993. Thallium poisoning. Diagnosis may be elusive but alopecia is the clue. Br. Med. J. 306 (6891), 1527–1529. https://doi.org/10.1136/bmj.306.6891.1527.
Mulkey, J.P., Oehme, F.W., 1993. A review of thallium toxicity. Vet. Hum. Toxicol. 35 (5), 445–453.
Mulkey, J.P., Oehme, F.W., 1998. A safety study of thallium-containing ceramic superconductor material in rats. Vet. Hum. Toxicol. 40 (1), 11–14.
Mulkey, J.P., Oehme, F.W., 2000. Are 2,3-dimercapto-1-propanesulfonic acid or prussian blue beneficial in acute thallotoxicosis in rats? Vet. Hum. Toxicol. 42 (6), 325–329.
Munch, J.C., 1931. The pharmacology of thallium and its use in rodent control. United States Department of Agriculture. Tech. Bull. 238, 1–28.
Munch, J.C., 1934. Human thallotoxicosis. J. Am. Med. Assoc. 102, 1929–1942.
Munteanu, G.G., Munteanu, S.G., 2006. [Distribution of thallium (I) in the Dubossari reservoir of the r. Dniester]. Vodnye Resursy 33 (5), 637–640 (Russian). https://www.elibrary.ru/item.asp?id=9296264.
Muravyova, S.I., 1960. [Determination of small amounts of thallium in biological substrates (blood and urine)]. Occupation. Hygiene Occupation. Dis. Moscow (8), 53–54 (Russian).
Nachman, M., Hartley, P.L., 1975. Role of illness in producing learned taste aversions in rats: a comparison of several rodenticides. J. Comp. Physiol. Psychol. 89 (9), 1010–1018. https://doi.org/10.1037/h0077189.
Nadal, M., Bocio, A., Schuhmacher, M., Domingo, J.L. 2005. Monitoring metals in the population living in the vicinity of a hazardous waste incinerator: levels in hair of school children. Biol. Trace Elem. Res. 104(3): 203-213. https://doi.org/10.1385/BTER:104:3:203.
Nadel, H.R., 1993. Thallium-201 for oncologic imaging in children. Semin. Nucl. Med. 23 (3), 243–254. https://doi.org/10.1016/s0001-2998(05)80105-9.
Nasiri-Majd, M., Taher, M.A., Fazelirad, H., 2015. Synthesis and application of nano-sized ionic imprinted polymer for the selective voltammetric determination of thallium. Talanta 144, 204–209. https://doi.org/10.1016/j.talanta.2015.05.058.
Nehring, D., 1962. Experiments on the toxicological effect of thallium ions on fish and fish-food organisms. Z. Fischer. 11, 557–562.
Neizvestnova, E.M., Katsnelson, B.A., Davydova, V.I., Malysheva, L.G., Konovalova, N.E., 1992. [Toxicological Characterization and Regulation of a Number of Polymetallic Oxides]. Gigiena Truda i Professional'nye Zabolevaniia (Occupational Hygiene and Occupational Diseases), vol. 6, pp. 23–26 (Russian). https://www.journal-irioh.ru/jour/issue/archive.
Nielsen, S.G., Rehkämper, M., Norman, M.D., Halliday, A.N., Harrison, D., 2006. Thallium isotopic evidence for ferromanganese sediments in the mantle source of Hawaiian basalts. Nature 439 (7074), 314–317. https://doi.org/10.1038/nature04450.
Ning, Z., He, L., Xiao, T., Márton, L., 2015. High accumulation and subcellular distribution of thallium in green cabbage (*Brassica oleracea* L. Var. Capitata L.). Int. J. Phytoremediation 17 (11), 1097–1104. https://doi.org/10.1080/15226514.2015.1045133.
Oehme, F.M., 1978. Mechanisms of Heavy Metal Inorganic Toxicities. Toxicity of Heavy Metals in the Environment. Part 1. Marcel Dekker, New York, pp. 69–85.

Osytek, K.M., Blower, P.J., Costa, I.M., Smith, G.E., Abbate, V., Terry, S.Y.A., 2021. In vitro proof of concept studies of radiotoxicity from Auger electron-emitter thallium-201. EJNMMI Res. 11 (1), 63. https://doi.org/10.1186/s13550-021-00802-w.

OT, 2014. [Occupational Safety and Health]. http://ohrana-bgd.ru/gigien/gigien1_09.html.

Ouellet, J.D., Dubé, M.G., Niyogi, S., 2013. The influence of food quantity on metal bioaccumulation and reproduction in fathead minnows (*Pimephales promelas*) during chronic exposures to a metal mine effluent. Ecotoxicol. Environ. Saf. 91, 188–197. https://doi.org/10.1016/j.ecoenv.2013.02.002.

Papp, J.P., Gay, P.C., Dodson, V.N., Pollard, H.M., 1969. Potassium chloride treatment in thallotoxicosis. Ann. Intern. Med. 71 (1), 119–123. https://doi.org/10.7326/0003-4819-71-1-119.

Patterson, J.F., 1975. Chronic thallitoxicosis: treatment of the choreiform sequelae. South. Med. J. 68 (7), 923–925.

Pau, P.W., 2000. Management of thallium poisoning. Hong Kong Med. J. 6 (3), 316–318.

Paulet, M., 1863. Expériences sur faction physiologique des sels de thallium. C. R. Math. Acad. Sci. Paris. 57, 494–495.

Paulson, G., Vergara, G., Young, J., Bird, M., 1972. Thallium intoxication treated with dithizone and hemodialysis. Arch. Intern. Med. 129 (1), 100–103.

Pavlíčková, J., Zbíral, J., Smatanová, M., Habarta, P., Houserová, P., Kubáň, V., 2006. Uptake of thallium from artificially contaminated soils by kale (*Brassica oleracea* L. var. acephala). Plant Soil Environ. 52 (12), 544–549. https://doi.org/10.17221/3545-pse.

Pavlova, A.Z., Bogomolov, D.V., Larev, Z.V., Amanmuradov, A.K., 2012. [Hair as an object of research in case of poisoning with salts of heavy metals]. Sud. Med. Ekspert. 55 (6), 25–29 (Russian). https://www.elibrary.ru/item.asp?id=18758518.

Peele, D.B., Farmer, J.D., MacPhail, R.C., 1987. Conditioned flavor aversions: applications in assessing the efficacy of chelators in the treatment of heavy-metal intoxication. Toxicol. Appl. Pharmacol. 88 (3), 397–410.

Peele, D.B., MacPhail, R.C., Farmer, J.D., 1986. Flavor aversions induced by thallium sulfate: importance of route of administration. Neurobehav. Toxicol. Teratol. 8 (3), 273–277.

Pelclová, D., Urban, P., Ridzon, P., Senholdová, Z., Lukás, E., et al., 2009. Two-year follow-up of two patients after severe thallium intoxication. Hum. Exp. Toxicol. 28 (5), 263–272. https://doi.org/10.1177/0960327109106487.

Perrault, J.R., Buchweitz, J.P., Lehner, A.F., 2014. Essential, trace and toxic element concentrations in the liver of the world's largest bony fish, the ocean sunfish (*Mola mola*). Mar. Pollut. Bull. 79 (1–2), 348–353. https://doi.org/10.1016/j.marpolbul.2013.11.026.

Perrin, D.D., 1979. Stability Constants of Metal-Ion Complexes. Part B. Organic Ligands. Pergamon, Oxford, pp. 1–1263.

Peter, A.L., Viraraghavan, T., 2005. Thallium: a review of public health and environmental concerns. Environ. Int. 31 (4), 493–501. https://doi.org/10.1016/j.envint.2004.09.003.

Petersohn, K.L., 1960. Thallium poisoning in pregnancy. Arch. Toxikol. 18, 160–164.

Polyakova, M.M., Sidorova, N.V., Spiridonova, B.C., Shabalina, L.P., Shatalov, N.N., 1977. [State of health and occupational conditions of workers handling thallium and its compounds]. Gig. Tr. Prof. Zabol. (2), 14–16 (Russian. PMID: 838343).

Ponsoda, X., Jover, R., Núñez, C., Royo, M., Castell, J.V., Gómez-Lechón, M.J., 1995. Evaluation of the cytotoxicity of 10 chemicals in human and rat hepatocytes and in cell lines: correlation between in vitro data and human lethal concentration. Toxicol. Vitro 9 (6), 959–966. https://doi.org/10.1016/0887-2333(95)00053-4.

Ponton, D.E., Caron, A., Hare, L., Campbell, P.G., 2016. Hepatic oxidative stress and metal subcellular partitioning are affected by selenium exposure in wild yellow perch (*Perca flavescens*). Environ. Pollut. 214, 608–617. https://doi.org/10.1016/j.envpol.2016.04.051.

Prick, J.J.G., 1979. Thallium poisoning. Handbook of clinical neurology. In: Vinken, P.J., Bruyn, G.W. (Eds.), Intoxication of the Nervous System, vol. 36. North-Holland, New York, pp. 239–278.

Puschner, B., Basso, M.M., Graham, T.W., 2012. Thallium toxicosis in a dog consequent to ingestion of Mycoplasma agar plates. J. Vet. Diagn. Invest. 24 (1), 227–230. https://doi.org/10.1177/1040638711425941.

Qi, J., Lai, Y., Liang, C., Yan, S., Huang, K., et al., 2019. Prenatal thallium exposure and poor growth in early childhood: a prospective birth cohort study. Environ. Int. 123, 224–230. https://doi.org/10.1016/j.envint.2018.12.005.

Rade, J.E., Marafante, E., Sabbioni, E., Gregotti, C., Di Nucci, A., Manzo, L., 1982. Placental transfer and retention of ^{201}Tl-thallium in the rat. Toxicol. Lett. 11 (3–4), 275–280. https://doi.org/10.1016/0378-4274(82)90161-8.

Radić, S., Cvjetko, P., Glavas, K., Roje, V., Pevalek-Kozlina, B., Pavlica, M., 2009. Oxidative stress and DNA damage in broad bean (*Vicia faba* L.) seedlings induced by thallium. Environ. Toxicol. Chem. 28 (1), 189–196. https://doi.org/10.1897/08-188.1.

Ramsden, D.B., 2007. Thallium. In: Waring, R.H., Steventon, G.B., Mitchel, S.C. (Eds.), Molecules of Death (2nd). Imperial College Press, London, pp. 415–422.

Rana, S., Sharma, N., Ojha, H., Shivkumar, H.G., Sultana, S., Sharma, R.K., 2014. p-Tertbutylcalix[4]arene nanoemulsion: preparation, characterization and comparative evaluation of its decontamination efficacy against Technetium-99m, Iodine-131 and Thallium-201. Colloids Surf. B Biointerfaces 117, 114–121. https://doi.org/10.1016/j.colsurfb.2014.02.001.

Rauws, A.G., 1974. Thallium pharmacokinetics and its modification by Prussian blue. Naunyn-Schmiedeberg's Arch. Pharmacol. 284 (3), 294–306. https://doi.org/10.1007/BF00500348.

Rauws, A.G., ten Ham, M., Kamerbeek, H.H., 1969. Influence of the antidote dithiocarb on distribution and toxicity of thallium in the rat. Arch. Int. Pharmacodyn. Ther. 182 (2), 425–426.

Reed, D., Crawley, J., Faro, S.N., Pieper, S.J., Kurland, L.T., 1963. Thallotoxicosis. Acute manifestations and sequelae. J. Am. Med. Assoc. 183, 516–522.

Remy, H., 1956. Compounds of thallium. In: Kleinberg, J. (Ed.), Treatise on Inorganic Chemistry. Elsevier, Amsterdam, pp. 380–389.

Richelmi, P., Bono, F., Guardia, L., Ferrini, B., Manzo, L., 1980. Salivary levels of thallium in acute human poisoning. Arch. Toxicol. 43 (4), 321–325. https://doi.org/10.1007/bf00366188.

Rickwood, C.J., King, M., Huntsman-Mapila, P., 2015. Assessing the fate and toxicity of Thallium I and Thallium III to three aquatic organisms. Ecotoxicol. Environ. Saf. 115, 300–308. https://doi.org/10.1016/j.ecoenv.2014.12.024.

Rinck, P.A., Beckmann, H.O., 1982. Myocardial uptake of thallous ion in animals. Röfo 136 (6), 679–680. https://doi.org/10.1055/s-2008-1056129.

Ríos, C., Galván-Arzate, S., Tapia, R., 1989. Brain regional thallium distribution in rats acutely intoxicated with Tl_2SO_4. Arch. Toxicol. 63 (1), 34–37. https://doi.org/10.1007/BF00334631.

Rìos, C., Monroy-Noyola, A., 1992. D-penicillamine and prussian blue as antidotes against thallium intoxication in rats. Toxicology 74 (1), 69–76. https://doi.org/10.1016/0300-483x(92)90044-f.

Riyaz, R., Pandalai, S.L., Schwartz, M., Kazzi, Z.N., 2013. A fatal case of thallium toxicity: challenges in management. J. Med. Toxicol. 9 (1), 75–78. https://doi.org/10.1007/s13181-012-0251-1.

Rodríguez-Mercado, J.J., Altamirano-Lozano, M.A., 2013. Genetic toxicology of thallium: a review. Drug Chem. Toxicol. 36 (3), 369–383. https://doi.org/10.3109/01480545.2012.710633.

Roshchina, T.A., 1966. [Experimental data on toxicity of magnesium alloy containing thallium]. In: Hygienic Assessment of Chemical Environmental Factors. Proceedings 3st Scientific Conference of Hygienic Departments of the 1st. Moscow Medical Institute, Moscow, pp. 83–85 (Russian).

Roshchina, T.A., 1970. [Experimental evaluation of the toxicity of a condensation aerosol of a magnesium alloy containing thallium]. In: Organism and Environment. Hygienic Assessment of Chemical Environmental Factors, Part 1. 1st Moscow Medical Institute, Moscow, pp. 93–94 (Russian).

Rossi, F., Cazzola, M., De Santis, D., Lisa, M., Servodio, R., et al., 1987. Cardiovascular effects of thallium sulfate after intracerebroventricular administration in normotensive rats. Curr. Ther. Res. Clin. Exp. 42 (5), 778–789.

Ruhr, L.P., Andries, J.K., 1985. Thallium intoxication in a dog. JAVMA 186 (5), 498–499.

Rusyniak, D.E., Arroyo, A., Acciani, J., Froberg, B., Kao, L., Furbee, B., 2010. Heavy metal poisoning: management of intoxication and antidotes. Experientia 100, 365–396. https://doi.org/10.1007/978-3-7643-8338-1_11.

Rusyniak, D.E., Furbee, R.B., Kirk, M.A., 2002. Thallium and arsenic poisoning in a small midwestern town. Ann. Emerg. Med. 39 (3), 307–311. https://doi.org/10.1067/mem.2002.122008.

Rusyniak, D.E., Kao, L.W., Nanagas, K.A., Kirk, M.A., Furbee, R.B., et al., 2003. Dimercaptosuccinic acid and Prussian Blue in the treatment of acute thallium poisoning in rats. J. Toxicol. Clin. Toxicol. 41 (2), 137–142. https://doi.org/10.1081/clt-120019129.

Sabbioni, E., Gregotti, C., Edel, J., Marafante, E., Di Nucci, A., Manzo, L., 1982. Organ/tissue disposition of thallium in pregnant rats. Arch. Toxicol. 5 (Suppl. l), 225–230. https://doi.org/10.1007/978-3-642-68511-8_41.

Sabbioni, E., Goetz, L., Marafante, E., Gregotti, C., Manzo, L., 1980. Metabolic fate of different inorganic and organic species of thallium in the rat. Sci. Total Environ. 15 (2), 123–135. https://doi.org/10.1016/0048-9697(80)90003-0.

Sabbioni, E., Manzo, L., 1980. Metabolism and toxicity of thallium. In: Manzo, L. (Ed.), Advances in Neurotoxicology. Pergamon Press, Oxford, pp. 249–270.

Saljooghi, A.S., Fatemi, S.J., 2011. Removal of thallium by deferasirox in rats as biological model. J. Appl. Toxicol. 31 (2), 139–143. https://doi.org/10.1002/jat.1573.

Sanotskiy, I.V., 1961. [Study of the toxicity of thallium compounds (thallium carbonate, iodide and bromide)]. Toxicol. New Indus Chem. Moscow 2, 94–104 (Russian).

Schaller, K.H., Manke, G., Raithel, H.J., Bühlmeyer, G., Schmidt, M., Valentin, H., 1980. Investigations of thallium-exposed workers in cement factories. Int. Arch. Occup. Environ. Health 47 (3), 223–231. https://doi.org/10.1007/BF00381680.

Schoer, J., 1984. Thallium. In: Hutzinger, O. (Ed.), Handbook of Environmental Chemistry. Anthropogenic Compounds. Springer, New York, pp. 143–214.

Schwetz, B.A., O'Neil, P.V., Voelker, F.A., Jacobs, D.W., 1967. Effects of diphenylthiocarbazone and diethyldithiocarbamate on the excretion of thallium by rats. Toxicol. Appl. Pharmacol. 10 (1), 79–88. https://doi.org/10.1016/0041-008x(67)90130-5.

Sehweil, A.M., McKillop, J.H., Milroy, R., Wilson, R., Abdel-Dayem, H.M., Omar, Y.T., 1989. Mechanism of ^{201}Tl uptake in tumours. Eur. J. Nucl. Med. 15 (7), 376–379. https://doi.org/10.1007/BF00449228.
Senthilkumaran, S., Balamurugan, N., Jena, N.N., Menezes, R.G., Thirumalaikolundusubramanian, P., 2017. Acute alopecia: evidence to thallium poisoning. Int. J. Trichol. 9 (1), 30–32. https://doi.org/10.4103/ijt.ijt_82_16.
Sessions, H.K., Goren, S., 1947. Report of investigation of health hazards in connection with the industrial handling of thallium. U. S. Naval Med. Bull. 47 (3), 545–550.
Shabalina, L.P., Krillov, V.F., Spiridonova, V.S., Poliakova, M.M., Popova, O.L., Shamshinova, A.M., 1996. [Methods of diagnosis of thallium intoxication and its antidote therapy]. Med. Tr. Prom. Ekol. 10, 25–30 (Russian. PMID: 9064097).
Shabalina, L.P., Spiridonova, B.C., 1976. [Toxicity of thallium malonic formate)]. Gig. Tr. Prof. Zabol. 8, 35–37 (Russian. PMID: 1010373).
Shabalina, L.P., Spiridonova, B.C., Pavlenko, G.I., Fomenko, V.N., 1980. [Cutaneous-resorptive, local and mutagenic action of thallium carbonate]. Gig. Tr. Prof. Zabol. (4), 54–55 (Russian. PMID: 7409564).
Shakeri, M.T., Nezami, H., Nakhaee, S., Aaseth, J., Mehrpour, O., 2021. Assessing heavy metal burden among cigarette smokers and non-smoking individuals in Iran: cluster Analysis and principal component analysis. Biol. Trace Elem. Res. 199 (11), 4036–4044. https://doi.org/10.1007/s12011-020-02537-6.
Shamshinova, A.M., Ivanina, T.A., Yakovlev, A.A., Shabalina, L.P., Spiridonova, V.S., 1990. [Electroretinography in the diagnosis of thallium intoxication]. J. Hyg. Epidemiol. Microbiol. Immunol. 34 (2), 113–121 (Russian. PMID: 2212633).
Sharma, N., Ojha, H., Pathak, D.P., Goel, R., Sharma, R.K., 2017. Ex-vivo complexation, skin permeation, interaction and cytodermal toxicity studies of p-tertbutylcalix[4]arene nanoemulsion for radiation decontamination. Life Sci. 168, 65–76. https://doi.org/10.1016/j.lfs.2016.11.007.
Sheiman, B.S., Prodanchuk, N.G., Voloshina, N.A., Makarov, A.A., Urin, A.A., Itz, V.V., 2014. [The role of the renal elimination system in the treatment of acute thallium poisoning]. Emerg. Med. 4 (59), 52–67 (Russian). http://emergency.zaslavsky.com.ua.
Shelley, R., Kim, N.S., Parsons, P., Lee, B.K., Jaar, B., et al., 2012. Associations of multiple metals with kidney outcomes in lead workers. Occup. Environ. Med. 69 (10), 727–735. https://doi.org/10.1136/oemed-2012-100765.
Sheyman, B.S., Prodanchuk, N.G., Voloshina, N.A., Makarov, A.A., Urin, A.A., Its, V.V., 2014. Thallium toxicokinetics and endotoxemia parameters in acute thallium intoxication. Emerg. Med. 4 (59), 52–67 (Russian). http://emergency.zaslavsky.com.ua/article/view/83472.
Shiraishi, S., Sakamoto, F., Tsuda, N., Yoshida, M., Tomiguchi, S., et al., 2015. Prediction of left main or 3-vessel disease using myocardial perfusion reserve on dynamic thallium-201 single-photon emission computed tomography with a semiconductor gamma camera. Circ. J. 79 (3), 623–631. https://doi.org/10.1253/circj.CJ-14-0932.
Shvartsev, S.L., Gomonova, L.N., 1975. [Thallium in brines of the Angara-Lena artesian basin]. Doklady Acad. Sci. USSR. Earth Sci. Sect. 220 (5), 1183–1185 (Russian). http://www.ipgg.sbras.ru/ru/science/publications/article-journal-talliy-v-rassolakh-angaroleninskogo-artezialnogo-1975-004165.
Siegel, B.Z., Siegel, S.M., 1975. A selective role for potassium in the phytotoxicity of thallium. Bioinorg. Chem. 4 (2), 93–97. https://doi.org/10.1016/s0006-3061(00)81017-0.
Skelley, J.F., Gabriel, K.L., 1964. Thallium intoxication in the dog. Ann. N. Y. Acad. Sci. 111, 612–617. https://doi.org/10.1111/j.1749-6632.1964.tb53129.x.

Smeaton, C.M., Walshe, G.E., Fryer, B.J., Weisener, C.G., 2012. Reductive dissolution of Tl(I)-jarosite by Shewanella putrefaciens: providing new insights into Tl biogeochemistry. Environ. Sci. Technol. 46 (20), 11086–11094. https://doi.org/10.1021/es302292d.

Smith, D.H., Doherty, R.A., 1964. Thallitoxicosis: report of three cases in Massachusetts. Pediatrics 34, 480–490 (PMID: 14212463).

Solovyev, N.D., Ivanenko, N.B., Ivanenko, A.A., 2011. Whole blood thallium determination by GFAAS with high-frequency modulation polarization Zeeman effect background correction. Biol. Trace Elem. Res. 143 (1), 591–599. https://doi.org/10.1007/s12011-010-8865-0.

Spiridonova, B.C., Shabalina, L.P., 1978. [Toxicology of thallium and its compounds (Literature review)]. Gig. Tr. Prof. Zabol. 7, 41–43 (Russian. PMID: 352834).

Spiridonova, V.S., Shabalina, L.P., 1977a. [Toxicology of Thallium and its Compounds. Occupational Health Issues when Working with Them]. Moscow, pp. 1–68 (Russian).

Spiridonova, V.S., Shabalina, L.P., 1977b. Occupational hygiene and basic measures to prevent thallium intoxication. In: Tarasenko, N.Y. (Ed.), Physiology and Occupational Hygiene in the Scientific and Technical Process. Moscow, pp. 16–19 (Russian).

Stadion, V.M., 1865. Thallium in contact with an animal organism. Arch. Forensic Med. Public Hyg. 1 (1), 35–55 (Russian).

Stavinoha, W.B., Emerson, G.A., Nash, J.B., 1959. The effects of some sulfur compounds on thallotoxicosis in mice. Toxicol. Appl. Pharmacol. 1, 638–646. https://doi.org/10.1016/0041-008x(59)90068-7.

Sun, J., Zou, X., Xiao, T., Jia, Y., Ning, Z., et al., 2015. Biosorption and bioaccumulation of thallium by thallium-tolerant fungal isolates. Environ. Sci. Pollut. Res. 22 (21), 16742–16748. https://doi.org/10.1007/s11356-015-4859-y.

Sun, T.W., Xu, Q.Y., Zhang, X.J., Wu, Q., Liu, Z.S., et al., 2012. Management of thallium poisoning in patients with delayed hospital admission. Clin. Toxicol. 50 (1), 65–69. https://doi.org/10.3109/15563650.2011.638926.

Tabandeh, H., Thompson, G.M., 1993. Visual function in thallium toxicity. BMJ 307 (6899), 324.

Taber, P., 1964. Chronic thallium poisoning: rapid Diagnosis and treatment. J. Pediatr. 65 (3), 461–463. https://doi.org/10.1016/s0022-3476(64)80414-5.

Taheri, M., Mehrzad, J., Afshari, R., Gharaie, M.H., 2015. Geogenic thallium and lead pollution in soils and potential risk of toxicity: a report from Iran. J. Res. Med. Sci. 20 (4), 420–421.

Thompson, D.F., 1981. Management of thallium poisoning. Clin. Toxicol. 18 (8), 979–990. https://doi.org/10.3109/15563658108990328.

Thyresson, N., 1950. The influence of dietary factors, especially brewer's yeast, cystine and B vitamins, on the course of chronic thallium poisoning in the rat. Acta Derm. Venereol. 30 (1), 9–26.

Thyresson, N., 1951. Experimental investigation on thallium poisoning in the rat. Distribution of thallium, especially in the skin, and excretion of thallium under different experimental conditions. A study with the use of the radioactive isotope Tl^{204}. Acta Derm. Venereol. 31 (1), 3–27.

Thyresson, N., 1952. Effect of thallium on the growth of tactile hair in the white rat. Acta Derm. Venereol. Suppl. 32 (29), 370–375.

Tikhova, T.S., 1967. [Thallium and its compounds]. In: [New Data on the Toxicology of Rare Metals and Their Compounds]. Moscow, pp. 24–34 (Russian).

Treshchinsky, A.I., Shlapak, I.P., Chepkiy, L.P., Dmitrieva, M.B., Kharchenko, L.A., et al., 1989. [Clinical picture and treatment of acute thallium poisoning]. Vrach. Delo 3, 100–103 (Russian). https://www.ncbi.nlm.nih.gov/nlmcatalog?term="Vrach+Delo.

Tromme, I., Van Neste, D., Dobbelaere, F., Bouffioux, B., Courtin, C., et al., 1998. Skin signs in the diagnosis of thallium poisoning. Br. J. Dermatol. 138 (2), 321–325. https://doi.org/10.1046/j.1365-2133.1998.02083.x.

Truhaut, R., 1960. The toxicology of thallium. J. Occup. Med. 2, 334–336. https://doi.org/10.1097/00043764-196007000-00008.

Tsai, Y.T., Huang, C.C., Kuo, H.C., Wang, H.M., Shen, W.S., et al., 2006. Central nervous system effects in acute thallium poisoning. Neurotoxicology (Little Rock) 27 (2), 291–295. https://doi.org/10.1016/j.neuro.2005.10.009.

Ushakova, V.G., 2009. [Environmental factors affecting the general level of health of the population of mountainous areas (on the example of the Altai Republic)]. Reg. Environ. (3), 75–79. Russian). https://www.ecoregion.ru/index.php?razdel=pre.

Uspenskaya, G.S., 1940. [On the distribution of thallium in the body]. Bull. Exp. Biol. Med. 9 (5), 329–331 (Russian).

Varão, A.M., Silva, J.D.S., Amaral, L.O., Aleixo, L.L.P., Onduras, A., et al., 2021. Toxic effects of thallium acetate by acute exposure to the nematode *C. elegans*. J. Trace. Elem. Med. Biol. 68, 126848. https://doi.org/10.1016/j.jtemb.2021.126848.

Vilavert, L., Nadal, M., Schuhmacher, M., Domingo, J.L., 2012. Concentrations of metals in soils in the neighborhood of a hazardous waste incinerator: assessment of the temporal trends. Biol. Trace Elem. Res. 149 (3), 435–442. https://doi.org/10.1007/s12011-012-9441-6.

Vodyanitskii, Y.N., 2012. [Standards for the contents of heavy metals and metalloids in soils]. Eurasian Soil Sci. 45 (3), 321–328. https://doi.org/10.1134/S1064229312030131 (Russian).

Volmer, P.A., Merola, V., Osborne, T., Bailey, K.L., Meerdink, G., 2006. Thallium toxicosis in a pit bull terrier. J. Vet. Diagn. Invest. 18 (1), 134–137. https://doi.org/10.1177/104063870601800124.

Vorobyev, N.V., 2003. [Clinical and Pathogenetic Features of Acute Thallium Poisoning]. The Dissertation Abstract of the Candidate of Medical Sciences. St. Petersburg, Russian.

Vrij, A.A., Cremers, H.M., Lustermans, F.A., 1995. Successful recovery of a patient with thallium poisoning. Neth. J. Med. 47 (3), 121–126. https://doi.org/10.1016/0300-2977(95)00006-9.

Wang, C., Chen, Y., Liu, J., Wang, J., Li, X., et al., 2013. Health risks of thallium in contaminated arable soils and food crops irrigated with wastewater from a sulfuric acid plant in western Guangdong province, China. Ecotoxicol. Environ. Saf. 90, 76–81. https://doi.org/10.1016/j.ecoenv.2012.12.014.

Wanger, T., Scheich, H., Ohl, F.W., Goldschmidt, J., 2012. The use of thallium diethyldithiocarbamate for mapping CNS potassium metabolism and neuronal activity: Tl^+ -redistribution, Tl^+ -kinetics and Tl^+ -equilibrium distribution. J. Neurochem. 122 (1), 106–114. https://doi.org/10.1111/j.1471-4159.2012.07757.x.

Waters, C.B., Hawkins, E.C., Knapp, D.W., 1992. Acute thallium toxicosis in a dog. J. Am. Vet. Med. Assoc. 201 (6), 883–885.

Weinig, E., Zink, P., 1967. On the quantitative mass spectrometry of standartsal thallium content in the human organism (in German). Arch. Toxicol. 22 (4), 255–274.

WHO, 1996. Thallium: Environmental Health Criteria. World Health Organization, Geneva, Switzerland. http://www.inchem.org/documents/ehc/ehc/ehc182.htm.

Woods, J.S., Fowler, B.A., 1986. Alteration of hepatocellular structure and function by thallium chloride: ultrastructural, morphometric, and biochemical studies. Toxicol. Appl. Pharmacol. 83 (2), 218–229. https://doi.org/10.1016/0041-008x(86)90299-1.

Worthley, L.I., 2002a. Clinical toxicology: part II. Diagnosis and management of uncommon poisonings. Crit. Care Resusc. 4 (3), 216–230.

Worthley, L.I., 2002b. Clinical toxicology: part I. Diagnosis and management of common drug overdosage. Crit. Care Resusc. 4 (3), 192–215.

Xiao, T., Guha, J., Boyle, D., Liu, C.Q., Chen, J., 2004a. Environmental concerns related to high thallium levels in soils and thallium uptake by plants in southwest Guizhou. Sci. Total Environ. 318 (1–3), 223–244. https://doi.org/10.1016/S0048-9697(03)00448-0.

Xiao, T., Guha, J., Boyle, D., Liu, C.Q., Zheng, B., et al., 2004b. Naturally occurring thallium: a hidden geoenvironmental health hazard? Environ. Int. 30 (4), 501–507. https://doi.org/10.1016/j.envint.2003.10.004.

Xu, H., Luo, Y., Wang, P., Zhu, J., Yang, Z., Liu, Z., 2019. Removal of thallium in water/wastewater: a review. Water Res. 165, 114981. https://doi.org/10.1016/j.watres.2019.114981.

Yang, Y., Faustino, P.J., Progar, J.J., Brownell, C.R., Sadrieh, N., et al., 2008. Quantitative determination of thallium binding to ferric hexacyanoferrate: prussian blue. Int. J. Pharm. 353 (1–2), 187–194. https://doi.org/10.1016/j.ijpharm.2007.11.031.

Yudin, A.L., Vasiliev, P.V., Pokrovsky, Y.A., 2010. [Modern methods of tomographic diagnostics for evaluating the effectiveness of treatment for cancer of the larynx and laryngopharynx]. Bull. Otorhinolaryngol. 3, 47–49 (Russian). https://www.elibrary.ru/item.asp?id=22793163.

Yumoto, T., Tsukahara, K., Naito, H., Iida, A., Nakao, A., 2017. A successfully treated case of criminal thallium poisoning. J. Clin. Diagn. Res. 11 (4), OD01–OD02. https://doi.org/10.7860/JCDR/2017/24286.9494.

Zasukhina, G.D., Krasovsky, G.N., Vasilieva, I.M., Sdirkova, N.I., Sokolovsky, V.V., et al., 1980. [Molecular biological effect of thallium carbonate]. Bull. Exp. Biol. Med. 90 (12), 723–725 (Russian. PMID: 6258670).

Zavaliy, L.B., Petrikov, S.S., Simonova, A.Y., Potskhveriya, M.M., Zaker, F., et al., 2021. Diagnosis and treatment of persons with acute thallium poisoning. Toxicol. Rep. 8, 277–281. https://doi.org/10.1016/j.toxrep.2021.01.013.

Zaychenkov, E.P., 1970. [Treatment of certain diseases of the peripheral nervous system with thallium applications]. Medi. Radiol. 15 (9), 86–88. https://medradiol.fmbafmbc.ru.

Zaychenkov, E.P., 1971. [The Action of Radioactive Thallium on the Nervous System in the Application Method]. The Dissertation Abstract of the Candidate of Medical Sciences. Saratov, Russian Federation. Russian.

Zhao, Z., Xiong, Y., Cheng, X., Hou, X., Yang, Y., et al., 2020. Adsorptive removal of trace thallium(I) from wastewater: a review and new perspectives. J. Hazard Mater. 393, 122378. https://doi.org/10.1016/j.jhazmat.2020.122378.

Zhou, Y., He, H., Wang, J., Liu, J., Lippold, H., et al., 2021. Stable isotope fractionation of thallium as novel evidence for its geochemical transfer during lead-zinc smelting activities. Sci. Total Environ. 803, 150036. https://doi.org/10.1016/j.scitotenv.2021.150036.

Ziskoven, R., Achenbach, C., Schulten, H.R., Roll, R., 1983. Thallium determinations in fetal tissues and maternal brain and kidney. Toxicol. Lett. 19 (3), 225–231. https://doi.org/10.1016/0378-4274(83)90123-6.

Zitko, V., 1975. Toxicity and pollution potential of thallium. Sci. Total Environ. 4 (2), 185–192.

Zitko, V., Carson, W.V., 1975. Accumulation of thallium in clams and mussels. Bull. Environ. Contam. Toxicol. 14 (5), 530–533. https://doi.org/10.1007/BF01683366.

Zitko, V., Carson, W.V., Carson, W.G., 1975. Thallium: occurrence in the environment and toxicity to fish. Bull. Environ. Contam. Toxicol. 13 (1), 23–30. https://doi.org/10.1007/BF01684859.

Zook, B.C., Gilmore, C.E., 1967. Thallium poisoning in dogs. J. Am. Vet. Med. Assoc. 151 (2), 206–217.

Zook, B.C., Holzworth, J., Thornton, G.W., 1968. Thallium poisoning in cats. J. Am. Vet. Med. Assoc. 153 (3), 285–299.

CHAPTER 2

Toxicology of thallium

List of abbreviations

ALT	alanine aminotransferase
AST	aspartate aminotransferase
ATP	adenosine triphosphate
DNP	2,4-dinitrophenol
NMR	nuclear magnetic resonance
ROS	reactive oxygen species
$\Delta\Psi_{mito}$	inner membrane potential

2.1 The toxic effect of thallium on the human body and animals

2.1.1 The toxic effect of thallium on the human body

Penetrating the body through the skin, respiratory, or digestive systems, thallium damages the cardiovascular, central nervous, and renal systems, as well as the gastrointestinal tract, skin, and causes hair loss (Hologgitas et al., 1980; Clarkson, 1987; Mulkey and Oehme, 1998, 2000; Leung and Ooi, 2000; Goel and Aggarwal, 2007; Al Hammouri et al., 2011; Rodríguez-Mercado and Altamirano-Lozano, 2013). Thallium causes an inflammatory reaction on contact with mucous membranes (Prick, 1979). After poisoning a person with inorganic thallium salts, at first, its maximum concentration is found in the kidneys and large intestine, while it is not as concentrated in the muscles, heart, bones, nails, lungs, liver, and brain (Weinig and Zink, 1967; Leung and Ooi, 2000; Hoffman, 2003). According to Davis et al., 1981, the maximum thallium concentration for people poisoned by its compounds appeared in the large intestine, left ventricle, and pituitary gland (Davis et al., 1981). However, significant amounts of this metal did find in the kidneys, liver, adrenal glands, and pancreas (Tanaka et al., 1978; Davis et al., 1981). Damage to the liver and muscles in thallium poisoning indicates by increased urinary excretion of coproporphyrin and uroporphyrin and creatinine clearance decrease. Increased blood nitrogen and noticeable urine protein excretion

Mitochondria as a Key Intracellular Target of Thallium Toxicity
ISBN 978-0-323-95531-7
https://doi.org/10.1016/B978-0-323-95531-7.00001-X

(proteinuria) characterize kidney damage (Prick, 1979). Poisoning of people with thallium compounds was accompanied by tachycardia, respiratory and renal failure, muscle pain and abdominal pain, decreased vision due to optic nerve disorders, hypertension due to impaired catecholamine metabolism, increased heart rate, sinus brachycardia, ventricular arrhythmia, extensive necrosis, damage and inflammation of the myocardium with thinning of myofibrils, vomiting, aspiration of stomach contents, development of intestinal obstruction, hair loss (alopecia), polyneuropathy, damage to the cranial axons, peripheral, gastrocnemius and vagus nerves with degeneration of myelin, as well as an increase in the volume of mitochondria and vacuoles, swelling and fragmentation of neurons of cranial nerves and spinal cord (Bank et al., 1972; Paulson et al., 1972; Tanaka et al., 1978; Hologgitas et al., 1980; Davis et al., 1981; Dumitru and Kalantri, 1990; Tabandeh and Thompson, 1993; Moore et al., 1993; Galván-Arzate, Santamaría, 1998; Mulkey and Oehme, 1998; Loyke, 2002; Ramsden, 2007; Pelclová et al., 2009; Huang et al., 2012; Riyaz et al., 2013; Curto-Barredo et al., 2015; Liu et al., 2015; Campbell et al., 2016). A 26-year-old man who took 10 g of thallium malonate died 40 hours later despite carrying out intensive hemoperfusion and hemodialysis (Aoyama et al., 1986). In this case, the heart tissue, but not tissue in the kidneys, muscles, testes, and brain, revealed the highest thallium concentrations.

When thallium enters the body, it distributes in the choroid plexus of the lateral ventricles of the heart, lacrimal and salivary glands, thyroid gland, myocardium, liver, spleen, visceral regions, kidneys, and testes (Repetto et al., 1998). The toxicity of thallium to humans is very high and complex, with a wide range of organs and tissues involved in this process. The severity of symptoms depends on time and level of exposure, absorption rate, age, and individual sensitivity (Repetto et al., 1998). The maximum thallium concentration did find in the colon wall, left ventricle, and pituitary gland (Davis et al., 1981). Simultaneously, a significant amount of thallium did discover in the adrenal glands, liver, kidneys, and pancreas (Tanaka et al., 1978; Davis et al., 1981).

2.1.2 General patterns of the toxic thallium effects on animals

The thallium distribution analysis in the animal body can calculate by the so-called three-part model (Leloux et al., 1987). The central compartments with the maximum affinity for thallium include kidneys, muscles, bones, bare skin, heart, tongue, and testes. The medium thallium affinity compartments embrace the small and large intestines, liver, spleen, and pancreas. The minimal thallium affinity compartments involve the brain, eyes, hair, ovaries, lungs, and adrenal glands (Leloux et al., 1987). The hepatoprotective and antioxidant effects against Tl-induced liver damage did research using garlic diallyl sulfide and curcumin in experiments with rats treated daily with thallium (6.4 mg/kg) for five days. Noticeable liver damage appeared there with increasing ALT and AST activity and concentrations of malondialdehyde and serum nitric oxide, and simultaneous decrease of liver glutathione concentration, hepatic catalase superoxide dismutase activity, and general antioxidant protection in the blood plasma of these animals (Appenroth and Winnefeld, 1999a; Abdel-Daim and Abdou, 2015). The simultaneous action of garlic diallyl sulfide and curcumin with thallium salts markedly reduced the toxic metal effects. Thallium did not affect the glutathione concentration in the kidneys of rats treated with 20 mg/kg Tl_2SO_4 (Appenroth and Winnefeld, 1999b). Thallium was shown to have not affected the growth of hair - follicles in rat pups, but at the same time, noticeably disrupted the process of normal differentiation and keratinization of their hair (Thyresson, 1951).

Experiments on mice exposed to thallium sulfate (20–2560 mg/kg) showed that this metal initially accumulated in the stomach. The maximum thallium concentration did find in the kidneys, heart, and liver, and after a while, the maximum thallium concentration did test in the brain tissues (Lie et al., 1960; Achenbach et al., 1980; Schoer, 1984). Chronic 5 ppm Tl showed an increase in damaged red muscle fibers in rats (Méndez-Armenta et al., 2011). The ^{204}Tl isotope use discovered that thallium first accumulated in the blood and then distributed in other organs of rats (Krasovsky et al., 1976). Similar thallium accumulation patterns (in the kidneys more than in other organs) did detect when studying the thallium accumulation in the body of gobies fed plants grown near cement plants in northwest Germany (Brockhaus et al., 1981; Schoer, 1984). Enterohepatic circulation of this metal may explain the high fecal and hepatic thallium concentrations long after poisoning (Thompson, 1981). A single injection of Tl_2SO_4 (5–20 mg/kg) in male mice caused physiological dysfunction of vital organs. It stimulated disturbances in the intestinal flora, membrane metabolites, energy, and protein metabolism (Tyagi et al., 2011). Tl^+ accumulated more actively than K^+ in the intestinal mucosa of goldfish (Ellory et al., 1973).

2.1.3 Teratogenic, mutagenic, and genotoxic effects of thallium compounds

Thallium compounds have genotoxic and mutagenic effects on human, animal, bacterial, yeast, and plant cells (Rodríguez-Mercado and Altamirano-Lozano, 2013; Rodríguez-Mercado et al., 2015). The teratogenic and mutagenic effect of thallium may be associated with this metal's ability to easily cross the placenta and accumulate in the fetal tissues, which causes poisoning for both the mother and the unborn child and manifests itself in the appearance of alopecia and Mees' lines both in the mother and in the born child (Erbslöh, 1960; Johnson, 1960; Petersohn, 1960; Oehme, 1978). These results did confirm by the data on the ^{201}Tl accumulation in female milk and experiments carried out on animals and birds. These findings showed that thallium poisoning leads to the fetus intrauterine development disorder and a decrease in its size and weight, the bone curvature, the impaired ossification of phalanges, vertebrae, skull bones, ribs, and limbs, the disruption in the development of cartilaginous and bone tissue, bone marrow, and even the embryo death (Gibson and Becker, 1970; Tarasenko et al., 1975; Roth, 1975; Achenbach et al., 1979; Ziskoven et al., 1980, 1983; Murphy et al., 1989; Mulkey and Oehme, 1993; Arbiser et al., 1997; Galván-Arzate, Santamaría, 1998; Hoffman, 2003; Álvarez-Barrera et al., 2019). The chronic experiments with zebrafish *Danio rerio* kept in water containing thallium (100–500 ng/L) showed a significant increase in mRNA expression, metallothionein (MT)2 synthesis, superoxide dismutase enzymatic activities, and Na^+/K^+-ATPase activity (Hou et al., 2017). The thallium genotoxicity and cytotoxicity are found in experiments in vivo with flies *Drosophila* subjected 600 μM Tl_2SO_4 with wing somatic mutation evaluation test (Reyes-Rodríguez et al., 2021). Thallium is recently found to shorten the telomere length in newborns and stunt growth in young children (Qi et al., 2019; Wu et al., 2021). Prenatal exposure to high thallium levels may cause underweight babies (Xia et al., 2016). A blood test of a patient who took 200 mg of Tl_2SO_4 revealed a sharp increase in the number of binucleated cells with micronuclei (Hantson et al., 1997). Thallium reduced the incorporation of amino acids into synthesized protein molecules and caused disaggregation of liver polyribosomes of mouses exposed to 80–200 μg/kg $TlNO_3$ (Hultin and Näslund, 1974). The mutagenic activity of thallium carbonate was higher than mercury chloride one (Zasukhina et al., 1983). Thallium damaged DNA and increased the number of chromosomal aberrations in human peripheral blood cells in experiments in vitro (Rodríguez-Mercado et al., 2015). Tl^{+} in human embryonic kidney cells showed the protein synthesis inhibition, eukaryotic initiation factor 2α phosphorylating, and the ribosome synthesis disturbance that resulted in cell growth inhibition and the cells' lethality (Chou and Lo, 2019).

The thallium isotopes (^{201}Tl and ^{204}Tl) used in experiments with pregnant mice and rats showed that this metal easily overcomes the placental barrier and noticeably accumulates in fetal tissues, especially in the brain (Olsen and Jonsen, 1982; Rade et al., 1982; Sabbioni et al., 1982). Within 15 minutes administration of $^{204}Tl_2SO_4$ (50 μCi) to white mice female at the 15th day of gestation revealed this metal accumulation in the fetal tissues and thallium even in the placenta more significant presence (Olsen and Jonsen, 1982). There was substantial damage to bone and cartilage tissue and a noticeable decrease in the weight of rat pups born to females receiving 32 mg/kg of thallium acetate (Barroso-Moguel et al., 1992). At the same time, this metal accumulated in the various parts of the brain of these rat pups (Barroso-Moguel et al., 1990). Thallium caused hair loss in young rat pups, but at the same time, it prevented the development of experimental cataracts in experiments with selenite (Thyresson, 1951b; Oštádalová and Babický, 1987). Thallium ingested with milk retarded the growth of rat pups has caused them to prolonged hair alopecia. In addition to that, this metal accumulated mainly in the kidneys, less in the liver, and noticeably lesser in the spleen, lungs, and brain (Downs et al., 1960). Newborn rat pups did treat with a single injection of thallium acetate. There was a noticeable decrease in large and medium fibers, axon degeneration, focal necrosis, small hemorrhages of the peroneal muscle, myopathy of muscle fibers, and the disappearance of striated fibers (Barroso-Moguel et al., 1996). A decrease in the glomerular filtration rate, a reduction in urine volume, and an increase in proteinuria and NO concentration in the blood were after a single administration of thallium sulfate (5–20 mg/kg) to adult rats and cubs (Appenroth et al., 1995, 1996; Appenroth and Winnefeld, 1999a).

Chronic administration of thallium carbonate orally to male albino rats increased the frequency of chromosomal aberrations and the mortality of embryos and pups born to intact females, caused changes in the seminal epithelium of the testes, reduced immunity and increased the production of nitric oxide, and disrupted the functional activity of leukocytes (Tarasenko et al., 1975; Popova et al., 1977; Shabalina et al., 1980; Zasukhina et al., 1980b; Vorobyev, 2003). Thallium caused noticeable disruption in the development of mouse embryos in experiments in vivo (Achenbach et al., 1979). Exposure to thallium sulfate resulted in bone curvature and beak reshaping in chick embryos, with many dead and damaged cells in cartilage tissue (Barclay et al., 1953; Landauer, 1960; Parker and Scheck, 1981). At the same time, ultrastructural changes observed in hypertrophied chondrocytes manifested themselves in the cytoplasmic structure violation and the cytoplasm appearance of lipid-like bodies with electron-dense granules, mineral inclusions, collagen, and mitochondria with signs of obvious degeneration (Parker and Scheck, 1981). If glucose, sorbose, arbidol, and imidazole were added into a medium simultaneously with thallium, the protective effect against thallium-induced teratogenic effects has been observed in experiments with chicken embryos (Landauer, 1960). Biotin, riboflavin, nicotinamide, and manganese have no protective effect in the same experiments with thallium.

Thallium reduced sperm motility in animals in experiments in vivo and in vitro (Formigli et al., 1986; Manzo and Sabbioni, 1988; Kamiljanov, 1993). Morphological studies in vitro of primary testicular cell cultures showed that 220 μM thallium caused after 48 hours, a significant loss of germ cells of rat spermatocytes and changes in the shape of Sertoli cells, which play a vital role in the spermatogenesis process (Gregotti et al., 1992). Thallium attenuated spermatogenesis in male rats and dogs and disrupted the sexual behavior of female rats, which showed a restoration after estrogen injection (Truhaut, 1960; Zook and Gilmore, 1967; Popova et al., 1977). In this case, the thallium carbonate effect was more substantial than the mercury chloride effect (Zasukhina et al., 1983). There were short-term experiments with thallium sulfate injected into pregnant rats' femoral veins and arteries. In this place, the thallium concentration after 32 minutes was in the fetal blood plasma 15 times less than that in the female plasma (Gibson and Becker, 1970). The 10–100 μM Tl_2SO_4 effect on cells of rat fibroblasts and embryonic cells in mice CBA line led to a decrease in the number of double-stranded DNA strands, did not cause breaks of single-stranded DNA strands and reduced the survival of vaccinia virus in these cells (Zasukhina et al., 1980a,b, 1983). Thallium reduced mitotic activity in rat hair follicle cells, and the result was general hyperplasia of skin cells with rarefaction and atrophy of hair follicles (Truhaut, 1960; Cavanagh et al., 1974; Cavanagh and Gregson, 1978). Chronic thallium poisoning of tadpoles inhibited their growth and blocked subsequent metamorphosis (Thyresson, 1950). Thallium accumulated in the gills of the fat-headed minnows *Pimephales promelas* and reduced the survival rate and weight of the fish fry (LeBlanc and Dean, 1984; Ouellet et al., 2013).

2.2 The toxic effect of thallium compounds on body systems

2.2.1 Thallium toxic effect on the nervous system

It should be stressed that the central and peripheral nervous system is the main critical organ in thallium intoxication (Peter and Viraraghavan, 2005). In particular, the human central nervous system is most sensitive to the toxic thallium effects, but the thallium accumulation in this one is "delayed" (Ponsoda et al., 1995; Sharma et al., 2004). If the thallotoxicosis treatment was delayed, neurological symptoms were observed even nine months after the penetration of thallium into the body in some cases (Pau, 2000). The nervous system diffuse damage and polyneuritis diagnosed by a laboratory assistant poisoned with thallium malonate (Fedorovskaya, 1971). The severity of these toxicity symptoms depended on the age, dose, and penetration method of thallium compounds into the body (Peter and Viraraghavan, 2005). Electrophysiological examination of a patient with thallium poisoning revealed severe sensorimotor polyneuropathy with predominantly motor axonal involvement (Lukács, 2003). The optic nerve damages can occur many months after poisoning with even relatively small thallium amounts and are signs of peripheral neuropathy (Cavanagh, 1979).

The thallium toxicity, particularly the thallium neurotoxicity, was higher than the toxicity of mercury and cadmium in experiments in vivo on rats (Krasovsky et al., 1976; Spiridonova and Shabalina, 1978). Lipid peroxidation and beta-galactosidase activity increased in brain tissues in experiments with rats receiving from 4 to 8 mg/kg thallium acetate (Brown et al., 1985). The brain of rats and mice turned out to be the least permeable to thallium due to the presence of the blood-brain barrier (Achenbach et al., 1979; Galván-Arzate and Ríos, 1994). However, Tl^{+} in the lipophilic dithiocarbamate complex easily penetrated this barrier (Goldschmidt et al., 2010; Wanger et al., 2012). Changes did find in the brains of animals exposed to thallium salts. These changes were the next: the concentration decrease of dopamine, serotonin, and catecholamine in the brain synapses various parts with a slowdown in the entry of these neurotransmitters into central neurons, a decrease in electrophysiological activity in central noradrenergic synapses, an increase in the spontaneous discharge rate of cerebellum Purkinje neurons, impaired adrenergic mechanisms, the monoxide activity increase, and lipid peroxidation stimulation (Hasan et al., 1978b; Marwaha et al., 1980; Hasan and Ali, 1981; Galván-Arzate, Santamaría, 1998). Thallium suppressed dendrite regrowth in experiments with the dorsal root neurons of the rat embryo isolated from the dorsal ganglion root (Windebank, 1986).

Significant changes in the concentrations of amino acids and neurotransmitters found in the brain various areas in acute and subacute experiments on rats injected intraperitoneally with 0–20 mg/kg thallium acetate (Ali et al., 1990). Critical thallium experiments in vivo stimulated the concentration increase of glutamate and taurine in the frontal cortex of rats. In contrast, there was a decrease in the hippocampus concentrations of taurine and aspartic acid. Subacute experiments showed a significant increase in amygdala concentrations of serotonin, 3,4-dihydroxyphenylacetic acid and dopamine, and hypothalamic dopamine concentrations (Ali et al., 1990). Lipid peroxidation and ROS production increase did detect in rats' brain tissues exposed to chronic and subchronic thallium salts (Galván-Arzate et al., 2000). Thallium-induced changes in the sustained impulse activity and spontaneously occurring inhibitory postsynaptic potentials were associated with the metal nonspecific effect on the intracellular metabolism of these cells.

Thallium reversibly reduced orthodromic propagating action potentials in pyramidal cells of the hippocampus sections of guinea pig and rat (Lohmann et al., 1989). The use of $^{86}Rb^{+}$ and $^{201}Tl^{+}$ isotopes injected into the mice nasal cavity showed that Tl^{+} could enter the axon through synaptic clefts and rapidly accumulate in the olfactory nerve (Kanayama et al., 2005). Thallium penetrated through potassium channels increased the excitability of neurons in the rat brain. This metal stimulated lipid peroxidation, reduced the concentration of reduced glutathione and glutathione peroxidase activity in experiments in vivo with rats receiving 32 mg/kg of thallium acetate (Osorio-Rico et al., 2015). All these thallium effects were reduced significantly in experiments with animals administered dizocilpine (N-methyl-D-aspartate receptor agonist) which may be associated with the glutamatergic system involvement in the development of Tl^{+}-induced toxicity in neurons.

Behavioral disturbances, disturbance of electrocortical potentials, and dose-dependent increase in blood pressure with the appearance of signs of very mild sinus bradycardia at high metal concentrations were the result of thallium sulfate administration from 4 to 385 μg/kg into the cerebral ventricles of rats (Nisticò et al., 1984; Rossi et al., 1987). At the same time, the use of various pharmacological agents (reserpine, propranolol, prazosin, clonidine, verapamil, diltiazem, atropine sulfate, hexamethonium, mecamylamine) eliminated the appearance of adverse reactions from the cardiovascular system before injection of Tl_2SO_4. The arterial hypertension appearance was concluded to be associated with the thallium activation of the sympathetic nervous system, and the sinus bradycardia appearance is a compensatory result of the vagus nerve discharge to the blood pressure increase (Rossi et al., 1987). On the other hand, the sodium channel permeability of the myelinated sciatic nerve of frog *Rana pipiens* showed the following series Na^+, $Li^+ > Tl^+ > K^+$ (Hille, 1973). The activity decrease of monoamine oxidase, acid phosphatase, succinate dehydrogenase, guanine deaminase, and protease (cathepsin) did discover in the brain striatum of rats treated with 5 mg/kg of thallium acetate (Hasan et al., 1977). At the same time, experiments with these animals showed the concentration decrease of glutathione and SH-groups in the cerebellum and the brain stem, and thallium neurotoxicity was associated with these facts (Hasan and Haider, 1989). Thallium-induced decrease in glutathione levels slowed down in different parts of the brain of white rats simultaneously receiving 5 mg/kg of thallium acetate and 8 mg/kg of thiolactone acetylhomocysteine (Hasan and Haider, 1989).

The research of spinal cell cultures from the dorsal roots' ganglia, spinal cord, mice peripheral nerves showed Tl^+-induced massive internal cavities in axons due to adhesion of adjacent vacuoles and Tl^+-stimulated mitochondrial swelling in axons of peripheral nerve fibers (Spencer et al., 1973). Acute potassium intoxication of rabbits caused degenerative lesions in the spinal cord white matter and led to the hind limbs' paralysis as a result (Truhaut, 1960). Tl^+ (1 mM) inhibited K^+-induced hyperpolarization of the rat demyelinated vagus nerve (Smith, 1979). Thallium increased the frequency but did not affect the amplitude of the potential of the endplates of mice diaphragmatic preparations. At the same time, the metal disrupted the mediator phase release up to the complete synapse blocking in these preparations (Wiegand et al., 1984b, 1986). Thallium also prevented calcium utilization by mitochondria and endoplasmic reticulum inside the presynaptic endings at the mediator release stage (Wiegand et al., 1984a, 1990; Lohmann and Wiegand, 1996). Low temperature and DNP reduced bidirectional axonal transport of $^{104}Tl^+$ in experiments in vitro and in vivo with the frog sciatic nerve after local application of this isotope to the nerve (Bergquist et al., 1983). The concentration of amino acids and other neurotransmitters showed significant changes in various brain areas of rats treated with up to 20 mg/kg of thallium acetate (Ali et al., 1990). Intracellular somatic mechanisms limit the toxic effect of thallium on the guinea pig hippocampus (Lohmann and Wiegand, 1996).

Thallium inhibited Rb^+ transport across the cell membrane of neurons in the mollusk Planorbarius corneus, which was associated with competition between Tl^+ and Rb^+ for a common carrier and with thallium inhibition of Na^+/K^+-ATPase (Skul'skiĭ and Pivovarova, 1988). Ouabain reduced $^{204}Tl^+$ accumulation in coil brain cells of mollusk *Planorbarius corneus* (Skul'skiĭ et al., 1976). At the same time, 0.1−0.2 mM Tl^+ activated Na^+ transport into these cells. The intracellular Na^+ concentration became stable in a medium containing above 1 mM Tl^+. The same Na^+ concentration was in a potassium-free ringer with 1.6 mM Tl^+. In addition, Tl^+ prevented Na^+ entry into the mollusk cells from the potassium-free ringer and reduced the already low K^+ concentration in these cells. Optimal Tl^+ concentrations, activating Na^+ transport, are 1/10 of similar K^+ concentrations (Skul'skiĭ et al., 1976).

2.2.2 Thallium toxic effect on muscle tissues

Thallium (1−100 μM) increased Purkinje fibers and interventricular preparations' contractility of the sheep heart muscle. Still, the contractility of these preparations weakened with an increase in the medium thallium concentration or decreased over time at the same thallium concentration (Wiemer et al., 1982; Ziskoven et al., 1982). At the same time, 0.1−100 μM thallium reduced the contractility of the preparations of the papillary muscles of the guinea pig. This effect of thallium on the contractile processes in the heart muscle may be associated with the metal effect on the slow incoming current (Wiemer et al., 1982; Ziskoven et al., 1982). High thallium concentrations caused irreversible damage to the frog's tailor muscle fibers (Mullins and Moore, 1960). The permeability of the potassium channels of the frog's tailor muscle fibers, the cell membrane of the neurons of the mollusk *Planorbarius corneus* and the giant squid, the Ranvier interception membrane of the single myelinated nerve of the frog *Rana pipiens* turned out to be higher for Tl^+ ions than for K^+ ions (Hagiwara et al., 1972; Hille, 1973; Skul'skiĭ and Malov, 1977; Ashcroft and Stanfield, 1983). At the same time, Tl^+ inhibited the incoming K^+ fluxes in solutions containing both Tl^+ and K^+. Still, it could not stoichiometrically replace K^+ in the frog sartorius muscle cells, which cannot distinguish between Tl^+ and K^+ ions (Mullins and Moore, 1960; Ling, 1977; Edelman, 1988). A similar inhibitory effect of Tl^+ on these channels did manifest in the presence of gramicidin ionophore, but Tl^+ did not affect the potassium conductance and the potential of the isolated muscle fiber *Rana esculenta* (Caffier and Shvinka, 1982; Shvinka and Caffier, 1983). The Tl^+ transport in mucous membrane preparations of the small, ileum and colon of rats proceeded in the direction from serosa to mucosa against an electric gradient with the participation of K^+-dependent transport systems and inhibited by ouabain, furosemide, or after removal of Na^+ and Cl^+ ions from the solution (Schäfer et al., 1981; Henning and Forth, 1982; Henning et al., 1982; Schäfer and Forth, 1987). No thallium secretion was found really in the study of rat stomach preparations (Henning and Forth, 1982). The addition of 30 μM Tl^+ or 150 μM K^+ reduced the contractions of the rat myometrium in potassium-free Ringer's solution by 50% (Grover et al., 1981). The concentrations of Tl^+ and K^+ half-activating nitrophenyl phosphatase in the granules of this myometrium were 0.47 and 1.6 mM, respectively. This result allowed concluding that Tl^+ can replace K^+ upon activation of Na^+/K^+-ATPase in the plasma membrane of rat myometrium.

The relative permeability of the end-plate channels of the frog muscle fibers was 2.5 times higher for Tl+ than one for K^+ and Na^+ (Adams et al., 1980; Talbot, 1992). In this case, the replacement of 2.5 mM K^+ in the solution by 2.5 mM Tl^+ stimulated the contractions' amplitude increase, and this effect did not depend on the increase in Ca^{2+} intracellular concentration. Blocking these thallium effects by ω-conotoxin suggested that Tl^+ can transport to the frog nerve endings via voltage-gated N-type calcium channels, and the metal does not cause a rise in intracellular Ca^{2+} cannot replace Ca^{2+} and Na^+ ions (Talbot, 1992). Ouabain markedly inhibited thallium transport into rat muscle fibers (Brismar, 1991). Stopped in Ringer's potassium-free solution, the isolated frog heart contractions resumed in the solutions containing 2 mM Tl^+ and 4 mM K^+ (Rusznyák et al., 1968). However, when the Ringer's concentration increased to 12 mM K^+ or 4 mM Tl^+, these contractions were blocked but restored when 4 mM Tl^+ did replace by 4 mM K^+. The thallium effect on the contraction of the isolated frog heart preparations was more substantial than the effect of potassium, which was associated with a faster penetration of

Tl^{+} into the frog tissue (Hughes et al., 1978). At the same time, thallium increased the heart rate and intermembrane space in the frog diaphragmatic preparations. It also blocked the phrenic nerve or the neuromuscular connection with the muscle fiber.

Experiments with sinoatrial and atria preparations of guinea pig and Purkinje fibers' and interventricular preparations of sheep cardiac muscle showed that thallium caused arrhythmia is associated with this metal action on the heart sinus node (Lameijer and van Zwieten, 1976; Achenbach et al., 1982). In general, the thallium effect on preparations' muscle contractions of warm-blooded animals is hypotensive (Lameijer and van Zwieten, 1976). Thallium reduced tone and slowed contraction with the simultaneous amplitude increase in rabbit preparations of the aorta, intestine, and ear or rat intestine preparations (Truhaut, 1960; Lameijer and van Zwieten, 1976). Diaphragm and left ventricle muscles found the maximum thallium concentrations in experiments with rabbits and rats receiving thallium salts 10 (mg/kg). At the same time, the thallium content in skeletal muscles was 2.5—4 times lower and reached a minimum in the brain (Rauws, 1974; Careaga-Olivares and Morales-Aguilera, 1993).

In vivo experiments with $^{201}Tl^{+}$ showed that thallium accumulates mainly in muscles and especially in the kidneys of rats (Ando et al., 1987). The ^{201}Tl using possibility in the myocardial infarction study previously investigated in experiments with dogs and rats (Chauncey et al., 1977). The Tl^{+} entry into the frog skeletal muscle cells led to membrane hyperpolarization and the rectification effect inactivation; however, K^{+} did not affect this muscle cell membrane (Stanfield et al., 1981). The hyperpolarization of rat diaphragm muscle fibers was more robust in experiments with Tl^{+} than with K^{+} (Den Hertog and Mooij, 1976). Thallium isotope is used as an agent for the myocardium scanning and diagnosing the lung, thyroid, liver, and brain tumors (Galván-Arzate and Santamaría, 1998). There were positive correlations between thallium concentrations and Chinese adults' cardiovascular disease mortality in the multiple-metals model with thallium and selenium (Shi et al., 2021).

2.2.3 Thallium effect on hemodynamics, blood, and kidneys

The level of renal function damage can serve as a good criterion for assessing the thallium intoxication degree in humans (Gozhenko et al., 1996). Intravenous administration of thallium sulfate (3—100 mg/kg) to rats resulted in a noticeable decrease in blood pressure in these animals

(Lameijer and van Zwieten, 1976). Exposure to thallium sulfate (20 mg/kg) in rats reduced the plasma concentration of amino acids and decreased the glomerular filtration rate with a markedly increased excretion of amino acids, and especially in adult animals (Fleck and Appenroth, 1996). There was an increase in plasma creatinine concentrations and protein concentration in urine after the administration of TlCl (5–50 mg/kg) to white male rats, while the rate of glomerular filtration, diuresis, and water reabsorption decreased (Gozhenko et al., 1996; Dikal, 2014). These effects were stopped mainly by melatonin administered 6 hours after thallium administration (Dikal, 2014). The Henle loops' epithelium necrosis, edematous degeneration, the epithelium swelling, and focal necrosis in the proximal convoluted tubules were detected using a light microscope in receiving thallium rats (Truhaut, 1960; Danilewicz et al., 1979). Thallium significantly reduced the excretion of K^+ ions by the kidneys of rats on the second day after a single administration of Tl_2SO_4 (5–20 mg/kg) (Appenroth et al., 1995). Increased creatinine, taurine, hippurate, and beta-hydroxybutyrate levels, along with decreased energy metabolites, trimethylamine, and choline did reveal using the method of high-resolution NMR spectroscopy in the urine metabolic analysis in Tl_2SO_4-exposed male mice (Tyagi et al., 2011).

2.2.4 Thallium toxic effect on frog skin

Thallium reduced the $^{86}Rb^+$ accumulation and the short-circuit current in frog skin. In addition to that, both Tl^+ and Cu^+ inhibited the Na^+ transport in frog skin cells (Maslova et al., 1971; Natochin and Skul'skiĭ, 1972; Skul'skiĭ et al., 1980a, 1982; Skul'skiĭ and Lapin, 1983, 1992; Zeiske and Van Driessche, 1986). Na^+ transport in frog skin stopped in Ringer's potassium-free solution, restored after adding K^+ to the solution, not restored after the 0.1–1.0 mM Tl_2CO_3 addition (Maslova et al., 1971). In the latter case, there was activation of Na^+/K^+-ATPase in the presence of 0.5–1.0 mM Tl_2CO_3 in the experiments with frog skin homogenates. At low Tl^+/K^+ ratios, thallium competitively inhibited K^+ current in frog skin, and the metal remained the only penetrating cation at higher Tl^+ ratios. Ba^{2+} inhibited K^+ not Tl^+ transport in frog skin (Zeiske and Van Driessche, 1983). Unlike K^+ ions, Tl^+ and Ba^{2+} ions added to potassium-free Ringer solution did not restore the active Na^+ transport in the frog skin (Natochin and Skul'skiĭ, 1972). One millimolar Tl^+ inhibited succinate dehydrogenase and alpha-ketoglutarate dehydrogenase by 70%–80% in the frog epithelial cells. But at the same time, it reduced the cells' Na^+/K^+-ATPase activity by 20%–30% (Skul'skiĭ et al., 1982).

2.2.5 Thallium accumulation in tumor tissues

onvolatile joint transport The thallium accumulation in tumor tissues is multifactorial. It is associated with blood flow, with tumor viability, with the Na^+/K^+-ATPase system, with a nonvolatile joint transport system, with a calcium channel system, with increased permeability of cell membranes, with internal rectification potassium channels being more permeable to Tl^+ than to K^+ (Sehweil et al., 1989; Abdel-Dayem et al., 1994; Galván-Arzate and Santamaría, 1998). However, there is a link between the thallium absorption and the tumor histological response to treatment. The ^{201}Tl accumulation dynamics in the human tumor tissues of lung and breast and lymphoma tumors parallel this isotope dynamics in these patients' myocardium (Sehweil et al., 1989). In this case, ^{201}Tl accumulation can be a helpful tool for tumor viability studying. Complexes of Tl^{3+} with pyridine-2,6-dicarboxylate derivatives have induced mitochondrial apoptosis in human melanoma A375 cells and showed selective anticancer activity to cause ROS production increase, p53 activation, cytochrome *c* release, $\Delta\Psi_{mito}$ decline, procaspase-9, and one-3 expression (Abdolmaleki et al., 2021). These authors suggested that the treatment of certain oncological diseases can develop drugs using these complexes.

2.3 Toxic effects of thallium on aquatic protozoa and unicellular organisms

The sensitivity of bacteria to thallium turned out to be very different. On the one hand, inhibition of the nitrification reaction in *Nitrobacter agilis* took 0.8–8 mg/L thallium acetate. However, up to 120 mg/L of this reagent is required for a noticeable effect on the *Staphylococcus aureus* bacterium (Schoer, 1984). Thallium acetate-induced transition from elongated to spherical shape in *E. coli* cells could result from less rigidity of the cell wall without losing cytoplasmic function (Nakamura and Kohguchi, 1971). Thallium is easier than potassium to penetrate bacterial cells *Streptococcus faecalis* and *Micrococcus lysodeikticus* (Skul'skiĭ et al., 1977). Thallium suppressed the ATPase activity and the potential of the bacterial plasma membrane *eutrophus* AE104 and *eutrophus* CH34 (Gruzina et al., 2000). Potassium present in the water reduced the accumulation of Tl^+ by freshwater bacteria, but it did not affect the collection of the organic derivative of this metal, dimethyltallium (I) (Twiss et al., 2004). The microalga *Chlorella* sp. or the cyanobacterium *Synechococcus leopoliensis* were applied as a model system in testing thallium toxicity. Wherein, a decrease in the toxic effects of Tl^+ (0–400 nM) on these unicellular organisms was found to achieve only at 40–160-fold excess of K^+ in water (Hassler et al., 2007). At the same time, K^+ inhibited the transport of Tl^+ into these cells. This transport did not change in the presence of Na^+, Mg^{2+}, or Ca^{2+}, and this one significantly enhanced with a lack of certain metals (Cu, Co, Mo, Mn, Fe, Zn) in water with a simultaneous weakening of the inhibition of this transport by K^+ ions. Ions of K^+ inhibited Tl^+-induced suppression of bacterial growth *Bacillus megaterium* and *Saccharomyces cerevisiae* (Norris et al., 1976). An assessment of the toxic effects of thallium on aquatic organisms developed using the algae *Pseudokirchneriella subcapitata, Ceriodaphnia dubia,* and black fathead *Pimephales promelas* (Rickwood et al., 2015).

Chlorella vulgaris algae completely bound thallium dissolved in water up to its concentration of 150 mg/L (Birungi and Chirwa, 2015). The accumulation of ^{204}Tl by the cells of the Black Sea alga *Ulva rigida* was more significant than the accumulation of ^{42}K (Skul'skiĭ et al., 1972a,b). This one has activated additionally in the daytime. Still, it is inhibited noticeably by NaF, which is associated with the intensity of metabolism and the participation of photosynthetic phenomena in this process. The accumulation of thallium in the cells of the marine microalga *Ulva lactuca* linearly depended on its concentration in water and took place with the participation of the Na^+, K^+, $2Cl^-$-cotransporter (Turner and Furniss, 2012). This metal continued to accumulate in algae even when potassium in the medium was 100,000 times that of thallium. The toxicity of Tl^{3+} for Daphnia was higher than the toxicity of Tl^+, and in this case, it was comparable to the toxicity of Hg^{2+} (Lan and Lin, 2005). Using the microalga *Chlorella* sp. as a model system for assessing the thallium toxicity found that a decrease in the toxic effects of Tl^+ could achieve only with a 40-160-fold excess of K^+ in the medium (Hassler et al., 2007). The half-life of thallium from these algae was 12–13.5 minutes. Thus, one can see that the presence of potassium in the medium significantly reduced the toxic effect of thallium on microalgae and bacteria (Norris et al., 1976; Hassler et al., 2007; Turner and Furniss, 2012).

References

Abdel-Daim, M.M., Abdou, R.H., 2015. Protective effects of diallyl sulfide and curcumin separately against thallium-induced toxicity in rats. Cell J. 17 (2), 379–388. https://doi.org/10.22074/cellj.2016.3752.

Abdel-Dayem, H.M., Scott, A.M., Macapinlac, H.A., El-Gazzar, A.H., 1994. In: Freeman, L.M. (Ed.), Role of ^{201}Tl Chloride and ^{99}mTc Sestamibi in Tumor Imaging. Nuclear Medicine Annual. Raven Press, New York, pp. 181–234.

Abdolmaleki, S., Ghadermazi, M., Aliabadi, A., 2021. Novel Tl(III) complexes containing pyridine-2,6-dicarboxylate derivatives with selective anticancer activity through inducing mitochondria-mediated apoptosis in A375 cells. Sci. Rep. 11 (1), 15699. https://doi.org/10.1038/s41598-021-95278-y.

Achenbach, C., Hauswirth, O., Heindrichs, C., Ziskoven, R., Köhler, F., et al., 1980. Quantitative measurement of time-dependent thallium distribution in organs of mice by field desorption mass spectrometry. J. Toxicol. Environ. Health 6 (3), 519–528. https://doi.org/10.1080/15287398009529870.

Achenbach, C., Wiemer, J., Ziskoven, R., Winter, U., 1982. Tl^+-ions: effects on the automaticity of sinoatrial tissue and dV/dtmax and Ik2 of cardiac Purkinje fibres. Z. Naturforsch C. J. Biosci. 37 (10), 1006–1014. https://doi.org/10.1515/znc-1982-1022.

Achenbach, C., Ziskoven, R., Koehler, F., Bahr, U., Schulten, H.R., 1979. Quantitative trace analysis of thallium in biological material. Angew. Chem. Int. Ed. Engl. 18 (11), 882–883. https://doi.org/10.1002/anie.197908821.

Adams, D.J., Dwyer, T.M., Hille, B., 1980. The permeability of endplate channels to monovalent and divalent metal cations. J. Gen. Physiol. 75 (5), 493–510. https://doi.org/10.1085/jgp.75.5.493.

Al Hammouri, F., Darwazeh, G., Said, A., Ghosh, R.A., 2011. Acute thallium poisoning: series of ten cases. J. Med. Toxicol. 7 (4), 306–311. https://doi.org/10.1007/s13181-011-0165-3.

Ali, S.F., Jairaj, K., Newport, G.D., Lipe, G.W., Slikker Jr., W., 1990. Thallium intoxication produces neurochemical alterations in rat brain. Neurotoxicology 11 (2), 381–390.

Álvarez-Barrera, L., Rodríguez-Mercado, J.J., Mateos-Nava, R.A., Vázquez-Martínez, Y., Altamirano-Lozano, M.A., 2019. Effect on the offspring of pregnant females CD-1 mice treated with a single thallium(I) application. Reprod. Toxicol. 90, 1–7. https://doi.org/10.1016/j.reprotox.2019.07.022.

Ando, A., Ando, I., Katayama, M., Sanada, S., Hiraki, T., et al., 1987. Biodistributions of ^{201}Tl in tumor bearing animals and inflammatory lesion induced animals. Eur. J. Nucl. Med. 12 (11), 567–572. https://doi.org/10.1007/BF00296099.

Aoyama, H., Yoshida, M., Yamamura, Y., 1986. Acute poisoning by intentional ingestion of thallous malonate. Hum. Toxicol. 5 (6), 389–392. https://doi.org/10.1177/096032718600500612.

Appenroth, D., Gambaryan, S., Winnefeld, K., Leiterer, M., Fleck, C., Bräunlich, H., 1995. Functional and morphological aspects of thallium-induced nephrotoxicity in rats. Toxicology 96 (3), 203–215. https://doi.org/10.1016/0300-483x(94)02919-l.

Appenroth, D., Tiller, S., Gambaryan, S., Winnefeld, K., Fleck, C., Bräunlich, H., 1996. Ontogenetic aspects of thallium-induced nephrotoxicity in rats. J. Appl. Toxicol. 16 (3), 235–243. https://doi.org/10.1002/(SICI)1099-1263(199605)16:3<235::AID-JAT338>3.0.CO;2-7.

Appenroth, D., Winnefeld, K., 1999a. Reconsideration of some hypotheses on the mechanism of thallium toxicity in rats with special respect to riboflavin and glutathione. Nephrol. Dial. Transplant. 14 (Suppl. 4), 16–18. https://doi.org/10.1093/oxfordjournals.ndt.a027948.

Appenroth, D., Winnefeld, K., 1999b. Is thallium-induced nephrotoxicity in rats connected with riboflavin and/or GSH?–reconsideration of hypotheses on the mechanism of thallium toxicity. J. Appl. Toxicol. 19 (1), 61–66. https://doi.org/10.1002/(sici)1099-1263(199901/02)19:1<61::aid-jat539>3.0.co;2-s.

Arbiser, J.L., Alani, R., Flynn, E., Wooten, V., Byers, H.R., 1997. Effects of thallium ion on cellular components of the skin. J. Dermatol. 24 (3), 147–155. https://doi.org/10.1111/j.1346-8138.1997.tb02762.x.

Ashcroft, F.M., Stanfield, P.R., 1983. The influence of the permeant ions thallous and potassium on inward rectification in frog skeletal muscle. J. Physiol. 343 (1), 407–428. https://doi.org/10.1113/jphysiol.1983.sp014900.

Bank, W.J., Pleasure, D.E., Suzuki, K., Nigro, M., Katz, R., 1972. Thallium poisoning. Arch. Neurol. 26 (5), 456–464. https://doi.org/10.1001/archneur.1972.00490110090009.

Barclay, R.K., Peacock, W.C., Karnofsky, D.A., 1953. Distribution and excretion of radioactive thallium in the chick embryo, rat, and man. J. Pharmacol. Exp. Therapeut. 107 (2), 178–187.

Barroso-Moguel, R., Mendez-Armenta, M., Villeda-Hernandez, J., Rios, C., Galvan-Arzate, S., 1996. Experimental neuromyopathy induced by thallium in rats. J. Appl. Toxicol. 16 (5), 385–389. https://doi.org/10.1002/(SICI)1099-1263(199609)16:5<385::AID-JAT359>3.0.CO;2-L.

Barroso-Moguel, R., Ríos-Castañeda, C., Villeda-Hernández, J., Méndez-Armenta, M., Galván-Arzate, S., 1990. Neurotoxicity of thallium: biochemical and morphological study of organic lesions. Arch. Invest. Med. 21 (2), 115–122.

Barroso-Moguel, R., Villeda-Hernández, J., Méndez-Armenta, M., Ríos, C., Galván-Arzate, S., 1992. Osteochondral lesions in developing rats intoxicated with thallium twenty four hours after birth. Arch. Med. Res. 23 (3), 129–133.

Bergquist, J.E., Edström, A., Hansson, P.A., 1983. Bidirectional axonal transport of thallium in frog sciatic nerve. Acta Physiol. Scand. 117 (4), 513–518. https://doi.org/10.1111/j.1748-1716.1983.tb07220.x.

Birungi, Z.S., Chirwa, E.M., 2015. The adsorption potential and recovery of thallium using green micro-algae from eutrophic water sources. J. Hazard Mater. 299, 67–77. https://doi.org/10.1016/j.jhazmat.2015.06.011.

Brismar, T., 1991. In vivo analysis of intracellular thallium-201 accumulation in skeletal muscle of the rat. Acta Physiol. Scand. 142 (4), 475–480. https://doi.org/10.1111/j.1748-1716.1991.tb09182.x.

Brockhaus, A., Dolgner, R., Ewers, U., Krämer, U., Soddemann, H., Wiegand, H., 1981. Intake and health effects of thallium among a population living in the vicinity of a cement plant emitting thallium containing dust. Int. Arch. Occup. Environ. Health 48 (4), 375–389. https://doi.org/10.1007/BF00378686.

Brown, D.R., Callahan, B.G., Cleaves, M.A., Schatz, R.A., 1985. Thallium induced changes in behavioral patterns: correlation with altered lipid peroxidation and lysosomal enzyme activity in brain regions of male rats. Toxicol. Ind. Health 1 (1), 81–98. https://doi.org/10.1177/074823378500100109.

Caffier, G., Shvinka, N.E., 1982. The effect of Rb^{+}, Cs^{+}, and $T1^{+}$ on the gramicidin A-induced conductance changes of the skeletal muscle cell membrane. Acta Biol. Med. Ger. 41 (11), 1087–1090.

Campbell, C., Bahrami, S., Owen, C., 2016. Anagen effluvium caused by thallium poisoning. JAMA Dermatol. 152 (6), 724–726. https://doi.org/10.1001/jamadermatol.2016.0194.

Careaga-Olivares, J., Morales-Aguilera, A., 1993. Differential accumulation of thallous ion by diverse rabbit and rat muscles. Bull. Environ. Contam. Toxicol. 51 (5), 764–771. https://doi.org/10.1007/BF00201656.

Cavanagh, J.B., 1979. The 'dying back' process. A common denominator in many naturally occurring and toxic neuropathies. Arch. Pathol. Lab Med. 103 (13), 659–664.

Cavanagh, J.B., Fuller, N.H., Johnson, H.R., Rudge, P., 1974. The effects of thallium salts, with particular reference to the nervous system changes. A report of three cases. Q. J. Med. 43 (170), 293–319.

Cavanagh, J.B., Gregson, M., 1978. Some effects of a thallium salt on the proliferation of hair follicle cells. J. Pathol. 125 (4), 179–191. https://doi.org/10.1002/path.1711250403.

Chauncey Jr., D.M., Schelbert, H.R., Halpern, S.E., Delano, F., McKegney, M.L., et al., 1977. Tissue distribution studies with radioactive manganese: a potential agent for myocardial imaging. J. Nucl. Med. 18 (9), 933–936.

Chou, Y.T., Lo, K.Y., 2019. Thallium(I) treatment induces nucleolar stress to stop protein synthesis and cell growth. Sci. Rep. 9 (1), 6905. https://doi.org/10.1038/s41598-019-43413-1.

Clarkson, T.W., 1987. Metal toxicity in the central nervous system. Environ. Health Perspect. 75 (1), 59–64. https://doi.org/10.1289/ehp.877559.

Curto-Barredo, L., Segura, S., Martín-Ezquerra, G., Lloveras, B., Gallardo, F., Pujol, R.M., 2015. Anagen effluvium due to thallium poisoning derived from the intake of Chinese herbal medicine and rodenticide containing thallium salts. J. Dermatol. 42 (10), 1027–1029. https://doi.org/10.1111/1346-8138.12997.

Danilewicz, M., Danilewicz, M., Kurnatowski, A., 1979. Histopathological changes in the kidney of experimental animals caused by lethal doses of thallium (in Polish). Med. Pr. 30 (4), 257–265.

Davis, L.E., Standefer, J.C., Kornfeld, M., Abercrombie, D.M., Butler, C., 1981. Acute thallium poisoning: toxicological and morphological studies of the nervous system. Ann. Neurol. 10 (1), 38–44. https://doi.org/10.1002/ana.410100108.

Den Hertog, A., Mooij, J.J., 1976. Potassium induced potential changes in rat diaphragm muscle. Pflügers Archiv. 362 (1), 69–79. https://doi.org/10.1007/BF00588683.

Dikal, M.V., 2014. Melatonin correction of renal functional disorders in rats under conditions of toxic action of thallium chloride. In: Current Problems and Achievements in Medicine. Proceedings of the International Interuniversity Scientific-Practical Conference, Samara, April 7, 2014. Innovative Center for the Development of Education and Science, Samara, pp. 103–105 (Russian).

Downs, W.L., Scott, J.K., Steadman, L.T., Maynard, E.A., 1960. Acute and sub-acute toxicity studies of thallium compounds. Am. Ind. Hyg. Assoc. J. 21 (5), 399–406. https://doi.org/10.1080/00028896009344093.

Dumitru, D., Kalantri, A., 1990. Electrophysiologic investigation of thallium poisoning. Muscle Nerve 13 (5), 433–437. https://doi.org/10.1002/mus.880130510.

Edelman, L., 1988. The cell water problem posed by electron microscopic studies of ion binding in muscles. Scanning Microsc. 2 (2), 851–865.

Ellory, J.C., Nibelle, J., Smith, M.W., 1973. The effect of salt adaptation on the permeability and cation selectivity of the goldfish intestinal epithelium. J. Physiol. 231 (1), 105–115. https://doi.org/10.1113/jphysiol.1973.sp010222.

Erbsloeh, J., 1960. Thallium poisoning in the second half of pregnancy (in Germany). Arch. Toxicol. 18, 156–159.

Fedorovskaya, L.A., 1971. On the issue of occupational thallium poisoning. In: Proceedings of the All-USSR Scientific Conference of Young Researchers on Occupational Hygiene and Occupational Pathology, Moscow, p. 50 (Russian).

Fleck, C., Appenroth, D., 1996. Renal amino acid transport in immature and adult rats during thallium-induced nephrotoxicity. Toxicology 106 (1–3), 229–236. https://doi.org/10.1016/0300-483x(95)03194-k.

Formigli, L., Scelsi, R., Poggi, P., Gregotti, C., Di Nucci, A., et al., 1986. Thallium-induced testicular toxicity in the rat. Environ. Res. 40 (2), 531–539. https://doi.org/10.1016/s0013-9351(86)80128-1.
Galván-Arzate, S., Martínez, A., Medina, E., Santamaría, A., Ríos, C., 2000. Subchronic administration of sublethal doses of thallium to rats: effects on distribution and lipid peroxidation in brain regions. Toxicol. Lett. 116 (1–2), 37–43. https://doi.org/10.1016/s0378-4274(00)00200-9.
Galván-Arzate, S., Ríos, C., 1994. Thallium distribution in organs and brain regions of developing rats. Toxicology 90 (1–2), 63–69. https://doi.org/10.1016/0300-483x(94)90205-4.
Galván-Arzate, S., Santamaría, A., 1998. Thallium toxicity. Toxicol. Lett. 99 (1), 1–13. https://doi.org/10.1016/s0378-4274(98)00126-x.
Gibson, J.E., Becker, B.A., 1970. Placental transfer, embryotoxicity, and teratogenicity of thallium sulfate in normal and potassium-deficient rats. Toxicol. Appl. Pharmacol. 16 (1), 120–132. https://doi.org/10.1016/0041-008x(70)90168-7.
Goel, A., Aggarwal, P., 2007. Pesticide poisoning. Natl. Med. J. India 20 (4), 182–191.
Goldschmidt, J., Wanger, T., Engelhorn, A., Friedrich, H., Happel, M., et al., 2010. High-resolution mapping of neuronal activity using the lipophilic thallium chelate complex TlDDC: protocol and validation of the method. Neuroimage 49 (1), 303–315. https://doi.org/10.1016/j.neuroimage.2009.08.012.
Gozhenko, A.I., Kukharchuk, A.L., Magalyas, V.N., Rasputnyak, S.G., Chalaya, E.N., 1996. Renal function in thallium intoxication. Med. Tr. Prom. Ekol. 11, 33–36 (Russian). PMID: 9213509.
Gregotti, C., Di Nucci, A., Costa, L.G., Manzo, L., Scelsi, R., et al., 1992. Effects of thallium on primary cultures of testicular cells. J. Toxicol. Environ. Health A. 36 (1), 59–69. https://doi.org/10.1080/15287399209531623.
Grover, A.K., Frederickson, M., Daniel, E.E., 1981. Thallous ion activation of p-nitrophenylphosphatase of rat myometrium plasma membrane. Can. J. Physiol. Pharmacol. 59 (11), 1180–1183. https://doi.org/10.1139/y81-182.
Gruzina, T.G., Balakina, M.N., Ul'berg, Z.R., 2000. Effect of ATP-ase activity and transmembrane potential in bacteria. Ukr. Biokhim. Zh. 72 (2), 72–76 (Russian). PMID: 10979585.
Hagiwara, S., Eaton, D.C., Stuart, A.E., Rosenthal, N.P., 1972. Cation selectivity of the resting membrane of squid axon. J. Membr. Biol. 9 (1), 373–384. https://doi.org/10.1007/BF01868063.
Hantson, P., Desoir, R., Léonard, E.D., Crutzen-Fayt, M.C., Léonard, A., Mahieu, P., 1997. Cytogenetic observations following thallium poisoning. J. Toxicol. Environ. Health A. 50 (2), 97–100. https://doi.org/10.1080/009841097160500.
Hasan, M., Ali, S.F., 1981. Effects of thallium, nickel, and cobalt administration of the lipid peroxidation in different regions of the rat brain. Toxicol. Appl. Pharmacol. 57 (1), 8–13. https://doi.org/10.1016/0041-008x(81)90019-3.
Hasan, M., Ali, S.F., Tariq, M., 1978. Levels of dopamine, norepinephrine and 5-hydroxytryptamine in different regions of the rat brain in thallium toxicosis. Acta Pharmacol. Toxicol. 43 (3), 169–173. https://doi.org/10.1111/j.1600-0773.1978.tb02251.x.
Hasan, M., Chandra, S.V., Bajpai, V.K., Ali, S.F., 1977. Electron microscopic effects of thallium poisoning on the rat hypothalamus and hippocampus: biochemica changes in the cerebrum. Brain Res. Bull. 2 (4), 255–261. https://doi.org/10.1016/0361-9230(77)90081-8.
Hasan, M., Haider, S.S., 1989. Acetyl-homocysteine thiolactone protects against some neurotoxic effects of thallium. Neurotoxicology 10 (2), 257–261.

Hassler, C.S., Chafin, R.D., Klinger, M.B., Twiss, M.R., 2007. Application of the biotic ligand model to explain potassium interaction with thallium uptake and toxicity to plankton. Environ. Toxicol. Chem. 26 (6), 1139–1145. https://doi.org/10.1897/06-315r.1.
Henning, C.H., Becker, G., Forth, W., 1982. Movement of thallium(I)-ions across the epithelium of intestinal segments in vitro of rats. Naunyn-Schmiedeberg's Arch. Pharmacol. 321 (2), 157–163. https://doi.org/10.1007/BF00518485.
Henning, C.H., Forth, W., 1982. The excretion of thallium (I)-ions into the gastrointestinal tract in situ of rats. Arch. Toxicol. 49 (2), 149–158. https://doi.org/10.1007/BF00332362.
Hille, B., 1973. Potassium channels in myelinated nerve. Selective permeability to small cations. J. Gen. Physiol. 61 (6), 669–686. https://doi.org/10.1085/jgp.61.6.669.
Hoffman, R.S., 2003. Thallium toxicity and the role of Prussian blue in therapy. Toxicol. Rev. 22 (1), 29–40. https://doi.org/10.2165/00139709-200322010-00004.
Hologgitas, J., Ullucci, P., Driscoll, J., Grauerholz, J., Martin, H., 1980. Thallium elimination kinetics in acute thallotoxicosis. J. Anal. Toxicol. 4 (2), 68–75. https://doi.org/10.1093/jat/4.2.68.
Hou, L.P., Yang, Y., Shu, H., Ying, G.G., Zhao, J.L., et al., 2017. Changes in histopathology, enzyme activities, and the expression of relevant genes in zebrafish (*Danio rerio*) following long-term exposure to environmental levels of thallium. Bull. Environ. Contam. Toxicol. 99 (5), 574–581. https://doi.org/10.1007/s00128-017-2176-5.
Huang, C.C., Shih, T.S., Liu, C.H., 2012. Acute thallium intoxication: an experience in Taiwan. Curr. Top. Toxicol. 8, 33–41.
Hughes, M.N., Man, W.K., Whaler, B.C., 1978. The toxicity of thallium(I) to cardiac and skeletal muscle. Chem. Biol. Interact. 23 (1), 85–97. https://doi.org/10.1016/0009-2797(78)90043-1.
Hultin, T., Näslund, P.H., 1974. Effects of thallium (I) on the structure and functions of mammalian ribosomes. Chem. Biol. Interact. 8 (5), 315–328. https://doi.org/10.1016/0009-2797(74)90010-6.
Johnson, W., 1960. A case of thallium poisoning during pregnancy. Med. J. Aust. 47 (2), 540–542. https://doi.org/10.5694/j.1326-5377.1960.tb68680.x.
Kamiljanov, A.K., 1993. Experimental substantiation of the maximum permissible concentration of thallium carbonate in the atmospheric air. Hyg. Sanit. (5), 8–10 (Russian).
Kanayama, Y., Enomoto, S., Irie, T., Amano, R., 2005. Axonal transport of rubidium and thallium in the olfactory nerve of mice. Nucl. Med. Biol. 32 (5), 505–512. https://doi.org/10.1016/j.nucmedbio.2005.03.009.
Krasovsky, G.N., Yurasova, O.I., Charyev, O.G., Skrypnikov, A.I., Kretova, L.G., 1976. Prediction of the toxicodynamic properties of heavy metals using their radioactive isotopes. In: Hygienic Aspects of Environmental Protection, 4, pp. 41–43 (Russian).
Lameijer, W., van Zwieten, P.A., 1976. Acute cardiovascular toxicity of thallium(I) ions. Arch. Toxicol. 35 (1), 49–61. https://doi.org/10.1007/BF00333985.
Lan, C.H., Lin, T.S., 2005. Acute toxicity of trivalent thallium compounds to Daphnia magna. Ecotoxicol. Environ. Saf. 61 (3), 432–435. https://doi.org/10.1016/j.ecoenv.2004.12.021.
Landauer, W., 1960. Experiments concerning the teratogenic nature of thallium: polyhydroxy compounds, histidine and imidazole as supplements. J. Exp. Zool. 143, 101–105. https://doi.org/10.1002/jez.1401430108.
LeBlanc, G.A., Dean, J.W., 1984. Antimony and thallium toxicity to embryos and larvae of fathead minnows (*Pimephales promelas*). Bull. Environ. Contam. Toxicol. 32 (5), 565–569. https://doi.org/10.1007/BF01607538.

Leloux, M.S., Nguyen, P.L., Claude, J.R., 1987. Experimental studies on thallium toxicity in rats. I-Localization and elimination of thallium after oral acute and sub-acute intoxication. J. Toxicol. Clin. Exp. 7 (4), 247–257.

Leung, K.M., Ooi, V.E., 2000. Studies on thallium toxicity, its tissue distribution and histopathological effects in rats. Chemosphere 41 (1–2), 155–159. https://doi.org/10.1016/s0045-6535(99)00404-x.

Lie, R., Thomas, R.G., Scott, J.K., 1960. The distribution and excretion of thallium-204 in the rat, with suggested MPC's and a bio-assay procedure. Health Psychol. 2, 334–340. https://doi.org/10.1097/00004032-195910000-00004.

Ling, G.N., 1977. Thallium and cesium in muscle cells compete for the adsorption sites normally occupied by K^+. Physiol. Chem. Phys. 9 (3), 217–225.

Liu, E.M., Rajagopal, R., Grand, M.G., 2015. Optic nerve atrophy and hair loss in a young man. JAMA Ophthalmol. 133 (12), 1469–1470. https://doi.org/10.1001/jamaophthalmol.2015.1957.

Lohmann, H., Csicsaky, M., Wiegand, H., 1989. The action of thallium on the excitability of CA1 pyramidal cells in hippocampal slices. Neurotoxicol. Teratol. 11 (6), 545–549. https://doi.org/10.1016/0892-0362(89)90035-4.

Lohmann, H., Wiegand, H., 1996. Reduced probability of orthodromically evoked action potential firing in CA1 pyramidal cells of Guinea pig hippocampal slices after acute thallium exposure. Arch. Toxicol. 70 (7), 430–439. https://doi.org/10.1007/s002040050295.

Loyke, H.F., 2002. Effects of elements in human blood pressure control. Biol. Trace Elem. Res. 85 (3), 193–209. https://doi.org/10.1385/BTER:85:3:193.

Lukács, M., 2003. Thallium poisoning induced polyneuropathy—clinical and electrophysiological data. Ideggyogyaszati Szle. 56 (11–12), 407–414.

Manzo, L., Sabbioni, E., 1988. Thallium. In: Seiler, H.G., Sigel, H., Sigel, A. (Eds.), Handbook on Toxicity of Inorganic Compounds. Marcel Dekker, New York, pp. 677–688.

Marwaha, J., Freedman, R., Hoffer, B., 1980. Electrophysiological changes at a central noradrenergic synapse during thallium toxicosis. Toxicol. Appl. Pharmacol. 56 (3), 345–352. https://doi.org/10.1016/0041-008x(80)90067-8.

Maslova, M.N., Natochin, Y.V., Skul'skiĭ, I.A., 1971. Inhibition of active transport of sodium and Na^+, K^+-dependent ATPase by thallium ions in frog skin. Biochemistry (Mosc.) 36 (4), 867–869 (Russian). https://pubmed.ncbi.nlm.nih.gov/4255389/.

Méndez-Armenta, M., Nava-Ruiz, C., Fernández-Valverde, F., Sánchez-García, A., Rios, C., 2011. Histochemical changes in muscle of rats exposed subchronically to low doses of heavy metals. Environ. Toxicol. Pharmacol. 32 (1), 107–112. https://doi.org/10.1016/j.etap.2011.03.001.

Moore, D., House, I., Dixon, A., 1993. Thallium poisoning. Diagnosis may be elusive but alopecia is the clue. British Med. J. 306 (6891), 1527–1529. https://doi.org/10.1136/bmj.306.6891.1527.

Mulkey, J.P., Oehme, F.W., 1993. A review of thallium toxicity. Vet. Hum. Toxicol. 35 (5), 445–453.

Mulkey, J.P., Oehme, F.W., 1998. A safety study of thallium-containing ceramic superconductor material in rats. Vet. Hum. Toxicol. 40 (1), 11–14.

Mulkey, J.P., Oehme, F.W., 2000. Are 2,3-dimercapto-1-propanesulfonic acid or prussian blue beneficial in acute thallotoxicosis in rats? Vet. Hum. Toxicol. 42 (6), 325–329.

Mullins, L.J., Moore, R.D., 1960. The movement of thallium ions in muscle. J. Gen. Physiol. 43, 759–773. https://doi.org/10.1085/jgp.43.4.759.

Murphy, P.H., Beasley, C.W., Moore, W.H., Stabin, M.G., 1989. Thallium-201 in human milk: observations and radiological consequences. Health Phys. 56 (4), 539–541.

Nakamura, M., Kohguchi, K., 1971. Morphological and biological changes induced by thallium acetate in *Escherichia coli*. Jpn. J. Microbiol. 15 (4), 285–294. https://doi.org/10.1111/j.1348-0421.1971.tb00583.x.
Natochin, Y.V., Skul'skiĭ, I.A., 1972. The difference in the effect of Tl^{+} and Ba^{2+} as partial analogues of K^{+} on the sodium pump in frog skin. Dokl. Akad. Nauk SSSR 203 (6), 1437–1440 (Russian). https://pubmed.ncbi.nlm.nih.gov/4537646/.
Nisticò, G., Faraone, V., Pujia, A., Persico, N., Marmo, E., 1984. Behavioural, electrocortical and spectrum power effects after intraventricular injection of thallium in rats. Zhongguo Yaoli Xuebao 5 (1), 1–4.
Norris, P., Man, W.K., Hughes, M.N., Kelly, D.P., 1976. Toxicity and accumulation of thallium in bacteria and yeast. Arch. Microbiol. 110 (23), 279–286. https://doi.org/10.1007/BF00690239.
Oehme, F.M., 1978. Mechanisms of Heavy Metal Inorganic Toxicities. Toxicity of Heavy Metals in the Environment. Part 1. Marcel Dekker, New York, pp. 69–85.
Olsen, I., Jonsen, J., 1982. Whole-body autoradiography of ^{204}Tl in embryos, fetuses and placentas of mice. Toxicology 23 (4), 353–358. https://doi.org/10.1016/0300-483x(82)90073-7.
Osorio-Rico, L., Villeda-Hernández, J., Santamaría, A., Königsberg, M., Galván-Arzate, S., 2015. The n-methyl-d-aspartate receptor antagonist MK-801 prevents thallium-induced behavioral and biochemical alterations in the rat brain. Int. J. Toxicol. 34 (6), 505–513. https://doi.org/10.1177/1091581815603936.
Oštádalová, I., Babický, A., 1987. Protective effect of thallium against the formation of selenium cataract in rats. Biol. Trace Elem. Res. 14 (1–2), 101–103. https://doi.org/10.1007/BF02795600.
Ouellet, J.D., Dubé, M.G., Niyogi, S., 2013. The influence of food quantity on metal bioaccumulation and reproduction in fathead minnows (*Pimephales promelas*) during chronic exposures to a metal mine effluent. Ecotoxicol. Environ. Saf. 91, 188–197. https://doi.org/10.1016/j.ecoenv.2013.02.002.
Parker, J., Scheck, M., 1981. Thallium chondrodystrophy in chick embryos: ultrastructural changes in the epiphyseal plate. Calcif. Tissue Int. 33 (1), 57–69. https://doi.org/10.1007/BF02409413.
Pau, P.W., 2000. Management of thallium poisoning. Hong Kong Med. J. 6 (3), 316–318.
Paulson, G., Vergara, G., Young, J., Bird, M., 1972. Thallium intoxication treated with dithizone and hemodialysis. Arch. Intern. Med. 129 (1), 100–103.
Pelclová, D., Urban, P., Ridzon, P., Senholdová, Z., Lukás, E., et al., 2009. Two-year follow-up of two patients after severe thallium intoxication. Hum. Exp. Toxicol. 28 (5), 263–272. https://doi.org/10.1177/0960327109106487.
Peter, A.L., Viraraghavan, T., 2005. Thallium: a review of public health and environmental concerns. Environ. Int. 31 (4), 493–501. https://doi.org/10.1016/j.envint.2004.09.003.
Petersohn, K.L., 1960. Thallium poisoning in pregnancy. Arch. Toxikol. 18, 160–164.
Ponsoda, X., Jover, R., Núñez, C., Royo, M., Castell, J.V., Gómez-Lechón, M.J., 1995. Evaluation of the cytotoxicity of 10 chemicals in human and rat hepatocytes and in cell lines: correlation between in vitro data and human lethal concentration. Toxicol. Vitro 9 (6), 959–966. https://doi.org/10.1016/0887-2333(95)00053-4.
Popova, O.Y., Silaev, A.A., Spiridonova, B.C., Strongina, O.M., Shabalina, L.P., Tsvetkova, R.P., 1977. The effect of some compounds of boron, barium, cadmium and thallium on reproductive function. In: Tarasenko, N.Y. (Ed.), Occupational Physiology and Hygiene in the Scientific and Technological Process, Moscow, pp. 46–50 (Russian).
Prick, J.J.G., 1979. Thallium poisoning. Handbook of clinical neurology. In: Vinken, P.J., Bruyn, G.W. (Eds.), Intoxication of the Nervous System, vol. 36. North-Holland, New York, pp. 239–278.

Qi, J., Lai, Y., Liang, C., Yan, S., Huang, K., et al., 2019. Prenatal thallium exposure and poor growth in early childhood: a prospective birth cohort study. Environ. Int. 123, 224–230. https://doi.org/10.1016/j.envint.2018.12.005.

Rade, J.E., Marafante, E., Sabbioni, E., Gregotti, C., Di Nucci, A., Manzo, L., 1982. Placental transfer and retention of ^{201}Tl-thallium in the rat. Toxicol. Lett. 11 (3–4), 275–280. https://doi.org/10.1016/0378-4274(82)90161-8.

Ramsden, D.B., 2007. Thallium. In: Waring, R.H., Steventon, G.B., Mitchel, S.C. (Eds.), Molecules of Death (2nd). Imperial College Press, London, pp. 415–422.

Rauws, A.G., 1974. Thallium pharmacokinetics and its modification by Prussian blue. Naunyn-Schmiedeberg's Arch. Pharmacol. 284 (3), 294–306. https://doi.org/10.1007/BF00500348.

Repetto, G., del Peso, A., Repetto, M., 1998. Human thallium toxicity. In: Nriagu, J.O. (Ed.), Thallium in the Environment. Wiley, New York, pp. 167–199.

Reyes-Rodríguez, M.L.Á., Santos-Cruz, L.F., García-Castro, C., Durán-Díaz, Á., Castañeda-Partida, L., et al., 2021. Genotoxicity and cytotoxicity evaluation of two thallium compounds using the *Drosophila* wing somatic mutation and recombination test. Heliyon 7 (5), e07087. https://doi.org/10.1016/j.heliyon.2021.e07087.

Rickwood, C.J., King, M., Huntsman-Mapila, P., 2015. Assessing the fate and toxicity of Thallium I and Thallium III to three aquatic organisms. Ecotoxicol. Environ. Saf. 115, 300–308. https://doi.org/10.1016/j.ecoenv.2014.12.024.

Riyaz, R., Pandalai, S.L., Schwartz, M., Kazzi, Z.N., 2013. A fatal case of thallium toxicity: challenges in management. J. Med. Toxicol. 9 (1), 75–78. https://doi.org/10.1007/s13181-012-0251-1.

Rodríguez-Mercado, J.J., Altamirano-Lozano, M.A., 2013. Genetic toxicology of thallium: a review. Drug Chem. Toxicol. 36 (3), 369–383. https://doi.org/10.3109/01480545.2012.710633.

Rodríguez-Mercado, J.J., Hernández-de la Cruz, H., Felipe-Reyes, M., Jaramillo-Cruz, E., Altamirano-Lozano, M.A., 2015. Evaluation of cytogenetic and DNA damage caused by thallium(I) acetate in human blood cells. Environ. Toxicol. 30 (5), 572–580. https://doi.org/10.1002/tox.21934.

Rossi, F., Cazzola, M., De Santis, D., Lisa, M., Servodio, R., et al., 1987. Cardiovascular effects of thallium sulfate after intracerebroventricular administration in normotensive rats. Curr. Ther. Res. Clin. Exp. 42 (5), 778–789.

Roth, M., 1975. Teratogenic micromelia in the chick embryo: osteo-vasculo-neural macrorelations. Folia Morphol. (Praha) 23 (3), 236–246.

Rusznyák, I., György, L., Ormai, S., Millner, T., 1968. On some potassium-like qualities of the thallium ion. Experientia 24 (8), 809–810. https://doi.org/10.1007/bf02144884.

Sabbioni, E., Gregotti, C., Edel, J., Marafante, E., Di Nucci, A., Manzo, L., 1982. Organ/tissue disposition of thallium in pregnant rats. Arch. Toxicol. 5 (Suppl. l), 225–230. https://doi.org/10.1007/978-3-642-68511-8_41.

Schäfer, S.G., Forth, W., 1987. Thallium(I) secretion across the isolated mucosa of rat descending colon. Arch. Toxicol. 59 (6), 413–420. https://doi.org/10.1007/BF00316207.

Schäfer, S.G., Nell, G., Henning, C.H., 1981. Movement of thallium (I) ions in vitro. Arch. Toxicol. 48 (4), 271–279. https://doi.org/10.1007/BF00319655.

Schoer, J., 1984. Thallium. In: Hutzinger, O. (Ed.), Handbook of Environmental Chemistry. Anthropogenic Compounds. Springer, New York, pp. 143–214.

Sehweil, A.M., McKillop, J.H., Milroy, R., Wilson, R., Abdel-Dayem, H.M., Omar, Y.T., 1989. Mechanism of ^{201}Tl uptake in tumours. Eur. J. Nucl. Med. 15 (7), 376–379. https://doi.org/10.1007/BF00449228.

Shabalina, L.P., Spiridonova, B.C., Pavlenko, G.I., Fomenko, V.N., 1980. Cutaneous-resorptive, local and mutagenic action of thallium carbonate. Gig. Tr. Prof. Zabol. (4), 54–55 (Russian). PMID: 7409564.

Sharma, A.N., Nelson, L.S., Hoffman, R.S., 2004. Cerebrospinal fluid analysis in fatal thallium poisoning: evidence for delayed distribution into the central nervous system. Am. J. Forensic Med. Pathol 25 (2), 156–158. https://doi.org/10.1097/01.paf.0000127382.31642.f1.

Shi, L., Yuan, Y., Xiao, Y., Long, P., Li, W., et al., 2021. Associations of plasma metal concentrations with the risks of all-cause and cardiovascular disease mortality in Chinese adults. Environ. Int. 157, 106808. https://doi.org/10.1016/j.envint.2021.106808.

Shvinka, N.E., Caffier, G., 1983. Effect of thallium ions on gramicidin-induced conductance of muscle fiber membrane. Biofizika 28 (6), 1006–1009 (Russian). PMID: 6197098.

Skul'skiĭ, I.A., Allakhverdov, B.L., Burovina, I.V., Lapin, A.V., 1980. X-ray local micro-analysis in cytology. Sh. Effect of thallium on the ionic composition and transport properties of frog skin epithelial cells. Cytology 22 (4), 409–414 (Russian).

Skul'skiĭ, I.A., Burovina, I.V., Glazunov, V.V., 1976. Investigation of the ionic selectivity of the mechanisms of transmembrane potassium transport in neurons of the mollusk *Planorbarius corneus*. Zh. Evolut. Biokhim. Fiziol. 12 (2), 105–112 (Russian). PMID: 941563.

Skul'skiĭ, I.A., Glazunov, V.V., Riabova, I.D., Gorneva, G.A., 1977. Selective permeability of bacterial membranes for monovalent thallium ions. Biokhimiia 42 (9), 1637–1641 (Russian). PMID: 410460.

Skul'skiĭ, I.A., Glazunov, V.V., Zesenko, A.I., Liubimov, A.A., 1972a. Accumulation of Tl^+ ions in cells of the Black Sea alga Ulva rigida. Biofizika 17 (5), 824–831 (Russian). PMID: 5086083.

Skul'skiĭ, I.A., Krestinskaya, T.V., Ivanova, T.I., Lapin, A.V., 1982. The influence of the Tallina on the activity of enzymes and unidirectional sodium transport in the skin of the frog. Cytology 24 (2), 188–193 (Russian). PMID: 6280346.

Skul'skiĭ, I.A., Lapin, A.V., 1983. The influence of Tallina on the transport of sodium and potassium in the skin frog. Cytology 25 (11), 1284–1288 (Russian). PMID: 6318410.

Skul'skiĭ, I.A., Lapin, A.V., 1992. Highly selective blockade of the frog skin sodium channels by monovalent copper cations. Biochim. Biophys. Acta 1112 (1), 27–28. https://doi.org/10.1016/0005-2736(92)90249-l.

Skul'skiĭ, I.A., Malov, A.S., 1977. Ionic selectivity of mollusk neuron membrane Planorbarius corneus. Zh. Evoliut. Biokh. Fiziol. 13 (5), 635–637 (Russian). PMID: 919907.

Skul'skiĭ, I.A., Pivovarova, N.B., 1988. Effect of inhibitors of energy metabolism on accumulation of potassium in neurons of the mollusk *Planorbarius corneus*. Zh. Evoliut. Biokh. Fiziol. 24 (6), 817–821 (Russian). PMID: 3245356.

Skul'skiĭ, I.A., Glazunov, V.V., Zesenko, A.I., Liubimov, A.A., 1972b. Selectivity of cells of the Black Sea alga *Ulva rigida* to Tl^+, Rb^+, Cs^+. Tsitologiia 14 (7), 849–856 (Russian). PMID: 4637817.

Smith, I.C., 1979. The electrogenic potential in rat C nerve fibres: some effects of lithium and thallium. J. Physiol. 294, 135–144. https://doi.org/10.1113/jphysiol.1979.sp012920.

Spencer, P.S., Peterson, E.R., Madrid, R., Raine, C.S., 1973. Effects of thallium salts on neuronal mitochondria in organotypic cord-ganglia-muscle combination cultures. J. Cell Biol. 58 (1), 79–95. https://doi.org/10.1083/jcb.58.1.79.

Spiridonova, B.C., Shabalina, L.P., 1978. Toxicology of thallium and its compounds (Literature review). Gig. Tr. Prof. Zabol. 7, 41–43 (Russian). PMID: 352834.

Stanfield, P.R., Ashcroft, F.M., Plant, T.D., 1981. Gating of a muscle K^+ channel and its dependence on the permeating ion species. Nature 289 (5797), 509–511. https://doi.org/10.1038/289509a0.
Tabandeh, H., Thompson, G.M., 1993. Visual function in thallium toxicity. BMJ 307 (6899), 324.
Talbot, P.A., 1992. Nature of increase in quantal release by the thallous ion at frog end plates with and without nerve stimulation. J. Gen. Physiol. 100 (5), 881–901. https://doi.org/10.1085/jgp.100.5.881.
Tanaka, J., Yonezawa, T., Ueyama, M., 1978. Acute thallotoxicosis: neuropathological and spectrophotometric studies on an autopsy case. J. Toxicol. Sci. 3 (4), 325–334. https://doi.org/10.2131/jts.3.325.
Tarasenko, N.Y., Vorobyeva, R.S., Zyuzyukin, Y.V., Popova, O.Y., Pronin, O.A., et al., 1975. Occupational hygiene and toxicology of alkaline earth and other metals. In: New in Research Methods for Diagnosis, Treatment and Prevention of Major Diseases. I.M. Sechenov First Moscow State Medical Institute, Moscow, pp. 40–47 (Russian).
Thompson, D.F., 1981. Management of thallium poisoning. Clin. Toxicol. 18 (8), 979–990. https://doi.org/10.3109/15563658108990328.
Thyresson, N., 1950. The influence of dietary factors, especially brewer's yeast, cystine and B vitamins, on the course of chronic thallium poisoning in the rat. Acta Derm. Venereol. 30 (1), 9–26.
Thyresson, N., 1951. Experimental investigation on thallium poisoning; effect of thallium on the growth and differentiation of hairs and epidermis, studied in young rats and in tissue cultures of embryonic rat skin. Acta Derm. Venereol. 31 (2), 133–146.
Truhaut, R., 1960. The toxicology of thallium. J. Occup. Med. 2, 334–336. https://doi.org/10.1097/00043764-196007000-00008.
Turner, A., Furniss, O., 2012. An evaluation of the toxicity and bioaccumulation of thallium in the coastal marine environment using the macroalga, *Ulva lactuca*. Mar. Pollut. Bull. 64 (12), 2720–2724. https://doi.org/10.1016/j.marpolbul.2012.09.023.
Twiss, M.R., Twining, B.S., Fisher, N.S., 2004. Bioconcentration of inorganic and organic thallium by freshwater phytoplankton. Environ. Toxicol. Chem. 23 (4), 968–973. https://doi.org/10.1897/02-643.
Tyagi, R., Rana, P., Khan, A.R., Bhatnagar, D., Devi, M.M., et al., 2011. Study of acute biochemical effects of thallium toxicity in mouse urine by NMR spectroscopy. J. Appl. Toxicol. 31 (7), 663–670. https://doi.org/10.1002/jat.1617.
Vorobyev, N.V., 2003. Clinical and Pathogenetic Features of Acute Thallium Poisoning The Dissertation Abstract of the Candidate of Medical Sciences. St. Petersburg, Russian.
Wanger, T., Scheich, H., Ohl, F.W., Goldschmidt, J., 2012. The use of thallium diethyldithiocarbamate for mapping CNS potassium metabolism and neuronal activity: Tl^+ -redistribution, Tl^+ -kinetics and Tl^+ -equilibrium distribution. J. Neurochem. 122 (1), 106–114. https://doi.org/10.1111/j.1471-4159.2012.07757.x.
Weinig, E., Zink, P., 1967. On the quantitative mass spectrometry of standards thallium content in the human organism (in German). Arch. Toxicol. 22 (4), 255–274.
Wiegand, H., Csicsaky, M., Krämer, U., 1984a. The action of thallium acetate on neuromuscular transmission in the rat phrenic nerve-diaphragm preparation. Arch. Toxicol. 55 (1), 55–58. https://doi.org/10.1007/BF00316586.
Wiegand, H., Lohmann, H., Chandra, S.V., 1986. The action of thallium acetate on phasic transmitter release in the mouse neuromuscular junction. Arch. Toxicol. 58 (4), 265–270. https://doi.org/10.1007/BF00297118.
Wiegand, H., Papadopoulos, R., Csicsaky, M., Krämer, U., 1984b. The action of thallium acetate on spontaneous transmitter release in the rat neuromuscular junction. Arch. Toxicol. 55 (4), 253–257. https://doi.org/10.1007/BF00341021.

Wiegand, H., Uhlig, S., Gotzsch, U., Lohmann, H., 1990. The action of cobalt, cadmium and thallium on presynaptic currents in mouse motor nerve endings Neurotoxicol. Teratology 12 (4), 313–318. https://doi.org/10.1016/0892-0362(90)90049-i.

Wiemer, J., Ziskoven, R., Achenbach, C., 1982. Tl^+-ions: comparison on the effects on the slow inward current and contractility of ventricular tissues. Z. Naturforsch C. J. Biosci. 37 (10), 1015–1022. https://doi.org/10.1515/znc-1982-1023.

Windebank, A.J., 1986. Specific inhibition of myelination by lead in vitro; comparison with arsenic, thallium, and mercury. Exp. Neurol. 94 (1), 203–212. https://doi.org/10.1016/0014-4886(86)90283-9.

Wu, M., Wang, L., Song, L., Liu, B., Liu, Y., et al., 2021. The association between prenatal exposure to thallium and shortened telomere length of newborns. Chemosphere 265, 129025. https://doi.org/10.1016/j.chemosphere.2020.129025.

Xia, W., Du, X., Zheng, T., Zhang, B., Li, Y., et al., 2016. A case-control study of prenatal thallium exposure and low birth weight in China. Environ. Health Perspect. 124 (1), 164–169. https://doi.org/10.1289/ehp.1409202.

Zasukhina, G.D., Krasovsky, G.N., Vasilieva, I.M., Sdirkova, N.I., Sokolovsky, V.V., et al., 1980b. Molecular biological effect of thallium carbonate. Bull. Exp. Biol. Med. 90 (12), 723–725 (Russian). PMID: 6258670.

Zasukhina, G.D., Vasilieva, I.M., Sdirkova, N.I., 1980a. Approach to estimation of the mutagenic potential of the environmental pollutants using determination of the mutagenic effect of thallium carbonate as an example. Dokl. Akad. Nauk SSSR 250 (3), 766–768 (Russian). PMID: 7353468.

Zasukhina, G.D., Vasilyeva, I.M., Sdirkova, N.I., Krasovsky, G.N., Vasyukovich, L.Y., et al., 1983. Mutagenic effect of thallium and mercury salts on rodent cells with different repair activities. Mutat. Res. 124 (2), 163–173. https://doi.org/10.1016/0165-1218(83)90176-3.

Zeiske, W., van Driessche, W., 1983. The interaction of "K^+-like" cations with the apical K^+ channel in frog skin. J. Membr. Biol. 76 (1), 57–72. https://doi.org/10.1007/BF01871453.

Zeiske, W., Van Driessche, W., 1986. Impairment of Na^+ transport across frog skin by Tl^+: effects on turnover, area density and saturation kinetics of apical Na^+ channels. Pflügers Archiv. 407 (2), 145–152. https://doi.org/10.1007/BF00580666.

Ziskoven, R., Achenbach, C., Bahr, U., Schulten, H.R., 1980. Quantitative investigations on the diaplacental transfer of thallium by field desorption mass spectrometry. Z. Naturforsch C. J. Biosci. 35 (11–12), 902–906. https://doi.org/10.1515/znc-1980-11-1205.

Ziskoven, R., Achenbach, C., Schulten, H.R., Roll, R., 1983. Thallium determinations in fetal tissues and maternal brain and kidney. Toxicol. Lett. 19 (3), 225–231. https://doi.org/10.1016/0378-4274(83)90123-6.

Ziskoven, R., Achenbach, C., Wiemer, J., Winter, U., 1982. Tl^+-ions: influence on cardiac contractility. Z. Naturforsch C. J. Biosci. 37 (10), 995–1005. https://doi.org/10.1515/znc-1982-1021.

Zook, B.C., Gilmore, C.E., 1967. Thallium poisoning in dogs. J. Am. Vet. Med. Assoc. 151 (2), 206–217.

CHAPTER 3

Toxic thallium effects on cells

List of abbreviations

5-HD	5-hydroxyodecanoate
ADP	adenosine diphosphate
ATP	adenosine triphosphate
CsA	cyclosporin A
DMSO	dimethyl sulfoxide
DNP	2,4-dinitrophenol
EDTA	ethylenediaminetetraacetic acid
EGTA	ethylene guanine tetraacetic acid
IMM	inner mitochondrial membrane
$\mathbf{K_{ATP}}$	mitochondrial ATP dependent channel
MPTP	mitochondrial permeability transition pore
NADPH	nicotinamide adenine dinucleotide phosphate
RLM	rat liver mitochondria
ROS	reactive oxygen species
$\mathbf{\Delta\Psi_{mito}}$	inner membrane potential

3.1 Molecular mechanisms of thallium transport across cell membranes: use of isotopes and ion channel inhibitors

Monovalent thallium ions can be used as a probe to study biological functions and transport of alkali metal ions such as K^+ and Na^+ (Williams, 1970; Skul'skiĭ, 1991; Nriagu, 1998; Goldschmidt et al., 2010; Wanger et al., 2012). Numerous studies have shown that the main pathways of Tl^+ transport into cells are Na^+/K^+-ATPase, Na^+-K^+-$2Cl^-$-cotransporter, and potassium channels of cell membranes (Mulkey and Oehme, 1993; Nriagu, 1998; Zhang et al., 2010; Blain and Kazantzis, 2015). The affinity of most cellular potassium channels to Tl^+ turned out to be noticeably higher than one to K^+ (Nriagu, 1998). One of Na^+/K^+-ATPase of human and rabbit erythrocytes was higher for Tl^+ than for K^+ (Lishko et al., 1973; Skulskii et al., 1973). Moreover, low concentrations of Tl^+ activated this ATPase, while high concentrations, on the contrary, reduced its activity. A similar regularity did find in reacting Tl^+ with other K^+ dependent enzymes: yeast

Mitochondria as a Key Intracellular Target of Thallium Toxicity
ISBN 978-0-323-95531-7
https://doi.org/10.1016/B978-0-323-95531-7.00005-7

aldehyde dehydrogenase, homoserine dehydrogenase, pyruvate kinase, L-threonine dehydrogenase, AMP deaminase, and vitamin B12-dependent diol dehydratase (Zitko, 1975; Bostian et al., 1982). Chronic thallium poisoning decreased Mg^{2+}-ATPase and Na^+/K^+-ATPase activity of human erythrocytes (Schoer, 1984).

Studies of the active center of Na^+/K^+-ATPase in yeast cells have shown that the most critical molecular groups determining Na^+/K^+-ATPase binding to Tl^+ and K^+ ions are the next: carboxamide groups of asparagine-776, methyl groups of threonine-772 and threonine-774, hydroxyl groups of serine-775 and threonine-771, as well as carboxylate groups of glutamic acid and asparagine (Nielsen et al., 1998; Pedersen et al., 1998). The permeability of the potassium channels of frog and starfish oocytes was noticeably higher to ions of Tl^+ than ones of K^+ (Lopatin et al., 1998; Choe et al., 2000, 2001). Interacting with thiol and carbonyl groups, Tl^+ changed the potassium channel structure of frog oocytes *Xenopus laevis*. As a result, the channel stabilized in an open state, and the ability of Ba^{2+} to inhibit this channel decreased significantly (Lu et al., 2001). The affinity of Tl^+ for the potassium channel (ROMK2, isolated from rat kidney and incorporated into frog oocytes) was greater than that of univalent cations K^+, Rb^+, and NH_4^+ (Chepilko et al., 1995). A patch-clamp study of the voltage-dependent channel of photoreceptor cells in tiger salamander showed that the affinity of this channel for Tl^+ was 1.5 times greater than that for K^+ (Wollmuth and Hille, 1992). Tl^+ blocked transport of K^+ via this channel (Wollmuth, 1995). However, this effect of thallium decreased with an increase in its concentration in the washing solution. Perhaps this was because thallium, in this case, increased nonspecific permeability of cell membrane, similar to the detected activation by thallium in passive potassium permeability of the IMM. In the presence of inorganic phosphate, Tl^+ ions are more active than Rb^+ ions in uptaking by yeast cells (Roomans and Borst-Pauwels, 1977). We found a similar activation of RLM swelling in a medium with $TlNO_3$ and inorganic phosphate (Korotkov et al., 2008). More recently, Tl^+ and a new fluorescent dye Thallos (TEFLabs) were used to study K_{ATP}-dependent channel in HEK-293 cells in the presence of new modulator of new Kir6.2/SUR1 activator VU063 (Kharade et al., 2016). Participation of Na^+-K^+-$2Cl^-$-cotransporter in Tl^+ transport to LLC-PK1 cells was studied using Fluorescent dye FluxOR (Carmosino et al., 2013).

High-throughput screening of more than 200,000 substances in studying Tl^+ transport into HEK293 cells in the presence of fluorescent dye

BTC-AM made it possible to find many new inhibitors and modulators of Ca^{2+}-activated K^+ channel in a cell membrane, including a potent inhibitor of this channel, NS6180 (Jørgensen et al., 2013). Similar studies of voltage-gated potassium channels in human osteosarcoma cells (U2-OS) were previously made using Tl^+ and more sensitive fluorescent dye, FluxOR (Beacham et al., 2010). Tetrapine Q (a stable variant of bee venom toxin) significantly suppressed the K_{ATP}-dependent channel of rat heart mitochondria and Tl^+ transport into permeabilized H9C2 cells and RomK knockdown cells (Foster et al., 2012). We used another inhibitor of the mitoKATP channel (5-HD) in our studies of the MPTP opening in media containing $TlNO_3$ (Korotkov et al., 2015a).

Fluorescence studies using BTC-AM dye and patch-clamp experiments revealed that Tl^+ transports into HEK293 cells and cultured rat muscle cells through voltage-dependent or ligand-gated potassium channels, as well as low-conductive and high-conductive Ca^{2+}-activated K^+ channels (Blatz and Magleby, 1984; Weaver et al., 2004; Hougaard et al., 2007; Jørgensen et al., 2008; Schroeder, Hansen, 2008). Tl^+ more strongly than K^+ activated the voltage-dependent BK potassium channels of HEK293 cells, and their permeability for Tl^+ was higher than that of K^+ (Piskorowski and Aldrich, 2006). The patch-clamp study of thallium accumulation in human osteosarcoma cells (HTB-96) revealed the cardiotoxicity of many substances taken from 1408 synthesized compounds inhibited voltage-dependent hERG potassium channels (Xia et al., 2011). The use of models to evaluate thallium transport into HEK293 cells transfected with human and rat potassium Kir1.1 channels and electrophysiological recordings of channel activity found a marked correlation between the effects of various compounds. This finding allowed for successful screening libraries of multiple substances and provided essential assistance in developing research carried out in medicinal chemistry (Felix et al., 2012).

^{201}Tl after intravenous injection mainly distributed in rat myocardium mitochondria, sarcoplasmic reticulum, nuclei, and myofibrils (Llaurado et al., 1979, 1983; Fukumoto et al., 1997, 1998). These results refer to the importance of studying the poorly studied mechanisms of toxic effects of thallium on mitochondria. Thallium isotopes are a helpful tool in studying cellular and bacterial K^+ transport mechanisms and radiation-induced apoptotic processes in cells (Bakker, 1978; Bakker-Grunwald, 1979; Brismar et al., 1994, 1995; Khoda et al., 2012). At the same time, it is necessary to emphasize the significance of using various isotopes ($^{204}Tl^+$, $^{86}Rb^+$, $^{42}K^+$, $^{22}Na^+$) when studying transport pathways of Tl^+ ions into erythrocytes and the effect of this toxic metal on the transport of K^+ and Na^+ via membrane of algal cells, snail ganglia and the inner mitochondrial membrane (Skul'skiĭ, 1991). Channel blockers (Cs^+, bupivacaine) and Ba^{2+} inhibited potassium channels of internal rectification and depolarized glial cells (Brismar et al., 1995). Perhaps this result is associated with the high affinity of Tl^+ ions to various active groups of molecules that form the structure of potassium channels.

Transport of thallium isotopes (^{201}Tl and ^{204}Tl) into cardiomyocyte culture cells inhibited visibly in the presence of oxidative phosphorylation uncoupler 2,4-dinitrophenol (DNP), 1–4 mM KCN, or 0.5 mM ouabain (Gachon et al., 1986; McCall et al., 1986; Chen et al., 2000). Monocarboxylic ionophore grysorixin, respiratory chain inhibitor cyanide, and glycolysis inhibitor iodoacetate inhibited ^{201}Tl uptake in myocardial cultures' cells (Maublant al., 1985, 1993). These data indicate the vital role of mitochondria supplying necessary energy for Tl^+ transport in cells. The use of thallium isotopes ($^{201}Tl^+$ and $^{204}Tl^+$), patch-clamp method, or fluorescent dyes revealed that the main parts of thallium transport into cells are Na^+/K^+-ATPase and Na^+-K^+-$2Cl^-$-cotransporter. This conclusion resulted in experiments with various cellular systems, which include: rat myocytes, oocytes of lamprey *(Lampetra fluviatilis)*, cells of small-cell adrenal cortex carcinoma (SW-13), mouse colon cancer cells (LS180), human leukemia cells (K562), Ehrlich carcinoma cells, rat myocardial culture cells, dog nephrons, smooth muscle cells, yeast cells, intact giant squid axons, human, rat, rabbit and lamprey erythrocytes, human embryonic kidney cells and rabbit heart Purkinje cells (Gehring and Hammond, 1964; Skulskii et al., 1973, 1990, 1992a,b; Landowne, 1975; Kashket, 1979; Delano et al., 1985; Favari and Mourelle, 1985; McCall et al., 1985; Sands et al., 1985; Sessler et al., 1986; Glynn et al., 1987; Sherstobitov et al., 1990, 2003, 2010, 2011; Bielen et al., 1993; Brismar et al., 1994, 1995; Nakamura et al., 1996; Arbab et al., 1997, 1998; Peluffo and Berlin, 1997; Mori et al., 1998; Nielsen et al., 1998; Chen et al., 2000; Zhang et al., 2010). In many cases, isotopes ($^{201}Tl^+$ and $^{204}Tl^+$) uptake was notably higher than absorption of $^{42}K^+$ and $^{86}Rb^+$. Interestingly, the involvement of Na^+/K^+-ATPase and Na^+-K^+-$2Cl^-$-cotransporter in accumulating $^{204}Tl^+$ by lamprey oocytes was the same until March (Sherstobitov et al., 2011). However, when oocytes become active in the prespawning period, the involvement of Na, K-pump increases significantly.

In animals exposed to thallium salts, this metal concentration in erythrocytes was higher than in blood plasma (Leloux et al., 1987). In this regard, the transport of thallium into erythrocytes can be one way to deliver this metal to different compartments of a body (Leloux et al., 1987). Passive transport of $^{204}Tl^{+}$ across membranes of human and lamprey erythrocytes was maximal in the presence of anions forming strong ion pairs with this cation (Cl^{-}, Br^{-}, I^{-}, HCO_3^{-}). On the contrary, it was minimal in a medium containing anions (NO_3^{-}, ClO_4^{-}, F^{-}, CH_3COO^{-}, SO_4^{2-}), which have a weak association with Tl^{+} (Sherstobitov et al., 1990; Skulskii et al., 1992a,b). Lamprey erythrocyte membrane was found to have potassium channels permeable to Tl^{+} (Sherstobitov et al., 1991). Ba^{2+}, quinine, and amiloride blocked these channels. The transport rate of $^{204}Tl^{+}$ into rat erythrocytes was several times lower than that in lamprey erythrocytes (Sherstobitov et al., 1990, 1991).

An increase in the concentration of potassium ions in the external environment reduced transport of $^{204}Tl^{+}$ or $^{201}Tl^{+}$ into erythrocytes of *Escherichia coli*, human, rabbit, and lamprey, and rat myocardial culture cells (Gehring and Hammond, 1964; Skulskii et al., 1978b; McCall et al., 1985; Sands et al., 1985; Sherstobitov et al., 1991). Adding 10 mM K^{+} or 10 mM Tl^{+} or ouabain into an external medium stimulated $^{22}Na^{+}$ release from human erythrocytes and inhibited $^{42}K^{+}$ entry into these organelles (Sands et al., 1985). Accumulation of $^{204}Tl^{+}$ by human erythrocytes decreased in increasing concentration of nonradioactive thallium from 0.005 to 0.2 mM (Sherstobitov et al., 1990). This result may indicate a rapid saturation of the carrier by Tl^{+} ions. The membrane hyperpolarization in the presence of valinomycin increased the uptake of $^{204}Tl^{+}$ in human erythrocytes (Sherstobitov et al., 1990). It is interesting to note that an increase in the concentration of Na^{+} in erythrocytes prevented a decrease in the activity of Na^{+}/K^{+}-ATPase in the presence of 5–10 mM Tl^{+} (Skulskii et al., 1973). Permeability of erythrocyte membrane for Tl^{+} was significantly higher than that for K^{+} (Skulskii et al., 1978). Tl^{+} transport through membranes of erythrocytes and bacteria was assumed to be carried out by the electrophoretic uniport mechanism (Bakker, 1978; Skulskii and Manninen, 1981). Associated with these data, it is interesting to mention exciting studies of Japanese scientists to find the relationship between thallium concentration and activity of Na^{+}/K^{+}-ATPase. The essence of these experiments was as follows; after a single administration of thallium sulfate (25 mg/kg) to mice, thallium concentration in liver and kidneys increased significantly, while the activity of Na^{+}/K^{+}-ATPase, on the contrary, significantly decreased (Yoshida et al., 1997). However, after releasing in part of accumulated thallium, the activity of this enzyme increased in comparison with the control.

3.2 Thallium toxicity mechanisms on mammalian cells

To assess toxic thallium effects with various antioxidants present, biochemical and histopathological changes in the liver (most important detoxifying tissue) with rats exposed to thallium salts researched (Herman and Bensch, 1967; Mourelle et al., 1988; Leung and Ooi, 2000). Thallium intoxication resulted in damage of intracellular membranes, including the mitochondrial membranes of the liver and kidneys, that was associated with massive swelling of mitochondria. Penetrated the cytoplasm, Tl^+ showed a toxic effect on some cytoplasmic enzymes.

Tl^+ was found to have inhibited the transport of K^+ ions into rat myocardial culture cells (McCall et al., 1985). Tl^+ penetrated easily through a cation-selective channel of rabbit sarcoplasmic reticulum and blocked this channel in the presence of other penetrating ions (Fox, 1983). Thallium damaged neurons in the cerebral cortex and striatum (Hasan et al., 1977b; Huang et al., 2012). Thallium lethal dose for isolated human and rat hepatocytes is 1.12 ± 0.43 and 0.72 ± 0.11 mM of Tl_2SO_4, respectively (Ponsoda et al., 1995). Thallium-poisoned humans or thallium-exposed rodents showed a significant violation of the structural integrity of the plasma and mitochondrial membranes. Intense swelling and vacuolization of mitochondria and axons, the intermembrane space increase, the outer and inner membrane disintegration, the cristae damage, the endoplasmic or sarcoplasmic reticulum degeneration, the destruction of myofilaments in hepatocytes and cardiomyocytes, the lower extremities in skeletal muscle cells, renal tubules' epithelial cells, brain striatum cells, glial cells, and neurons accompanied this thallium toxicity (Herman and Bensch, 1967; Bank et al., 1972; Deshimaru et al., 1977; Hasan et al., 1977a,b; Woods and Fowler, 1986; Leung and Ooi, 2000; Kiliç and Kutlu, 2010; Rodríguez-Mercado and Altamirano-Lozano, 2013; Li et al., 2015).

Thallium uncoupled oxidative phosphorylation in mitochondria of mammalian cells (Douglas et al., 1990; Repetto et al., 1994; Puga Molina and Verstraeten, 2008; Rodríguez-Mercado and Altamirano-Lozano, 2013). High electron density granules did find in experiments with rats exposed to thallium. These granules were in cytoplasm and mitochondria of oligodendrocytes, brain perikaryon, cerebellum, hypothalamus, hippocampus, kidneys, and liver (Herman and Bensch, 1967; Hasan et al., 1977a,b, 1978a; Kiliç and Kutlu, 2010; Li et al., 2015). Isolated hepatocytes added into a medium with 100 μM TlCl showed the increased cytoplasm concentration of Ca^{2+}, Na^{+}, and inorganic phosphate and the decreased K^{+} concentration (Zierold, 2000). We recently showed that Ca^{2+} uptake into neonatal rat cardiomyocytes increased by 1.5 mM Tl^{+} (Korotkov et al., 2020). This effect of Tl^{+} decreased by thallium ionophore nonactin and showed elimination after injection into a medium of EGTA. It recently found that thallium damaged the isolated rat hepatocytes, isolated synaptosomal mitochondrial fractions P2, and neuronal cells of PC12 and HN9.10e lines were in experiments in vitro. This damage resulted in the following phenomena: apoptosis of these cells and their proteolysis, depletion of glutathione pool, $\Delta\Psi_{mito}$ decrease, formation of ROS and H_2O_2, lipid peroxidation increase, and membrane fluidity rise (Hanzel and Verstraeten, 2006, 2009; Verstraeten, 2006; Pourahmad et al., 2010; Eskandari et al., 2011; Pino et al., 2017; Maya-López et al., 2018; Bramanti et al., 2019). These thallium cytotoxic effects were reduced markedly by MPTP inhibitors (CsA and carnitine) (Pourahmad et al., 2010; Eskandari et al., 2011). There is no study of thallium effects' made together with the principal matrix cation K^{+} in experiments with isolated mitochondria. The toxicity of thallium turned out to be lower than that of mercury and cadmium in experiments with isolated intestinal epithelial cells that may be due to features of these cells (Keogh et al., 1994).

Thallium-inhibited acetylcholinesterase activity in isolated cells of mouse neuroblastoma markedly. However, succinate dehydrogenase activation was observed due to the active transport of Tl^{+} into mitochondria of these cells (Repetto et al., 1994). Thallium quinolinate activated respiration of rat mast cell suspension in the presence of 10 mM sodium succinate (Tomilets et al., 1980a,b). In other cases, succinate dehydrogenase activity in mouse skin homogenates did not change upon exposure to 500 mM Tl^{+} (Flesch and Goldstone, 1950). Considering these contradictions, we investigated the effects of various concentrations of Tl^{+} on RLM energized by succinate. Thallium promoted selenium retention in the liver and kidney of rats and did not protect these cells from chronic selenosis (Levander and Argrett, 1969). On the other hand, thallium accumulated noticeably more strongly in the liver, kidneys, and bones of rats, exposed to cooperative action of thallium (20–30 mg/kg) and Na_2SeO_4 (10 mg/kg) (Rusiecki and Brzeziński, 1966). Thallium damaged keratinocytes, melanocytes, and endothelial rat cells in experiments in vivo. And it has exerted a pleiotropic effect on the proliferation, shape, and motility of several types of cells (Arbiser et al., 1997). Thallium disrupted keratin synthesis in experiments in vitro with cell cultures of keratinocytes isolated from the skin of rat embryos (Thyresson, 1951b).

A convincing assumption did make previously that the toxic effect of thallium on cells of living organisms may be associated with the proximity of crystal-chemical radii of K^+ and Tl^+ ions as well as with the easy polarizability of the latter (Korenman, 1960; Williams, 1970; Douglas et al., 1990; Skul'skiĭ, 1991). This circumstance allows Tl^+ compared to K^+ to form stronger chemical bonds with reactive electron-donor groups of biological molecules ($R-CO_2^-$, $R-OPO_3^{2-}$, R—SH In turn, the isomorphism of K^+ and Tl^+ ions makes it possible to use Tl^+ and Rb^+ as molecular probes to study various potassium-dependent biological processes (Douglas et al., 1990; Skul'skiĭ, 1991). It was found earlier that Tl^+ and Tl^{3+} ions induced apoptotic processes in Jurkat, C6 glioma, and PC12 cells. These effects manifested in endosomal acidification and the formation of proapoptotic proteins (Bragadin et al., 2003; Chia et al., 2005; Hanzel and Verstraeten, 2009; Hanzel et al., 2012; Pino et al., 2017).

One of the reasons for the toxic effect of Tl^+ on living organisms is the ability of Tl^+ to inhibit the activity of Na^+/K^+-ATPase, to replace K^+ in K^+-dependent intracellular enzymes, transporters, and biochemical processes requiring potassium in the cytoplasm, and it accumulates in this way in places with a high potassium content (nerves, muscles, liver). In addition, thallium damages the Langerhans islets in the pancreas, and prevents the participation of K^+ ions in myocardial contraction mechanisms, replaces K^+ in ribosomes, and depolarizes membranes in nerves and heart. At the same time, thallium disrupts carbohydrate metabolism in neurons, increases concentration of Ca^{2+} in cytoplasm, inactivates enzymes by forming ligand bonds with their thiol groups, blocks formation of disulfide bonds in keratin, creates compounds with riboflavin and riboflavin-containing cofactors, disrupts regulation of the cell cycle, promotes depletion of energy cell stores, inhibiting Mg^{2+}-regulated and K^+-dependent pyruvate kinase, disrupts glycolysis processes, interferes Krebs cycle and displaces intracellular potassium, thus reducing concentration of K^+ in cytoplasm (Thyresson, 1951a; Gehring and Hammond, 1967; Britten and Blank, 1968; Spencer et al., 1973; Hultin and Näslund, 1974; Barrera and Gomez-Puyou, 1975; Diwan and Lehrer, 1977; Ling, 1977; Innis and Moses, 1978; Prick, 1979; Skulskii et al., 1981;

Saddique, 1983; Schoer, 1984; Aoyama et al., 1988; Douglas et al., 1990; Skul'skiĭ, 1991; Mulkey and Oehme, 1993; Appenroth et al., 1995; Galván-Arzate and Santamaría, 1998; Appenroth and Winnefeld, 1999a; Gruzina et al., 2000; Zierold, 2000; Peter and Viraraghavan, 2005; Ramsden, 2007; Cvjetko et al., 2010; Wojtovich et al., 2010; Huang et al., 2012; Rodríguez-Mercado and Altamirano-Lozano, 2013; Gad and Pham, 2014; Blain and Kazantzis, 2015).

Tl^+ of 0.2–1.5 mM activated Na^+/K^+-ATPase of bovine brain microsomes (Inturrisi, 1969a). Experiments with vesicles from cells of the mucous membrane of pig's stomach found that Tl^+ at low concentrations activated gastric K^+/H^+-ATPase, but thallium (>3 mM) inhibited this enzyme (Rabon, 1981). At the same time, high permeability of the membrane lipid phase of the above vesicles for $TlNO_3$ and Tl^+ions was found that (in contrast to experiments with K^+ and Rb^+) led to reduction K^+/H^+-ATPase activity and a decrease in proton gradient on the vesicular membrane (Rabon, 1981). The affinity of K^+/H^+-ATPase and Na^+/K^+-ATPase of plasma membranes was 10 times higher for Tl^+ than for K^+ (Rabon, 1981; Mulkey and Oehme, 1993). Such a different effect (activation/inhibition) of Tl^+ ions on the activity of Na^+/K^+-ATPase and K^+/H^+-ATPase of cytoplasmic membranes is possibly related to the specific organization of this enzyme in different types of cells.

The formation of metallothioneins and the capture of free radicals by antioxidants (vitamins E and C) are known to reduce significantly the cellular toxicity of some heavy metals (zinc, cadmium, copper). In this regard, a similar mechanism has been proposed previously for thallium poisoning (Appenroth and Winnefeld, 1998; Kiliç and Kutlu, 2010; Rodríguez-Mercado and Altamirano-Lozano, 2013; Krizkova et al., 2016). However, according to other sources, unlike other heavy metals, thallium did not stimulate the formation of metallothioneins in cells (Zierold, 2000). Thallium forms an insoluble complex with riboflavins. However, the relationship between thallium nephrotoxicity and riboflavin-dependent processes, including, after effect of riboflavin on animals in experiments in vivo, has not been found (Appenroth and Winnefeld, 1999a,b).

Other causes of thallium toxicity to living organisms can be the next: the formation of ligand bonds Tl^+ with amino, imino, and SH groups of proteins of mitochondrial and cell membranes, the decrease in the activity of both hepatic NADPH-cytochrome c (P-450) reductase and microsomal oxidases with mixed function, lipid peroxidation, the body antioxidants defense decrease, free from the depletion of glutathione, and increased production of ROS and hydrogen peroxide (Herman and Bensch, 1967; Hasan and Ali, 1981; Woods and Fowler, 1984, 1986; Chandler and Scott, 1986; Aoyama et al., 1988; Douglas et al., 1990; Mulkey and Oehme, 1993; Delvalls et al., 1999; Zierold, 2000; Villaverde et al., 2004; Galván-Arzate et al., 2005; Hanzel and Verstraeten, 2006, 2009; Verstraeten, 2006; Puga Molina and Verstraeten, 2008; Cvjetko et al., 2010; Kiliç and Kutlu, 2010; Pourahmad et al., 2010; Méndez-Armenta et al., 2011; Osorio-Rico et al., 2015). Associated with the assumption that the formation of Tl^{+} bonds with SH groups of mitochondrial membrane proteins may be one of the reasons for its toxic effect on mitochondria. In this regard, it seemed relevant to study the reaction of this metal with mitochondrial proteins, which was noted in this work when describing and discussing our studies (see Chapter 4). Tl^{+}-induced ROS formation and thallotoxicosis manifestations were significantly reduced in experiments with selenium compounds (Mulkey and Oehme, 1993). It is known that Tl^{+} can replace K^{+} and Na^{+} (to a lesser extent) during the glutamate electrogenic transport into a cell (Tao et al., 2008; Teichman et al., 2009).

Finally, it should be noted that a wide range of thallium compounds are used to study the toxic effect of this metal on cells and isolated mitochondria. For example, in experiments conducted at a temperature of 37°C with isolated human and rat hepatocytes, rat liver mitochondria, and PC12 and 3T3 cell lines, 2–350 μM thallium concentrations were used (Ponsoda et al., 1995; Zierold, 2000; Villaverde et al., 2004; Hanzel and Verstraeten, 2006, 2009; Pourahmad et al., 2010; Eskandari et al., 2015). Thallium salts from 2 to 150 mM are used to study rates of oxygen uptake and potassium fluxes in experiments with isolated mitochondria (Barrera and Gomez-Puyou, 1975; Melnick et al., 1976; Diwan and Lehrer, 1977; Wojtovich et al., 2010), DMSO-induced water channels in the membrane of CHO–K1 cells (He et al., 2012), potassium channels of frog and starfish oocytes (Hagiwara et al., 1977; Lopatin et al., 1998; Choe et al., 2000, 2001; Lu et al., 2001; Shieh and Lee, 2001), rat cardiomyocytes (Foster et al., 2012), HEK293 line cells (Schroeder and Hansen, 2008), and membranes of proximal renal tubules (Mauerer et al., 1998). Media with 250 mM $TlNO_3$ used in patch-clamp studies of gating models in charoh algae cells (Hansen et al., 2003). Compounds of trivalent thallium (Tl^{3+}) do not exist in nature, but it makes sense to analyze scientific literature on the effects of Tl^{3+} on various biological processes in vivo and in vitro. The fact is that we cannot exclude some possibility in formatting Tl (III) compounds in a body under certain conditions or their release into an environment with industrial effluents (Korotkov et al., 2015b).

It was found that organic compounds Tl^{3+} have reacted actively with lactate dehydrogenase and phosphoglycerate kinase of rabbit muscles (Bunni and Douglas, 1984; Bunni et al., 1987). Reducing Tl^{3+} to Tl^{+} is relatively slow (Delvalls et al., 1999; Harris and Messori, 2002). Toxic effects of Tl^{3+} on PC12 cell proliferation were higher than Tl^{+} (Pino and Verstraeten, 2015). In addition, Tl^{3+} increased nitric oxide production by PC12 cells (Pino et al., 2017). Tl^{+}- and Tl^{3+}-induced apoptotic processes in PC12 cells were recently shown to be associated with activation of MAPK-dependent cascade of protein kinases ERK1/2, JNK, p38, and p53 of the Bcl-2 family (Pino et al., 2017). The epidermal growth factor partially inhibited the unfavorable effects of Tl^{+} and Tl^{3+} on PC12 cells (Pino et al., 2017). The toxicity of trivalent metal cations and antitumor activity for various lines of tumor cells in rats and mice increased in the order of $Al^{3+} < Ga^{3+} < In^{3+} < Tl^{3+}$ (Hart and Adamson, 1971). However, all these metals did not have a notable effect on tumor cells of ascites leukemia, plasmacytoma, and Ehrlich carcinoma (Adamson et al., 1975). Experiments with brain liposomes from phosphatidylcholine and phosphatidylserine showed that Tl^{3+} compared to Tl^{+} was more intensive: increased the membrane surface potential, reduced the fluidity of anionic regions of the membrane in combination with increased fluidity in its cationic areas, and promoted more actively rearrangement of lipids in their membrane (Villaverde and Verstraeten, 2003). The toxicity of Tl^{3+} for many aquatic organisms was thousands of times higher than that of Tl^{+} and Cd^{2+} (Ralph and Twiss, 2002; Peter and Viraraghavan, 2005). Tl^{+} can be actively oxidized to Tl^{3+} in the cells of freshwater planktonic bacteria (Twining et al., 2003). Organic compounds of trivalent thallium (analogs of organomercurials) in comparison to Tl acetate reacted more actively with thiouridine tRNA from *Escherichia coli* (Baindur and Douglas, 1987).

3.2.1 Intracellular mechanisms of thallium toxicity

Many researchers in the mechanism of the toxic effects of thallium on living organisms assign a unique role to thallium-inhibited pyruvate kinase, a key enzyme of glycolysis. The affinity of this enzyme for Tl^{+} was 50 times higher than that for K^{+} (Kayne, 1971; Melnick et al., 1976; Schoer, 1984; Hoffman, 2003). Thallium-induced increase in intracellular Ca^{2+} concentration can occur due to inhibition of Na^{+}/K^{+}-ATPase and H^{+}-ATP synthase, uncoupling of oxidative phosphorylation, and disruption of cellular antioxidants processes (Mulkey and Oehme, 1993). Thallium neurotoxicity can also be associated with glycolysis inhibition in the nervous system cells, which get the most ATP through glycolysis (Prick, 1979). Thallium inhibits enzymes containing SH groups involved in the synthesizing of protoporphyrin-9 and subsequent transformation of coproporphyrins, which are necessary for synthesizing the heme, a precursor of hemoglobin (Prick, 1979). According to some reports, Tl^+ activated alkaline phosphatase isolated from the bull brain (Inturrisi, 1969b).

Experiments with ^{205}Tl showed Tl^+ ions possess a high affinity for Ca^{2+} sites of bovine activated plasma protein C (Hill et al., 1987). ATPase from amino acid accumulating granules of the adrenal medulla and splenic nerves was inhibited by 5–30 μM Tl^{3+}, while Tl^+ was ineffective (Burger and Starke, 1969). Complexation constants of Tl^+ with molecules having vicinal thiol groups turned out to be two orders of magnitude lower than that for ions of bivalent heavy metals (Perrin, 1979). Thallium(I) acetate activated monoamine oxidase in rat liver and rat brain mitochondria (Woods and Fowler, 1986; Osorio-Rico et al., 1995). On the other hand, mitochondrial respiratory enzymes (NADH-, succinate- and malate-dehydrogenases) pure affected by Tl^+ of millimolar concentrations (Flesch and Goldstone, 1950; Melnick et al., 1976; Woods and Fowler, 1986; Mulkey and Oehme, 1993). However, significant swelling of liver and kidney mitochondria of rats treated with thallium acetate in experiments in vivo indicates an essential role of mitochondria in toxic mechanisms of thallium action on living organisms (Herman and Bensch, 1967). Thus, overcoming the existing contradictions in the above studies required more detailed research of mechanisms in the toxic effect of thallium on mitochondria and, in particular, a more detailed examination of the reaction of Tl^+ with thiol groups of mitochondria. On the other hand, Tl^+ ions can use as a suitable tool to analyze different ion channels' activity (Weaver, 2018). The neuronal K^+/Cl^- cotransporter 2 activity was researched by Tl^+ uptake assay in HEK-293 cells (Lee et al., 2021). The dimethyl rhodol Tl^+ sensor showed the best result in studying potassium channel activity with cellular imaging (Dutter et al., 2018).

3.3 Interaction of Tl^+ and Tl^{3+} with molecular thiol groups

It was previously assumed that the toxicity of thallium is based on its ability to form potent ligand (chelate) complexes with intracellular and the inner mitochondrial membrane enzymes with vicinal thiol groups, which can change mitochondria's spatial and structural integrity. However, this effect of Tl^+ does not relate to direct inhibition of mitochondrial respiratory enzymes (Flesch and Goldstone, 1950; Truter, 1975; Melnick et al., 1976; Oehme, 1978; Prick, 1979; Woods and Fowler, 1986; Mulkey and Oehme, 1993; Huang et al., 2012). The formation of thallium complexes with SH groups of keratin prevents the formation of S—S bonds necessary for its structure synthesis (Thyresson, 1951a; Prick, 1979). This prevention may be one of the reasons observed in thallotoxicosis of hair growth disorders (alopecia) or nails (Mees' line).

Currently, it is widely believed that the toxic effects of thallium (like bivalent heavy metals) are associated with the ability of Tl^+ to firmly bind to the thiol groups of cellular and mitochondrial proteins. As a result, inhibition of enzymatic reactions, the free glutathione level decrease, and the lipid peroxidation increase can take place, which together would lead to cell damage (Chandler and Scott, 1986; Mulkey and Oehme, 1993; Repetto et al., 1994; Aoyama et al., 1988; Galván-Arzate and Santamaría, 1998; Goyer and Clarkson, 2001; Hoffman, 2003; Peter and Viraraghavan, 2005; Ramsden, 2007; Cvjetko et al., 2010; Kiliç and Kutlu, 2010; Huang et al., 2012; Rodríguez-Mercado and Altamirano-Lozano, 2013). One to 25 μM $Tl(OH)_3$ significantly reduced the content of reduced glutathione and inhibited the activity of glutathione peroxidase in cytosolic fractions of the rat brain (Hanzel et al., 2005). Tl^+ accumulated in yeast mitochondria was oxidized to Tl^{3+} and excreted into a protoplast in the form of insoluble oxide Tl_2O_3 (Lindegren and Lindegren, 1971). The interaction of Tl^+ with molecular thiol groups was much weaker than that found for other heavy metals (Ag^+, Hg^{2+}, Cd^{2+}, Pb^{2+}, Zn^{2+}) (Perrin, 1979). At the same time, Tl^+ decreased the content of reduced glutathione in aged rat liver mitochondria, which had been previously incubated for an hour in a medium with $TlNO_3$ at 37°C (Eskandari et al., 2015). Even with this, Tl acetate taken in fivefold excess did not affect titer of 0.1 M glutathione solution with Elman's reagent, unlike Tl^{3+} and its organic compounds, which oxidized glutathione (Bunni et al., 1987; Khan et al., 2018).

Experiments were performed with rat liver mitochondria (37°C), PC12 cells, isolated rat hepatocytes, and rats exposed to mono- and trivalent thallium salts. These experiments showed that Tl^+ and Tl^{3+} ions induced the production of H_2O_2 and oxygen- and nitrogen-containing radicals, enhanced lipid peroxidation, and decreased the reduced glutathione cytoplasmic concentration (Sabbioni et al., 1980b; Villaverde et al., 2004; Galván-Arzate et al., 2000, 2005; Hanzel, Verstraeten, 2006, 2009; Kiliç and Kutlu, 2010; Pourahmad et al., 2010; Eskandari et al., 2015; Osorio-Rico et al., 2017; Puga Molina et al., 2017). It should be emphasized that these experiments did not reveal any significant differences between the effects of Tl^+ and Tl^{3+} (Sabbioni et al., 1980b; Rodríguez-Mercado and Altamirano-Lozano, 2013). It was concluded that thallium's oxidative state does not affect the metabolism of this metal and its distribution in animal organisms in experiments in vivo (Sabbioni et al., 1980b; Rodríguez-Mercado and Altamirano-Lozano, 2013).

To reduce toxic effects of bivalent heavy metals did use sulfur-containing chelators (2,3-dimercaptopropanol, N-acetylcysteine, methionine), ion exchange resins (Dowex 50), thiomerin diuretic, EDTA, dimercaptopropanesulfonic acid, dithiocarb, sodium thiosulfate, and cystine. However, the use of these reagents was not effective in the treatment of people and animals poisoned with thallium, or even the use increased the toxicity of thallium due to its redistribution in the central nervous system (Thyresson, 1950; Lund, 1956; Rauws et al., 1969; Thompson, 1981; Schoer, 1984; Hoffman, 2003; Ramsden, 2007; Rusyniak et al., 2010). On the other hand, the toxicity of thallium salts showed visible weakening when animals did subject to betaine, cystine, methionine, S-aminoethylisothiouronium, and S-aminopropylisothiuronium and thioacetamide bromides (Gross et al., 1948; Joszpa and Batko, 1957; Meyer and Tal, 1957; Stavinoha et al., 1959). But according to other sources, cystine and cysteine did not reduce thallium toxicity (Lund, 1956). The use of L-methionine or L-cysteine, regardless of the presence of Prussian blue, which binds Tl^+ ions, increased mortality in rats exposed to thallium acetate (Montes et al., 2007, 2011). Thus, the presence of contradictions in the ability of sulfur-containing compounds to reduce the toxic effects of thallium led to the need for a thorough study of mechanisms of interaction of Tl^+ with mitochondrial thiol groups.

The increase in thallium secretion did leave the human body or with animals in experiments with dimercaptosuccinic acid, dithizone, 2,3-dimercaptopropanol, diethyldithiocarbamate, KCl, and furosemide. However, a rise in the cytoplasm and nervous system thallium concentration was a side effect in this case (Lund, 1956; Schwetz et al., 1967; Kamerbeek et al., 1971a; Lameijer, 1979; Schoer, 1984; Chandler and Scott, 1986; Adams et al., 2009; Goldschmidt et al., 2010; Wanger et al., 2012). Interaction of Tl^{3+} with membranes of artificial vesicles containing bovine brain phosphatidylcholine and bovine heart cardiolipin resulted in the alteration of the membranes' physical properties, lipid oxidation, cardiolipin hydrolysis, and a decrease in cytochrome C binding with these membranes. These effects can partly explain mitochondrial dysfunction of thallium-exposed cells in vivo (Puga Molina and Verstraeten, 2008). The orally administered chelators deferasirox and deferiprone significantly decreased Tl^{3+}-induced toxic effects in experiments in vivo with rats (Salehi et al., 2017).

References

Adams, J.B., Baral, M., Geis, E., Mitchell, J., Ingram, J., et al., 2009. Safety and efficacy of oral DMSA therapy for children with autism spectrum disorders: Part A - medical results. BMC Clin. Pharmacol. 9 (16), 1–22. https://doi.org/10.1186/1472-6904-9-16.

Adamson, R.H., Canellos, G.P., Sieber, S.M., 1975. Studies on the antitumor activity of gallium nitrate (NSC-15200) and other group IIIa metal salts. Cancer Chemother. Rep. 59 (3), 599–610.

Aoyama, H., Yoshida, M., Yamamura, Y., 1988. Induction of lipid peroxidation in tissues of thallous malonate-treated hamster. Toxicology 53 (1), 11–18. https://doi.org/10.1016/0300-483x(88)90232-6.

Appenroth, D., Gambaryan, S., Winnefeld, K., Leiterer, M., Fleck, C., Bräunlich, H., 1995. Functional and morphological aspects of thallium-induced nephrotoxicity in rats. Toxicology 96 (3), 203–215. https://doi.org/10.1016/0300-483x(94)02919-l.

Appenroth, D., Winnefeld, K., 1998. Vitamin E and C in the prevention of metal nephrotoxicity in developing rats. Exp. Toxicol. Pathol. 50, 391–396. https://doi.org/10.1016/S0940-2993(98)80024-9.

Appenroth, D., Winnefeld, K., 1999a. Reconsideration of some hypotheses on the mechanism of thallium toxicity in rats with special respect to riboflavin and glutathione. Nephrol. Dial. Transplant. 14 (Suppl. 4), 16–18. https://doi.org/10.1093/oxfordjournals.ndt.a027948.

Appenroth, D., Winnefeld, K., 1999b. Is thallium-induced nephrotoxicity in rats connected with riboflavin and/or GSH?—reconsideration of hypotheses on the mechanism of thallium toxicity. J. Appl. Toxicol. 19 (1), 61–66. https://doi.org/10.1002/(sici)1099-1263(199901/02)19:1<61::aid-jat539>3.0.co;2-s.

Arbab, A.S., Koizumi, K., Toyama, K., Arai, T., Araki, T., 1997. Ion transport systems in the uptake of ^{99}Tcm-tetrofosmin, ^{99}Tcm-MIBI and ^{201}Tl in a tumour cell line. Nucl. Med. Commun. 18 (3), 235–240. https://doi.org/10.1097/00006231-199703000-00007.

Arbab, A.S., Koizumi, K., Toyama, K., Arai, T., Araki, T., 1998. Technetium-99m-tetrofosmin, technetium-99m-MIBI and thallium-201 uptake in rat myocardial cells. J. Nucl. Med. 39 (2), 266–271.

Arbiser, J.L., Alani, R., Flynn, E., Wooten, V., Byers, H.R., 1997. Effects of thallium ion on cellular components of the skin. J. Dermatol. 24 (3), 147–155. https://doi.org/10.1111/j.1346-8138.1997.tb02762.x.

Baindur, S.R., Douglas, K.T., 1987. Organothallium(III) reagents for modification of biomacromolecules: interaction of thallium derivatives with transfer RNA. Biochim. Biophys. Acta. 923 (1), 66–73. https://doi.org/10.1016/0304-4165(87)90127-9.

Bakker, E.P., 1978. Accumulation of thallous ions Tl^+ as a measure of the electrical potential difference across the cytoplasmic membrane of bacteria. Biochemistry 17 (14), 2899–2904. https://doi.org/10.1021/bi00607a031.

Bakker-Grunwald, T., 1979. Movement of thallous ion across the ascites cell membrane. J. Membr. Biol. 47 (2), 171–183. https://doi.org/10.1007/BF01876115.

Bank, W.J., Pleasure, D.E., Suzuki, K., Nigro, M., Katz, R., 1972. Thallium poisoning. Arch. Neurol. 26 (5), 456–464. https://doi.org/10.1001/archneur.1972.00490110090009.

Barrera, H., Gomez-Puyou, A., 1975. Characteristics of the movement of K^+ across the mitochondrial membrane and the inhibitory action of Tl^+. J. Biol. Chem. 250 (14), 5370–5374.

Beacham, D.W., Blackmer, T., O' Grady, M., Hanson, G.T., 2010. Cell-based potassium ion channel screening using the FluxOR assay. J. Biomol. Screen. 15 (4), 441–446. https://doi.org/10.1177/1087057109359807.

Bielen, F.V., Glitsch, H.G., Verdonck, F., 1993. Na^+ pump current-voltage relationships of rabbit carDiac Purkinje cells in Na^+-free solution. J. Physiol. 465 (1), 699–714. https://doi.org/10.1113/jphysiol.1993.sp019701.

Blain, R., Kazantzis, G., 2015. Thallium. In: Nordberg, G.F., Fowler, B.A., Nordberg, M. (Eds.), Handbook on the Toxicology of Metals, Specific Metals, Vol. II. Elsevier, Amsterdam, pp. 1229–1240.

Blatz, A.L., Magleby, K.L., 1984. Ion conductance and selectivity of single calcium-activated potassium channels in cultured rat muscle. J. Gen. Physiol. 84 (1), 1–23. https://doi.org/10.1085/jgp.84.1.1.

Bostian, K.A., Betts, G.F., Man, W.K., Hughes, M.N., 1982. Multiple binding of thallium and rubidium to potassium-activated yeast aldehyde dehydrogenase. Influences on tertiary structure, stability and catalytic activity. Biochem. J. 207 (1), 73–80. https://doi.org/10.1042/bj2070073.

Bragadin, M., Toninello, A., Bindoli, A., Rigobello, M.P., Canton, M., 2003. Thallium induces apoptosis in Jurkat cells. Ann. N. Y. Acad. Sci. 1010, 283–291. https://doi.org/10.1196/annals.1299.049.

Bramanti, E., Onor, M., Colombaioni, L., 2019. Neurotoxicity induced by low thallium doses in living hippocampal neurons: evidence of early onset mitochondrial dysfunction and correlation with ethanol production. ACS Chem. Neurosci. 10 (1), 451–459. https://doi.org/10.1021/acschemneuro.8b00343.

Brismar, T., Anderson, S., Collins, V.P., 1995. Mechanism of high K^+ and Tl^+ uptake in cultured human glioma cells. Cell. Mol. Neurobiol. 15 (3), 351–360. https://doi.org/10.1007/BF02089945.

Brismar, T., Lefvert, A., Jondal, M., 1994. A method for analysis of cellular K-transport mechanisms through thallium (^{201}Tl) uptake in human lymphocytes. Acta Physiol. Scand. 150 (3), 299–303. https://doi.org/10.1111/j.1748-1716.1994.tb09690.x.

Britten, J.S., Blank, M., 1968. Thallium activation of the (Na^+-K^+)-activated ATPase of rabbit kidney. Biochim. Biophys. Acta 159 (1), 160–166. https://doi.org/10.1016/0005-2744(68)90254-4.

Bunni, M.A., Baindur, S.R., Douglas, K.T., Drake, A.F., 1987. Organothallium(III) reagents for modification of biomacromolecules: irreversible labelling of phosphoglycerate kinase from rabbit muscle. Biochim. Biophys. Acta 923 (3), 421–430. https://doi.org/10.1016/0304-4165(87)90050-x.

Bunni, M.A., Douglas, K.T., 1984. Arylthallium(III) reagents for protein modification. Inhibition of lactate dehydrogenase from various sources by o-carboxyphenylthallium(III) bistrifluoroacetate. Biochem. J. 217 (2), 383–390. https://doi.org/10.1042/bj2170383.

Burger, A., Starke, K., 1969. Influencing by thallium of the ATPase of amino-storing granules in adrenal medulla and splenic nerves. Experientia 25 (6), 578–579. https://doi.org/10.1007/BF01896519.

Carmosino, M., Rizzo, F., Torretta, S., Procino, G., Svelto, M., 2013. High-throughput fluorescent-based NKCC functional assay in adherent epithelial cells. BMC Cell Biol. 14 (16), 1–9. https://doi.org/10.1186/1471-2121-14-16.

Chandler, H.A., Scott, M., 1986. A review of thallium toxicology. J. R. Nav. Med. Serv. 72 (2), 75–79.

Chen, Y.W., Yeh, J.L., Huang, Y.C., Chen, I.J., Huang, Y.F., Liu, G.C., 2000. The interference of uptake of thallium-201 in cultured rat myocardial cells with existence of potassium related pharmaceuticals—a preliminary report. Kaohsiung J. Med. Sci. 16 (3), 134–140.

Chepilko, S., Zhou, H., Sackin, H., Palmer, L.G., 1995. Permeation and gating properties of a cloned renal K^+ channel. Am. J. Physiol. 268 (2), C389–C401. https://doi.org/10.1152/ajpcell.1995.268.2.C389. Part. 1.

Chia, C.F., Chen, S.C., Chen, C.S., Shih, C.M., Lee, H.M., Wu, C.H., 2005. Thallium acetate induces C6 glioma cell apoptosis. Ann. N. Y. Acad. Sci. 1042, 523–530. https://doi.org/10.1196/annals.1338.064.

Choe, H., Sackin, H., Palmer, L.G., 2000. Permeation properties of inward-rectifier potassium channels and their molecular determinants. J. Gen. Physiol. 115 (4), 391–404. https://doi.org/10.1085/jgp.115.4.391.

Choe, H., Sackin, H., Palmer, L.G., 2001. Gating properties of inward-rectifier potassium channels: effects of permeant ions. J. Membr. Biol. 184 (1), 81–89. https://doi.org/10.1007/s00232-001-0076-3.

Cvjetko, P., Cvjetko, I., Pavlica, M., 2010. Thallium toxicity in humans. Arh. Hig. Rada Toksikol. Croatia. 61 (1), 111–119. https://doi.org/10.2478/10004-1254-61-2010-1976.

Delano, M.L., Sands, H., Gallagher, B.M., 1985. Transport of $^{42}K^+$, $^{201}Tl^+$ and $[^{99m}Tc(dmpe)_2 \cdot Cl_2]^+$ by neonatal rat myocyte cultures. Biochem. Pharmacol. 34 (18), 3377–3380. https://doi.org/10.1016/0006-2952(85)90361-2.

Delvalls, T.A., Saenz, V., Arias, A.M., Blasco, J., 1999. Thallium in the marine environment: first ecotoxicological assessments in the Guadalquivir 500 estuary and its potential adverse effect on the Donana European natural reserve after the Aznalcollar mining spill. Cienc. Mar. 25 (2), 161–175.

Deshimaru, M., Miyakawa, T., Sumiyoshi, S., Yasuoka, F., Kawano, K., 1977. Electron microscopic study of experimental thallotoxicosis. Folia Psychiatr. Neurol. Jpn. 31 (2), 269–275. https://doi.org/10.1111/j.1440-1819.1977.tb02728.x.

Diwan, J.J., Lehrer, P.H., 1977. Inhibition of mitochondrial potassium ion flux by thallous ions. Biochem. Soc. Trans. 5 (1), 203–205. https://doi.org/10.1042/bst0050203.

Douglas, K.T., Bunni, M.A., Baindur, S.R., 1990. Thallium in biochemistry. Int. J. Biochem. 22 (5), 429–438. https://doi.org/10.1016/0020-711x(90)90254-z.

Dutter, B.F., Ender, A., Sulikowski, G.A., Weaver, C.D., 2018. Rhodol-based thallium sensors for cellular imaging of potassium channel activity. Org. Biomol. Chem. 16 (31), 5575–5579. https://doi.org/10.1039/c8ob01098f.

Eskandari, M.R., Mashayekhi, V., Aslani, M., Hosseini, M.J., 2015. Toxicity of thallium on isolated rat liver mitochondria: the role of oxidative stress and MPT pore opening. Environ. Toxicol. 30 (2), 232–241. https://doi.org/10.1002/tox.21900.

Eskandari, M.R., Pourahmad, J., Daraei, B., 2011. Thallium(I) and thallium(III) induce apoptosis in isolated rat hepatocytes by alterations in mitochondrial function and generation of ROS. Toxicol. Environ. Chem. 93 (1), 145–156. https://doi.org/10.1080/02772248.2010.505826.

Favari, L., Mourelle, M., 1985. Thallium replaces potassium in activation of the (Na^+,K^+)-ATPase of rat liver plasma membranes. J. Appl. Toxicol. 5 (1), 32–34. https://doi.org/10.1002/jat.2550050106.

Felix, J.P., Priest, B.T., Solly, K., Bailey, T., Brochu, R.M., et al., 2012. The inwardly rectifying potassium channel Kir1.1: development of functional assays to identify and characterize channel inhibitors. Assay Drug Dev. Technol. 10 (5), 417–431. https://doi.org/10.1089/adt.2012.462.

Flesch, P., Goldstone, S.B., 1950. Effect of thallium on sulphydryl compounds *in vitro*. J. Invest. Dermatol. 15 (5), 345–347. https://doi.org/10.1038/jid.1950.111.

Foster, D.B., Ho, A.S., Rucker, J., Garlid, A.O., Chen, L., et al., 2012. Mitochondrial ROMK channel is a molecular component of mitoK_{ATP}. Circ. Res. 111 (4), 446–454. https://doi.org/10.1161/CIRCRESAHA.112.266445.

Fox, J.A., 1983. Thallous ion permeation through the cation-selective channel of the sarcoplasmic reticulum. Anomalous mole fraction dependence. Biochim. Biophys. Acta 736 (2), 241–245. https://doi.org/10.1016/0005-2736(83)90290-0.

Fukumoto, M., Kurohara, A., Yoshimura, N., Yoshida, D., Akagi, N., Yoshida, S., 1998. Relationship between ATP synthesis and ^{201}Tl uptake in transformed and non-transformed cell lines. Nucl. Med. Commun. 19 (12), 1169–1175. https://doi.org/10.1097/00006231-199812000-00009.

Fukumoto, M., Yoshida, D., Yoshida, S., 1997. Subcellular distribution of thallium: morphological and quantitative study in rat myocardium. Ann. Nucl. Med. 11 (4), 291–297. https://doi.org/10.1007/BF03165296.

Gachon, P., Moins, N., Maublant, J.C., Ross, M.R., Davidson, W.D., Mena, I., 1986. Relationship between ^{201}Tl uptake and oxidative metabolism in cultured myocardial cells. Nucl. Med. Commun. 7 (1), 59–64. https://doi.org/10.1097/00006231-198601000-00008.

Gad, S.C., Pham, T., 2014. Thallium. In: Wexler, P. (Ed.), Encyclopedia of Toxicology, third ed. Academic Press, pp. 527–529.

Galván-Arzate, S., Martínez, A., Medina, E., Santamaría, A., Ríos, C., 2000. Subchronic administration of sublethal doses of thallium to rats: effects on distribution and lipid peroxidation in brain regions. Toxicol. Lett. 116 (1–2), 37–43. https://doi.org/10.1016/s0378-4274(00)00200-9.

Galván-Arzate, S., Pedraza-Chaverrí, J., Medina-Campos, O.N., Maldonado, P.D., Vázquez-Román, B., et al., 2005. Delayed effects of thallium in the rat brain: regional changes in lipid peroxidation and behavioral markers, but moderate alterations in antioxidants, after a single administration. Food Chem. Toxicol. 43 (7), 1037–1045. https://doi.org/10.1016/j.fct.2005.02.006.

Galván-Arzate, S., Santamaría, A., 1998. Thallium toxicity. Toxicol. Lett. 99 (1), 1–13. https://doi.org/10.1016/s0378-4274(98)00126-x.

Gehring, P.J., Hammond, P.B., 1964. The uptake of thallium by rabbit erythrocytes. J. Pharmacol. Exp. Therapeut. 145 (2), 215–221.

Gehring, P.J., Hammond, P.B., 1967. The interrelationship between thallium and potassium in animals. J. Pharmacol. Exp. Therapeut. 155 (1), 187–201.

Glynn, I.M., Hara, Y., Richards, D.E., Steinberg, M., 1987. Comparison of rates of cation release and of conformational change in dog kidney Na, K-ATPase. J. Physiol. 383 (1), 477–485. https://doi.org/10.1113/jphysiol.1987.sp016422.

Goldschmidt, J., Wanger, T., Engelhorn, A., Friedrich, H., Happel, M., et al., 2010. High-resolution mapping of neuronal activity using the lipophilic thallium chelate complex TlDDC: protocol and validation of the method. Neuroimage 49 (1), 303–315. https://doi.org/10.1016/j.neuroimage.2009.08.012.

Goyer, R.A., Clarkson, T.W., 2001. Toxic effects of metals. Thallium. In: Klaasen, C.D. (Ed.), Casarett and Doull's Toxicology: The Basic Science of Poisons, sixth ed. Macmillan Publishing Co., Inc., New York, pp. 855–856.

Gross, P., Runne, E., Wilson, J.W., 1948. Studies on the effect of thallium poisoning of the rat; the influence of cystine and methionine on alopecia and survival periods. J. Invest. Dermatol. 10 (3), 119–134. https://doi.org/10.1038/jid.1948.20.

Gruzina, T.G., Balakina, M.N., Ul'berg, Z.R., 2000. Effect of ATP-ase activity and transmembrane potential in bacteria. Ukr. Biokhim. Zh. 72 (2), 72–76. Russian. PMID: 10979585.

Hagiwara, S., Miyazaki, S., Krasne, S., Ciani, S., 1977. Anomalous permeabilities of the egg cell membrane of a starfish in K^+-Tl^+ mixtures. J. Gen. Physiol. 70 (3), 269–281. https://doi.org/10.1085/jgp.70.3.269.

Hansen, U.P., Cakan, O., Abshagen-Keunecke, M., Farokhi, A., 2003. Gating models of the anomalous mole-fraction effect of single-channel current in Chara. J. Membr. Biol. 192 (1), 45–63. https://doi.org/10.1007/s00232-002-1063-z.

Hanzel, C.E., Almeira Gubiani, M.F., Verstraeten, S.V., 2012. Endosomes and lysosomes are involved in early steps of Tl(III)-mediated apoptosis in rat pheochromocytoma (PC12) cells. Arch. Toxicol. 86 (11), 1667–1680. https://doi.org/10.1007/s00204-012-0878-3.

Hanzel, C.E., Verstraeten, S.V., 2006. Thallium induces hydrogen peroxide generation by impairing mitochondrial function. Toxicol. Appl. Pharmacol. 216 (3), 485–492. https://doi.org/10.1016/j.taap.2006.07.003.

Hanzel, C.E., Verstraeten, S.V., 2009. Tl(I) and Tl(III) activate both mitochondrial and extrinsic pathways of apoptosis in rat pheochromocytoma (PC12) cells. Toxicol. Appl. Pharmacol. 236 (1), 59–70. https://doi.org/10.1016/j.taap.2008.12.029.

Hanzel, C.E., Villaverde, M.S., Verstraeten, S.V., 2005. Glutathione metabolism is impaired in vitro by thallium(III) hydroxide. Toxicology 207 (3), 501–510. https://doi.org/10.1016/j.tox.2004.11.002.

Harris, W.R., Messori, L., 2002. A comparative study of aluminum(III), gallium(III), indium(III), and thallium(III) binding to human serum transferring. Coord. Chem. Rev. 228, 237–262.

Hart, M.M., Adamson, R.H., 1971. Antitumor activity and toxicity of salts of inorganic group 3a metals: aluminum, gallium, indium, and thallium. Proc. Natl. Acad. Sci. U.S.A. 68 (7), 1623–1626. https://doi.org/10.1073/pnas.68.7.1623.

Hasan, M., Ali, S.F., 1981. Effects of thallium, nickel, and cobalt administration of the lipid peroxidation in different regions of the rat brain. Toxicol. Appl. Pharmacol. 57 (1), 8–13. https://doi.org/10.1016/0041-008x(81)90019-3.

Hasan, M., Ashraf, I., Bajpai, V.K., 1978. Electron microscopic study of the effects of thallium poisoning on the rat cerebellum. Forensic Sci. 11 (2), 139–146. https://doi.org/10.1016/s0379-0738(78)80008-5.

Hasan, M., Bajpai, V.K., Shipstone, A.C., 1977a. Electron microscope study of thallium-induced alterations in the oligodendrocytes of the rat area postrema. Exp. Pathol. (Jena) 13 (6), 338–345. https://doi.org/10.1016/s0014-4908(77)80021-5.

Hasan, M., Chandra, S.V., Dua, P.R., Raghubir, R., Ali, S.F., 1977b. Biochemical and electrophysiologic effects of thallium poisoning on the rat corpus striatum. Toxicol. Appl. Pharmacol. 41 (2), 353–359. https://doi.org/10.1016/0041-008x(77)90036-9.

He, F., Liu, W., Zheng, S., Zhou, L., Ye, B., Qi, Z., 2012. Ion transport through dimethyl sulfoxide (DMSO) induced transient water pores in cell membranes. Mol. Membr. Biol. 29 (3–4), 107–113. https://doi.org/10.3109/09687688.2012.687460.

Herman, M.M., Bensch, K.G., 1967. Light and electron microscopic studies of acute and chronic thallium intoxication in rats. Toxicol. Appl. Pharmacol. 10 (2), 199–222. https://doi.org/10.1016/0041-008x(67)90104-4.

Hill, K.A., Steiner, S.A., Castellino, F.J., 1987. $^{205}Tl^{+}$ as a spectroscopic probe of the monovalent cation binding sites of bovine plasma activated protein C and des-1-41-light-chain-activated protein C. J. Biol. Chem. 262 (15), 7098–7104.

Hoffman, R.S., 2003. Thallium toxicity and the role of Prussian blue in therapy. Toxicol. Rev. 22 (1), 29–40. https://doi.org/10.2165/00139709-200322010-00004.

Hougaard, C., Eriksen, B.L., Jørgensen, S., Johansen, T.H., Dyhring, T., et al., 2007. Selective positive modulation of the SK3 and SK2 subtypes of small conductance Ca^{2+}-activated K^{+} channels. Br. J. Pharmacol. 151 (5), 655–665. https://doi.org/10.1038/sj.bjp.0707281.

Huang, C.C., Shih, T.S., Liu, C.H., 2012. Acute thallium intoxication: an experience in Taiwan. Curr. Top. Toxicol. 8, 33–41.

Hultin, T., Näslund, P.H., 1974. Effects of thallium (I) on the structure and functions of mammalian ribosomes. Chem. Biol. Interact. 8 (5), 315–328. https://doi.org/10.1016/0009-2797(74)90010-6.

Innis, R., Moses, H., 1978. Clinical conferences at the Johns Hopkins hospital. Thallium poisoning. Johns Hopkins Med. J. 142 (1), 27–31.

Inturrisi, C.E., 1969a. Thallium-induced dephosphorylation of a phosphorylated intermediate of the (sodium plus thallium-activated) ATPase. Biochim. Biophys. Acta 178 (3), 630–633. https://doi.org/10.1016/0005-2744(69)90233-2.

Inturrisi, C.E., 1969b. Thallium activation of K^{+}-activated phosphatases from beef brain. Biochim. Biophys. Acta 173 (3), 567–569. https://doi.org/10.1016/0005-2736(69)90022-4.

Jørgensen, S., Dyhring, T., Brown, D.T., Strøbæk, D., Christophersen, P., Demnitz, J., 2013. A high-throughput screening campaign for detection of Ca^{2+}-activated K^{+} channel activators and inhibitors using a fluorometric imaging plate reader-based Tl^{+}-influx assay. Assay Drug Dev. Technol. 11 (3), 163–172. https://doi.org/10.1089/adt.2012.479.

Jørgensen, S., Johansen, T.H., Dyhring, T., 2008. Fluorescence-based Tl^{+}-influx assays as a novel approach for characterization of small-conductance Ca^{2+}-activated K^{+} channel modulators. Methods Mol. Biol. 491, 257–266. https://doi.org/10.1007/978-1-59745-526-8_20.

Joszpa, O., Batko, B., 1957. Two cases of severe thallium poisoning treated with thioacetamide. Pol. Tyg. Lek. 12 (5), 179–181 (Polish).

Kamerbeek, H.H., Rauws, A.G., ten Ham, M., van Heijst, A.N., 1971. Dangerous redistribution of thallium by treatment with sodium diethyldithiocarbamate. Acta Med. Scand. 189 (3), 149–154. https://doi.org/10.1111/j.0954-6820.1971.tb04356.x.

Kashket, E.R., 1979. Active transport of thallous ions by Streptococcus lactis. J. Biol. Chem. 254 (17), 8129–8131.

Kayne, F.J., 1971. Thallium (I) activation of pyruvate kinase. Arch. Biochem. Biophys. 143 (1), 232–239. https://doi.org/10.1016/0003-9861(71)90204-9.

Keogh, J.P., Steffen, B., Siegers, C.P., 1994. Cytotoxicity of heavy metals in the human small intestinal epithelial cell line I-407: the role of glutathione. J. Toxicol. Environ. Health A. 43 (3), 351–359. https://doi.org/10.1080/15287399409531926.

Khan, M.J., Mukhtiar, M., Qureshi, M.M., Jan, S.U., Ullah, I., et al., 2018. Spectrophotometric investigation of Glutathione modulation by Thallium chloride in aqueous medium. Pak. J. Pharm. Sci. 31 (4), 1463–1467.

Kharade, S.V., Nichols, C., Denton, J.S., 2016. The shifting landscape of K_{ATP} channelopathies and the need for 'sharper' therapeutics. Future Med. Chem. 8 (7), 789–802. https://doi.org/10.4155/fmc-2016-0005.

Khoda, M., Utsunomiya, K., Ha-Kawa, S., Kanno, S., Kono, Y., Sawada, S., 2012. An investigation of the early detection of radiation induced apoptosis by ^{99}mTc-Annexin V and ^{201}thallium-chloride in a lung cancer cell line. J. Radiat. Res. Tokyo 53 (3), 361–367. https://doi.org/10.1269/jrr.11177.

Kiliç, G.A., Kutlu, M., 2010. Effects of exogenous metallothionein against thallium-induced oxidative stress in rat liver. Food Chem. Toxicol. 48 (3), 980–987. https://doi.org/10.1016/j.fct.2010.01.013.

Korenman, I.M., 1960. Analytical Chemistry of Thallium. Moscow. Russian. https://studizba.com/files/show/djvu/2342-1-i-m-korenman–analiticheskaya-himiya.html.

Korotkov, S.M., Brailovskaya, I.V., Shumakov, A.R., Emelyanova, L.V., 2015a. Closure of mitochondrial potassium channels favors opening of the Tl^{+}-induced permeability transition pore in Ca^{2+}-loaded rat liver mitochondria. J. Bioenerg. Biomembr. 47 (3), 243–254. https://doi.org/10.1007/s10863-015-9611-2.

Korotkov, S.M., Emelyanova, L.V., Konovalova, S.A., Brailovskaya, I.V., 2015b. Tl^{+} induces the permeability transition pore in Ca^{2+}-loaded rat liver mitochondria energized by glutamate and malate. Toxicol. in Vitro 29 (5), 1034–1041. https://doi.org/10.1016/j.tiv.2015.04.006.

Korotkov, S.M., Emel'yanova, L.V., Yagodina, O.V., 2008. Inorganic phosphate stimulates the toxic effects of Tl^{+} in rat liver mitochondria. J. Biochem. Mol. Toxicol. 22 (3), 148–157. https://doi.org/10.1002/jbt.20215.

Korotkov, S.M., Nesterov, V.P., Belostotskaya, G.B., Brailovskaya, I.V., Novozhilov, A.V., Sobol, C.V., 2020. Influence of Tl^{+} on the movement of Ca^{2+} and Na^{+} ions across membranes of rat neonatal cardiomyocytes and rat heart mitochondria. Ukrainian Biochem. J. 92 (1), 41–55.

Krizkova, S., Kepinska, M., Emri, G., Rodrigo, M.A., Tmejova, K., et al., 2016. Microarray analysis of metallothioneins in human diseases. A review. J. Pharm. Biomed. Anal. 117, 464–473. https://doi.org/10.1016/j.jpba.2015.09.031.

Lameijer, W., 1979. Accelerated renal elimination of thallium in the rat due to treatment with furosemide or potassium ions. Arch. Toxicol. 2 (Suppl), 365–366.

Landowne, D.A., 1975. A comparison of radioactive thallium and potassium fluxes in the giantaxon of the squid. J. Physiol. 252 (1), 79–96. https://doi.org/10.1113/jphysiol.1975.sp011135.

Lee, K.L., Abiraman, K., Lucaj, C., Ollerhead, T.A., Brandon, N.J., et al., 2021. Inhibiting with-no-lysine kinases enhances K^{+}/Cl^{-} cotransporter 2 activity and limits status epilepticus. Brain. https://doi.org/10.1093/brain/awab343.

Leloux, M.S., Nguyen, P.L., Claude, J.R., 1987. Experimental studies on thallium toxicity in rats. I-Localization and elimination of thallium after oral acute and sub-acute intoxication. J. Toxicol. Clin. Exp. 7 (4), 247–257.

Leung, K.M., Ooi, V.E., 2000. Studies on thallium toxicity, its tissue distribution and histopathological effects in rats. Chemosphere 41 (1–2), 155–159. https://doi.org/10.1016/s0045-6535(99)00404-x.

Levander, O.A., Argrett, L.C., 1969. Effects of arsenic, mercury, thallium, and lead on selenium metabolism in rats. Toxicol. Appl. Pharmacol. 14 (2), 308–314. https://doi.org/10.1016/0041-008x(69)90112-4.

Li, S., Huang, W., Duan, Y., Xing, J., Zhou, Y., 2015. Human fatality due to thallium poisoning: autopsy, microscopy, and mass spectrometry assays. J. Forensic Sci. 60 (1), 247–251. https://doi.org/10.1111/1556-4029.12623.

Lindegren, C.C., Lindegren, G., 1971. Thallium as a vital stain for yeast mitochondria. Nature 234 (5327), 297. https://doi.org/10.1038/234297a0.

Ling, G.N., 1977. Thallium and cesium in muscle cells compete for the adsorption sites normally occupied by K^+. Physiol. Chem. Phys. 9 (3), 217–225.

Lishko, V.K., Kolchinskaya, L.I., Parkhomenko, N.T., 1973. Thallium and the erythrocyte sodium pump. Ukr. Biochim. Zh. 45 (1), 42–46. Russian. PMID: 4266781.

Llaurado, J.G., Madden, J.A., Smith, G.A., 1983. Effects of dietary magnesium deficiency on thallium-201 kinetics and distribution in rat myocardium: concise communication. J. Nucl. Med. 24 (5), 402–407.

Llaurado, J.G., Smith, G.A., Madden, J.A., Meade, R.C., 1979. Partition of thallium-201 in isolated myocardial tissue of rats previously injected at rest or after exercise. J. Nucl. Med. 20 (11), 1136–1141.

Lopatin, A.N., Makhina, E.N., Nichols, C.G., 1998. A novel crystallization method for visualizing the membrane localization of potassium channels. Biophys. J. 74 (5), 2159–2170. https://doi.org/10.1016/S0006-3495(98)77925-1.

Lu, T., Wu, L., Xiao, J., Yang, J., 2001. Permeant ion-dependent changes in gating of Kir2.1 inward rectifier potassium channels. J. Gen. Physiol. 118 (5), 509–521. https://doi.org/10.1085/jgp.118.5.509.

Lund, A., 1956. The effect of various substances on the excretion and the toxicity of thallium in the rat. Acta Pharmacol. Toxicol. 12 (3), 260–268. https://doi.org/10.1111/j.1600-0773.1956.tb01386.x.

Maublant, J.C., Moins, N., Gachon, P., Renoux, M., Zhang, Z., Veyre, A., 1993. Uptake of technetium-99m-teboroxime in cultured myocardial cells: comparison with thallium-201 and technetium-99m-sestamibi. J. Nucl. Med. 34 (2), 255–259.

Maublant, J.C., Moins, N., Gachon, P., Ross, M.R., Davidson, W.D., Mena, I., 1985. Effects of grisorixin on the distribution of thallium-201 and on the oxidative metabolism in cultured myocardial cells. J. Nucl. Med. 26 (6), 626–629.

Mauerer, U.R., Boulpaep, E.L., Segal, A.S., 1998. Properties of an inwardly rectifying ATP-sensitive K^+ channel in the basolateral membrane of renal proximal tubule. J. Gen. Physiol. 111 (1), 139–160. https://doi.org/10.1085/jgp.111.1.139.

Maya-López, M., Mireles-García, M.V., Ramírez-Toledo, M., Colín-González, A.L., Galván-Arzate, S., et al., 2018. Thallium-induced toxicity in rat brain crude synaptosomal/mitochondrial fractions is sensitive to anti-excitatory and antioxidants agents. Neurotox. Res. 33 (3), 634–640. https://doi.org/10.1007/s12640-017-9863-1.

McCall, D., Zimmer, L.J., Katz, A.M., 1985. Kinetics of thallium exchange in cultured rat myocardial cells. Circ. Res. 56 (3), 370–376. https://doi.org/10.1161/01.res.56.3.370.

McCall, D., Zimmer, L.J., Katz, A.M., 1986. Effect of ischemia-related metabolic factors on thallium exchange in cultured rat myocardial cells. Can. J. Cardiol. 2 (3), 176–183.

Melnick, R.L., Monti, L.G., Motzkin, S.M., 1976. Uncoupling of mitochondrial oxidative phosphorylation by thallium. Biochem. Biophys. Res. Commun. 69 (1), 68–73. https://doi.org/10.1016/s0006-291x(76)80273-2.

Méndez-Armenta, M., Nava-Ruiz, C., Fernández-Valverde, F., Sánchez-García, A., Rios, C., 2011. Histochemical changes in muscle of rats exposed subchronically to low doses of heavy metals. Environ. Toxicol. Pharmacol. 32 (1), 107–112. https://doi.org/10.1016/j.etap.2011.03.001.

Meyer, H., Tal, M., 1957. Studies on thallium poisoning. I. The detoxifying action of some compounds containing sulphur and labile methyl groups. J. Sci. Food Agric. 8, 516–519.

Montes, S., Pérez-Barrón, G., Rubio-Osornio, M., Ríos, C., Diaz-Ruíz, A., et al., 2011. Additive effect of DL-penicillamine plus Prussian blue for the antidotal treatment of thallotoxicosis in rats. Environ. Toxicol. Pharmacol. 32 (3), 349–355. https://doi.org/10.1016/j.etap.2011.07.002.

Montes, S., Soriano, L., Ríos, C., Monroy-Noyola, A., 2007. Endogenous thiols enhance thallium toxicity. Arch. Toxicol. 81 (10), 683–687. https://doi.org/10.1007/s00204-007-0203-8.

Mori, K., Yamaguchi, T., Maeda, M., 1998. Mechanism of 201-thallium-chloride uptake in tumor cells and its relationship to potassium channels. Neurol. Res. 20 (1), 19–22. https://doi.org/10.1080/01616412.1998.11740479.

Mourelle, M., Favari, L., Amezcua, J.L., 1988. Protection against thallium hepatotoxicity by silymarin. J. Appl. Toxicol. 8 (5), 351–354. https://doi.org/10.1002/jat.2550080503.

Mulkey, J.P., Oehme, F.W., 1993. A review of thallium toxicity. Vet. Hum. Toxicol. 35 (5), 445–453.

Nakamura, K., Sammiya, T., Hashimoto, J., Ishibashi, R., Matsumoto, K., Kubo, A., 1996. Comparison of cationic myocardial perfusion agents: characteristics of accumulation in cultured smooth muscle cells. Ann. Nucl. Med. 10 (4), 375–381. https://doi.org/10.1007/BF03164797.

Nielsen, J.M., Pedersen, P.A., Karlish, S.J., Jorgensen, P.L., 1998. Importance of intramembrane carboxylic acids for occlusion of K^+ ions at equilibrium in renal Na,K-ATPase. Biochemistry 37 (7), 1961–1968. https://doi.org/10.1021/bi972524q.

Nriagu, J.O., 1998. Thallium in the Environment. In: Wiley Series in Advances in Environmental Science and Technology. Wiley, New York, pp. 1–284.

Oehme, F.M., 1978. Mechanisms of Heavy Metal Inorganic Toxicities. In: Toxicity of Heavy Metals in the Environment. Part 1. Marcel Dekker, New York, pp. 69–85.

Osorio-Rico, L., Galván-Arzate, S., Ríos, C., 1995. Thallium increases monoamine oxidase activity and serotonin turnover rate in rat brain regions. Neurotoxicol. Teratol. 17 (1), 1–5. https://doi.org/10.1016/0892-0362(94)00047-h.

Osorio-Rico, L., Santamaria, A., Galván-Arzate, S., 2017. Thallium toxicity: general issues, neurological symptoms, and neurotoxic mechanisms. Adv. Neurobiol. 18, 345–353. https://doi.org/10.1007/978-3-319-60189-2_17.

Osorio-Rico, L., Villeda-Hernández, J., Santamaría, A., Königsberg, M., Galván-Arzate, S., 2015. The n-methyl-d-aspartate receptor antagonist MK-801 prevents thallium-induced behavioral and biochemical alterations in the rat brain. Int. J. Toxicol. 34 (6), 505–513. https://doi.org/10.1177/1091581815603936.

Pedersen, P.A., Nielsen, J.M., Rasmussen, J.H., Jorgensen, P.L., 1998. Contribution to Tl^+, K^+, and Na^+ binding of Asn776, Ser775, Thr774, Thr772, and Tyr771 in cytoplasmic part of fifth transmembrane segment in alpha-subunit of renal Na,K-ATPase. Biochemistry 37 (51), 17818–17827. https://doi.org/10.1021/bi981898w.

Peluffo, R.D., Berlin, J.R., 1997. Electrogenic K^+ transport by the Na^+-K^+ pump in rat carDiac ventricular myocytes. J. Physiol. 501 (Part 1), 33–40. https://doi.org/10.1111/j.1469-7793.1997.033bo.x.

Perrin, D.D., 1979. Stability Constants of Metal-Ion Complexes. Part B. Organic Ligands. Pergamon, Oxford, pp. 1–1263.

Peter, A.L., Viraraghavan, T., 2005. Thallium: a review of public health and environmental concerns. Environ. Int. 31 (4), 493–501. https://doi.org/10.1016/j.envint.2004.09.003.

Pino, M.T., Marotte, C., Verstraeten, S.V., 2017. Epidermal growth factor prevents thallium(I)- and thallium(III)-mediated rat pheochromocytoma (PC12) cell apoptosis. Arch. Toxicol. 91 (3), 1157–1174. https://doi.org/10.1007/s00204-016-1793-9.

Pino, M.T., Verstraeten, S.V., 2015. Tl(I) and Tl(III) alter the expression of EGF-dependent signals and cyclins required for pheochromocytoma (PC12) cell-cycle resumption and progression. J. Appl. Toxicol. 35 (8), 952–969. https://doi.org/10.1002/jat.3096.

Piskorowski, R.A., Aldrich, R.W., 2006. Relationship between pore occupancy and gating in BK potassium channels. J. Gen. Physiol. 127 (5), 557–576. https://doi.org/10.1085/jgp.200509482.

Ponsoda, X., Jover, R., Núñez, C., Royo, M., Castell, J.V., Gómez-Lechón, M.J., 1995. Evaluation of the cytotoxicity of 10 chemicals in human and rat hepatocytes and in cell lines: correlation between in vitro data and human lethal concentration. Toxicol. in Vitro. 9 (6), 959–966. https://doi.org/10.1016/0887-2333(95)00053-4.

Pourahmad, J., Eskandari, M.R., Daraei, B., 2010. A comparison of hepatocyte cytotoxic mechanisms for thallium (I) and thallium (III). Environ. Toxicol. 25 (5), 456–467. https://doi.org/10.1002/tox.20590.

Prick, J.J.G., 1979. Thallium poisoning. Handbook of clinical neurology. In: Vinken, P.J., Bruyn, G.W. (Eds.), Intoxication of the Nervous System, vol. 36. North-Holland, New York, pp. 239–278.

Puga Molina, L.C., Verstraeten, S.V., 2008. Thallium(III)-mediated changes in membrane physical properties and lipid oxidation affect cardiolipin-cytochrome c interactions. Biochim. Biophys. Acta 1778 (10), 2157–2164. https://doi.org/10.1016/j.bbamem.2008.04.013.

Puga Molina, L.C., Salvatierra Fréchou, D.M., Verstraeten, S.V., 2017. Early response of glutathione- and thioredoxin-dependent antioxidants defense systems to Tl(I)- and Tl(III)-mediated oxidative stress in adherent pheochromocytoma (PC12adh) cells. Arch. Toxicol. 92 (1), 195–211. https://doi.org/10.1007/s00204-017-2056-0.

Rabon, E.C., Sachs, G., 1981. Thallium interaction with the gastric (K, H)-ATPase. J. Membr. Biol. 62 (1–2), 19–27. https://doi.org/10.1007/BF01870196.

Ralph, L., Twiss, M.R., 2002. Comparative toxicity of thallium(I), thallium(III), and cadmium(II) to the unicellular alga Chlorella isolated from Lake Erie. Bull. Environ. Contam. Toxicol. 68 (2), 261–268. https://doi.org/10.1007/s001280247.

Ramsden, D.B., 2007. Thallium. In: Waring, R.H., Steventon, G.B., Mitchel, S.C. (Eds.), Molecules of Death, 2nd. Imperial College Press, London, pp. 415–422.

Rauws, A.G., ten Ham, M., Kamerbeek, H.H., 1969. Influence of the antidote dithiocarb on distribution and toxicity of thallium in the rat. Arch. Int. Pharmacodyn. Ther. 182 (2), 425–426.

Repetto, G., Sanz, P., Repetto, M., 1994. In vitro effects of thallium on mouse neuroblastoma cells. Toxicol. in Vitro 8 (4), 609–611. https://doi.org/10.1016/0300-483x(94)90173-2.

Rodríguez-Mercado, J.J., Altamirano-Lozano, M.A., 2013. Genetic toxicology of thallium: a review. Drug Chem. Toxicol. 36 (3), 369–383. https://doi.org/10.3109/01480545.2012.710633.

Roomans, G.M., Borst-Pauwels, G.W., 1977. Interaction of phosphate with monovalent cation uptake in yeast. Biochim. Biophys. Acta 470 (1), 84–91. https://doi.org/10.1016/0005-2736(77)90063-3.

Rusiecki, W., Brzeziński, J., 1966. The effect of sodium selenate on acute poisoning with thallium. Acta Pol. Pharm. 23 (1), 69–74.

Rusyniak, D.E., Arroyo, A., Acciani, J., Froberg, B., Kao, L., Furbee, B., 2010. Heavy metal poisoning: management of intoxication and antidotes. Experientia (Suppl) 100, 365–396. https://doi.org/10.1007/978-3-7643-8338-1_11.

Sabbioni, E., Marafante, E., Rade, J., Di Nucci, A., Gregotti, C., Manzo, L., 1980. Metabolic patterns of low and toxic doses of thallium in the rat. In: Holmstedt, B., Lauwerys, R., Mercier, M., Roberfroid, M. (Eds.), Mechanisms of Toxicity and Hazard Evaluation. Elsevier, New York, pp. 559–564.

Saddique, A., Peterson, C.D., 1983. Thallium poisoning: a review. Vet. Hum. Toxicol. 25 (1), 16–22.

Salehi, S., Saljooghi, A.S., Badiee, S., Moqadam, M.M., 2017. Chelation of thallium (III) in rats using combined deferasirox and deferiprone therapy. Toxicol. Res. 33 (4), 299–304. https://doi.org/10.5487/TR.2017.33.4.299.

Sands, H., Delano, M.L., Camin, L.L., Gallagher, B.M., 1985. Comparison of the transport of $^{42}K^{+}$, $^{22}Na^{+}$, $^{201}Tl^{+}$, and $[^{99m}Tc(dmpe)_2\ X\ Cl_2]^{+}$ using human erythrocytes. Biochim. Biophys. Acta 812 (3), 665–670. https://doi.org/10.1016/0005-2736(85)90259-7.

Schoer, J., 1984. Thallium. In: Hutzinger, O. (Ed.), Handbook of Environmental Chemistry. Anthropogenic Compounds. Springer, New York, pp. 143–214.

Schroeder, I., Hansen, U.P., 2008. Tl^{+}-induced micros gating of current indicates instability of the MaxiK selectivity filter as caused by ion/pore interaction. J. Gen. Physiol. 131 (4), 365–378. https://doi.org/10.1085/jgp.200809956.

Schwetz, B.A., O'Neil, P.V., Voelker, F.A., Jacobs, D.W., 1967. Effects of diphenylthiocarbazone and diethyldithiocarbamate on the excretion of thallium by rats. Toxicol. Appl. Pharmacol. 10 (1), 79–88. https://doi.org/10.1016/0041-008x(67)90130-5.

Sessler, M.J., Geck, P., Maul, F.D., Hör, G., Munz, D.L., 1986. New aspects of cellular thallium uptake: Tl^{+}-Na^{+}-$2Cl^{-}$-cotransport is the central mechanism of ion uptake. Nuklearmedizin 25 (1), 24–27.

Sherstobitov, A.O., Gusev, G.P., Skul'skiĭ, I.A., 1990. The transport of univalent thallium across the human erythrocyte membrane. Tsitologiia 32 (3), 239–244 (Russian). https://pubmed.ncbi.nlm.nih.gov/2171170/.

Sherstobitov, A.O., Gusev, G.P., Skul'skiĭ, I.A., 1991. Potassium transport in lamprey (*Lampetra fluviatilis*) erythrocytes: evidence for K^{+} channels. Tsitologiia 33 (11), 130–134. https://doi.org/10.1016/0300-9629(92)90510-W.

Sherstobitov, A.O., Lapin, A.A., Glazunov, V.V., Nikiforov, A.A., 2010. Transport of monovalent thallium across the membrane of oocyte of the lamprey *Lampetra fluviatilis*. J. Evol. Biochem. Physiol. 46 (3), 235–240. https://doi.org/10.1134/S0022093010030026.

Sherstobitov, A.O., Lapin, A.A., Glazunov, V.V., Nikiforov, V.A., Nikiforov, A.A., 2011. Accumulation of sodium and potassium ions in oocytes of the river lamprey *Lampetra fluviatilis* at the prespawning period. J. Evol. Biochem. Physiol. 47 (4), 328–332. https://doi.org/10.1134/S0022093011040037.

Sherstobitov, A.O., Lapin, A.V., Glazunov, V.V., 2003. Transport of monovalent thallium in oocytes of the lamprey *Lampetra fluviatilis*. J. Evol. Biochem. Physiol. 39 (3), 387–388. https://doi.org/10.1023/A:1026164329277.

Shieh, R.C., Lee, Y.L., 2001. Ammonium ions induce inactivation of Kir2.1 potassium channels expressed in Xenopus oocytes. J. Physiol. 535 (Pt 2), 359–370. https://doi.org/10.1111/j.1469-7793.2001.00359.x.

Skul'skiĭ, I.A., 1991. The isomorphism of univalent thallium and potassium in membrane transport processes. Tsitologiia 33 (11), 118–129 (Russian). PMID: 1668045.

Skul'skiĭ, I.A., Glazunov, V.V., Gusev, G.P., Sherstobitov, A.O., 1992. Influence of anions on passive transport of monovalent thallium ions through membrane of human erythrocytes. Biol. Membr. 9 (4), 376–381 (Russian).

Skul'skiĭ, I.A., Manninen, V., Glasunov, V.V., 1990. Thallium and rubidium permeability of human and rat erythrocyte membrane. Gen. Physiol. Biophys. 9 (1), 39–44. PMID: 2311912.

Skulskii, I.A., Gusev, G.P., Sherstobitov, A.O., Manninen, V., 1992. Anion-dependent transport of thallous ions through human erythrocyte membrane. J. Membr. Biol. 130 (3), 219–225. https://doi.org/10.1007/BF00240479.

Skulskii, I.A., Manninen, V., 1981. Effect of membrane potential on the passive transport of Tl^{+} in human red blood cells. Acta Physiol. Scand. 111 (3), 343–348. https://doi.org/10.1111/j.1748-1716.1981.tb06746.x.

Skulskii, I.A., Manninen, V., Järnefelt, J., 1973. Interaction of thallous ions with the cation transport mechanism in erythrocytes. Biochim. Biophys. Acta 298 (3), 702–709. https://doi.org/10.1016/0005-2736(73)90086-2.

Skulskii, I.A., Manninen, V., Järnefelt, J., 1978. Factors affecting the relative magnitudes of the ouabain-sensitive and the ouabain-insensitive fluxes of thallium ion in erythrocytes. Biochim. Biophys. Acta 506 (2), 233–241. https://doi.org/10.1016/0005-2736(78)90394-2.

Skulskii, I.A., Saris, N.E., Savina, M.V., Glasunov, V.V., 1981. Uptake of thallous ions by mitochondria is stimulated by nonactin but not by respiration alone. FEBS J. 120 (2), 263–266. https://doi.org/10.1111/j.1432-1033.1981.tb05698.x.

Spencer, P.S., Peterson, E.R., Madrid, R., Raine, C.S., 1973. Effects of thallium salts on neuronal mitochondria in organotypic cord-ganglia-muscle combination cultures. J. Cell Biol. 58 (1), 79–95. https://doi.org/10.1083/jcb.58.1.79.

Stavinoha, W.B., Emerson, G.A., Nash, J.B., 1959. The effects of some sulfur compounds on thallotoxicosis in mice. Toxicol. Appl. Pharmacol. 1, 638–646. https://doi.org/10.1016/0041-008x(59)90068-7.

Tao, Z., Gameiro, A., Grewer, C., 2008. Thallium ions can replace both sodium and potassium ions in the glutamate transporter excitatory amino acid carrier 1. Biochemistry 47, 12923–12930. https://doi.org/10.1021/bi8017174.

Teichman, S., Qu, S., Kanner, B.I., 2009. The equivalent of a thallium binding residue from an archeal homolog controls cation interactions in brain glutamate transporters. Proc. Natl. Acad. Sci. U.S.A. 106 (34), 14297–14302. https://doi.org/10.1073/pnas.0904625106.

Thompson, D.F., 1981. Management of thallium poisoning. Clin. Toxicol. 18 (8), 979–990. https://doi.org/10.3109/15563658108990328.

Thyresson, N., 1950. The influence of dietary factors, especially brewer's yeast, cystine and B vitamins, on the course of chronic thallium poisoning in the rat. Acta Derm. Venereol. 30 (1), 9–26.

Thyresson, N., 1951a. Experimental investigation on thallium poisoning in the rat. Distribution of thallium, especially in the skin, and excretion of thallium under different experimental conditions. A study with the use of the radioactive isotope Tl^{204}. Acta Derm. Venereol. 31 (1), 3–27.

Thyresson, N., 1951b. Experimental investigation on thallium poisoning; effect of thallium on the growth and differentiation of hairs and epidermis, studied in young rats and in tissue cultures of embryonic rat skin. Acta Derm. Venereol. 31 (2), 133–146.

Tomilets, V.A., Dontsov, V.I., Ado, V.A., Zakharova, I.A., 1980a. Toxic, anaphylactoid, and sensitizing properties of mercaptoquinolinates, metals of the 8th and 3d group. Bull. Exp. Biol. Med. 89 (3), 328–330 (Russian). PMID: 6155949.

Tomilets, V.A., Zakharova, I.A., Dontsov, V.I., Ado, V.A., 1980b. Comparative study of the toxic, anaphylactoid, and sensitizing properties of 5-sulfo-8-mercaptoquinolinates of metals of 8th and 3d groups of the periodic table. Bull. Exp. Biol. Med. 89 (4), 463–465 (Russian). PMID: 7388162.

Truter, M.R., 1975. Recognition of metal cations by biological systems. Philos. Trans. R. Soc. Lond. B Biol. Sci. 272 (915), 29–41. https://doi.org/10.1098/rstb.1975.0068.

Twining, B.S., Twiss, M.R., Fisher, N.S., 2003. Oxidation of thallium by freshwater plankton communities. Environ. Sci. Technol. 37 (12), 2720–2726. https://doi.org/10.1021/es026145i.

Verstraeten, S.V., 2006. Relationship between thallium(I)-mediated plasma membrane fluidification and cell oxidants production in Jurkat T cells. Toxicology 222 (1–2), 95–102. https://doi.org/10.1016/j.tox.2006.01.028.
Villaverde, M.S., Hanzel, C.E., Verstraeten, S.V., 2004. In vitro interactions of thallium with components of the glutathione-dependent antioxidants defence system. Free Radic. Res. 38 (9), 977–984. https://doi.org/10.1080/10715760400000950.
Villaverde, M.S., Verstraeten, S.V., 2003. Effect of thallium (I) and thallium (III) on liposome membrane physical properties. Arch. Biochem. Biophys. 417 (2), 235–243. https://doi.org/10.1016/s0003-9861(03)00366-7.
Wanger, T., Scheich, H., Ohl, F.W., Goldschmidt, J., 2012. The use of thallium diethyldithiocarbamate for mapping CNS potassium metabolism and neuronal activity: Tl^+ -redistribution, Tl^+ -kinetics and Tl^+ -equilibrium distribution. J. Neurochem. 122 (1), 106–114. https://doi.org/10.1111/j.1471-4159.2012.07757.x.
Weaver, C.D., 2018. Thallium flux assay for measuring the activity of monovalent cation channels and transporters. Methods Mol. Biol. 1684, 105–114. https://doi.org/10.1007/978-1-4939-7362-0_9.
Weaver, C.D., Harden, D., Dworetzky, S.I., Robertson, B., Knox, R.J., 2004. A thallium-sensitive, fluorescence-based assay for detecting and characterizing potassium channel modulators in mammalian cells. J. Biomol. Screen 9 (8), 671–677. https://doi.org/10.1177/1087057104268749.
Williams, R.J.V., 1970. The biochemistry of sodium, potassium, magnesium, and calcium. Q. Rev. Chem. Soc. 3, 331–365. https://doi.org/10.1039/qr9702400331.
Wojtovich, A.P., Williams, D.M., Karcz, M.K., Lopes, C.M., Gray, D.A., et al., 2010. A novel mitochondrial K_{ATP} channel assay. Circ. Res. 106 (7), 1190–1196. https://doi.org/10.1161/CIRCRESAHA.109.215400.
Wollmuth, L.P., 1995. Multiple ion binding sites in Ih channels of rod photoreceptors from tiger salamanders. Pflügers Archiv: Eur. J. Physiol. 430 (1), 34–43. https://doi.org/10.1007/BF00373837.
Wollmuth, L.P., Hille, B., 1992. Ionic selectivity of Ih channels of rod photoreceptors in tiger salamanders. J. Gen. Physiol. 100 (5), 749–765. https://doi.org/10.1085/jgp.100.5.749.
Woods, J.S., Fowler, B.A., 1986. Alteration of hepatocellular structure and function by thallium chloride: ultrastructural, morphometric, and biochemical studies. Toxicol. Appl. Pharmacol. 83 (2), 218–229. https://doi.org/10.1016/0041-008x(86)90299-1.
Woods, J.S., Fowler, B.A., Eaton, D.L., 1984. Studies on the mechanisms of thallium-mediated inhibition of hepatic mixed function oxidase activity. Correlation with inhibition of NADPH-cytochrome c (P-450) reductase. Biochem. Pharmacol. 33 (4), 571–576.
Xia, M., Shahane, S.A., Huang, R., Titus, S.A., Shum, E., et al., 2011. Identification of quaternary ammonium compounds as potent inhibitors of hERG potassium channels. Toxicol. Appl. Pharmacol. 252 (3), 250–258. https://doi.org/10.1016/j.taap.2011.02.016.
Yoshida, M., Igeta, S., Kawashima, R., Akama, Y., Yoshida, K., 1997. Changes in adenosine triphosphate (ATP) concentration and its activity in murine tissue after thallium administration. Bull. Environ. Contam. Toxicol. 59, 268–273. https://doi.org/10.1007/s001289900474.

Zhang, D., Gopalakrishnan, S.M., Freiberg, G., Surowy, C.S., 2010. A thallium transport FLIPR-based assay for the identification of KCC2-positive modulators. J. Biomol. Screen 15 (2), 177–184. https://doi.org/10.1177/1087057109355708.
Zierold, K., 2000. Heavy metal cytotoxicity studied by electron probe X-ray microanalysis of cultured rat hepatocytes. Toxicol. in Vitro. 14 (6), 557–563. https://doi.org/10.1016/s0887-2333(00)00049-7.
Zitko, V., 1975. Toxicity and pollution potential of thallium. Sci. Total Environ. 4 (2), 185–192.

CHAPTER 4

Effect of thallium on mitochondria

List of abbreviations

5-HD	5-hydroxyodecanoate
ADP	adenosine diphosphate
ANT	adenine nucleotide translocase
ATP	adenosine triphosphate
BKA	bongkrekic acid
CAT	carboxyatractyloside
CCP	m-chlorophenyl hydrazone
CsA	cyclosporin A
CyP-D	Cyclophyllin D
Dia	diazoxide
DIDS	4,4′-diisothiocyanostilbene-2,2′-disulfonate
DNP	2,4-dinitrophenol
EDTA	ethylenediaminetetraacetic acid
EGTA	ethylene guanine tetraacetic acid
ETC	electron transport chain
FCCP	carbonyl cyanide-trifluoromethoxyphenylhydrazone
IMM	inner mitochondrial membrane
mitoBK$_{Ca}$	mitochondrial calcium-activated BK-type potassium channel
mitoK$_{ATP}$	mitochondrial ATP-dependent channel
MPTP	mitochondrial permeability transition pore
MSL	mersalyl
NADH	reduced nicotinamide adenine dinucleotide
NEM	N-ethylmaleimide
PAO	phenylarsine oxide
Pax	paxilin
Pi	inorganic phosphate
PiC	phosphate symporter
Pin	pinacidil
RCR	respiratory control ratio
RHM	rat heart mitochondria
RLM	rat liver mitochondria
ROS	reactive oxygen species
RR	ruthenium red

Mitochondria as a Key Intracellular Target of Thallium Toxicity
ISBN 978-0-323-95531-7
https://doi.org/10.1016/B978-0-323-95531-7.00002-1

tBHP	tert-butyl hydroperoxide
VDAC	volt-dependent anion channel
$\Delta\Psi_{mito}$	inner membrane potential

4.1 Heavy metal effects on isolated mitochondria

We showed earlier that state 4 respiration of rat liver mitochondria (RLM) markedly stimulated due to Cd^{2+}-induced increase in the IMM ion permeability (Skul'skiĭ et al., 1988; Korotkov et al., 1998, 2008b; Belyaeva and Korotkov, 2003). At the same time, respiration of mitochondria being in states of 4, 3, or $3U_{DNP}$ (uncoupled DNP; Panov et al., 2009) decreased in the presence of 15–40 μM Cd^{2+} due to inhibition of the respiratory chain. On the other hand, 0.2–22 mM thallium acetate did not affect respiration of rat liver mitochondria in state 3, but at the same time, it stimulated respiration of these organelles in state 4 (Melnick et al., 1976; Bragadin et al., 2003). Considering these data imperfection, it became necessary to study the effect of higher (1–125 mM) concentrations of Tl^+ salts ($TlNO_3$ and Tl acetate) on RLM respiration in states 4_0, 4, 3, and $3U_{DNP}$ in the presence of Pi and ionophore nonactin that affect thallium transport (Korotkov et al., 2007, 2008a). Experiments in vivo with rats exposed to thallium acetate showed the damage of the inner membrane structure in liver and kidney mitochondria due to massive swelling of these organelles resulted (Herman and Bensch, 1967). The aforementioned results of experiments in vitro and in vivo come into some contradiction among themselves. As you know, cellular cytoplasm has a concentration of K^+ ions (about 130 mM). Taking these data into account, we used sucrose-supported 400 mOsm medium containing 25–75 mM $TlNO_3$ and 125 mM KNO_3 to evaluate the effects of Tl^+ on oxygen consumption rates of mitochondria being in the aforementioned states in the presence of K^+ close to intracellular concentration.

Thallium was found to have reduced the body's antioxidants defenses, deplete the intracellular pool of reduced glutathione, and increased the production of reactive oxygen species (ROS; Aoyama et al., 1988; Mulkey and Oehme, 1993; Zierold, 2000; Galván-Arzate et al., 2005; Hanzel and Verstraeten, 2009; Cvjetko et al., 2010; Kiliç and Kutlu, 2010; Méndez-Armenta et al., 2011; Osorio-Rico et al., 2015). Such action of thallium could be associated with its reaction with the thiol groups of cells and mitochondria. That, by analogy with the effects of the aforementioned heavy bivalent metals, could be one of the reasons for the toxic effect of thallium on living organisms (Herman and Bensch, 1967; Melnick et al., 1976; Saris et al., 1981; Woods and Fowler, 1986). At the same time, there are no data on the direct interaction of Tl^{+} with mitochondrial thiol groups. In this regard, it became necessary to directly assess the interaction of Tl^{+} with free mitochondrial thiol groups, which will be described in the corresponding section of experimental results.

Nonenergized mitochondria swelled weakly in a medium containing potassium, ammonium, or sodium nitrates. This result helped to reveal correspondingly low passive permeability of the inner membrane for monovalent cations of K^{+}, H^{+}, and Na^{+} (Mitchell and Moyle, 1969; Brierley et al., 1970, 1977). On the other hand, this permeability markedly increased with mitochondria exposed to bivalent heavy metals (Cd^{2+}, Zn^{2+}, Hg^{2+}, Pb^{2+}) and PAO, which have a high affinity to molecular SH groups (Perrin, 1979; Scott et al., 1971; Sanadi et al., 1981; Viarengo and Nicotera, 1991; Korotkov et al., 1998; Hernández-Esquivel et al., 2011). In addition, swelling of these mitochondria was very noticeable in media containing 20–80 mM $TlNO_3$, which was associated with high passive permeability of the IMM to Tl^{+} ions (Saris et al., 1981; Skul'skiĭ et al., 1984). Until now, the effects of Tl^{+} did not study on passive ionic permeability of the IMM to univalent cations (K^{+}, H^{+}, Na^{+}). To assess the impact of Tl^{+} on passive permeability of the IMM to K^{+}, H^{+}, Na^{+} ions, we investigated swelling of nonenergized rat liver and heart mitochondria in 400 mOsm nitrate media containing 25–75 mM $TlNO_3$ and respectively 125 mM nitrates (KNO_3, NH_4NO_3, $NaNO_3$).

4.1.1 Thallium effects on isolated mitochondria

Currently, researchers conclude that mitochondria are the critical organelle damaged in thallium poisoning (Osorio-Rico et al., 2017). Chronic thallium intoxication of rats resulted in damage of microvilli, endoplasmic reticulum, and mitochondrial membranes in kidneys, liver, muscles, brain, and other organs, which were due to massive mitochondrial swelling in these organelles (Herman and Bensch, 1967; Spencer et al., 1973; Danilewicz et al., 1979; Woods and Fowler, 1986). Electron microscopic studies of the organs of rats and mice treated with thallium salts have shown that mitochondria are one of the main cellular targets of thallium (Herman and Bensch, 1967; Spencer et al., 1973; Prick, 1979; Woods and Fowler, 1986). Intense swelling of nonenergized mitochondria in media with 20–40 mM $TlNO_3$ revealed high passive permeability of the inner membrane to Tl^+ (Skul'skiĭ et al., 1978a; Saris et al., 1981). Monovalent thallium damaged the structure of isolated rat liver mitochondria (Skul'skiĭ et al., 1984). Intracellular toxic effects of Tl^+ on living organisms were suggested to be based on its ability to easily penetrate the inner mitochondrial membrane and uncouple oxidative phosphorylation processes (Melnick et al., 1976; Saris et al., 1981).

High-amplitude swelling of nonenergized mitochondria did find in media containing millimolar concentrations of thallium salts ($TlNO_3$ and Tl-acetate) due to high permeability of the IMM to ions Tl^+ (Melnick et al., 1976; Skul'skiĭ, 1977; Saris et al., 1981; Skul'skiĭ et al., 1979b, 1981, 1984). On the other hand, swelling of nonenergized mitochondria in nitrate or chloride media increased in the presence of micromolar concentrations of bivalent heavy metals (Cd^{2+}, Pb^{2+}, Hg^{2+}, Cu^{2+}) and Ag^+ due to rise in the IMM passive ionic permeability, induced by these metals (Riley and Lehninger, 1964; Scott et al., 1971; Skul'skiĭ et al., 1988; Korotkov et al., 1998; Belyaeva et al., 2004a; Lee et al., 2005). Considering these data, it became necessary to research the effects of Tl^+ on the passive permeability to monovalent cations (K^+, H^+, Na^+, Li^+). For this purpose, we studied swelling of nonenergized mitochondria in 400 mOsm medium containing 25–75 mM $TlNO_3$ and 125 mM nitrate (KNO_3, NH_4NO_3, $NaNO_3$, $LiNO_3$). Tl^+ ions high reduced valinomycin-induced K^+ diffusion potential in nonenergized mitochondria (Skul'skiĭ et al., 1978a, 1979b). At the same time, transport of $^{204}Tl^+$ into the matrix of nonenergized mitochondria occurred when proton diffusion potential or potassium diffusion potential was present on the inner membrane. In the case of energized organelles, such transport became possible in the presence of ionophore nonactin (Skul'skiĭ, 1977, 1978a, 1979b, 1980b, 1981; Skul'skiĭ and Glazunov, 1981).

Succinate injection into suspension of rat liver mitochondria preswollen in a medium with $TlNO_3$ resulted in their powerful contraction, which a specific Tl^+/H^+ exchanger explains in the inner membrane (Saris et al., 1981). Nonenergized mitochondria preswollen in a medium with Tl-acetate swelled additionally after their energization (Melnick et al., 1976; Skul'skiĭ et al., 1979b, 1981, 1984; Saris et al., 1981; Korotkov et al., 2007). In this case, entry of Tl^+ into the mitochondrial matrix occurred by the mechanism of electrophoretic uniport (Melnick et al., 1976; Diwan and Lehrer, 1977; Skul'skiĭ et al., 1978a; Saris et al., 1981). Later there was direct evidence in participation of the K^+/H^+ exchanger in the transport of Tl^+ through the IMM (Diwan et al., 1990). Uptake of $^{86}Rb^+$ or $^{137}Cs^+$ by succinate-energized rat liver mitochondria in the presence of valinomycin decreased sharply after the addition of 50 nM nonactin and 1 μM $^{204}Tl^+$ into a medium (Skul'skiĭ et al., 1979a,b, 1981; Skul'skiĭ and Glazunov, 1981; Korotkov et al., 2005, 2007). This result can explain by the competition of these ions for a standard transport system.

The use of fluorescent dye BTC-AM showed that Tl^+ could transport into energized rat heart mitochondria (RHM) using ATP-sensitive (mitoK_{ATP}) or BK-type Ca^{2+} activated (mitoK_{Ca}) potassium channels (Wojtovich et al., 2010, 2013; Queliconi et al., 2011; Testai et al., 2013). MitoK_{ATP} blockers (glibenclamide and 5-HD) inhibited Tl^+ transport into succinate-energized rat liver mitochondria (Nikitina and Glazunov, 2003). Tl^+ ions inhibited the transport of K^+ and Mg^{2+} into mitochondria (Barrera and Gomez-Puyou, 1975; Diwan and Lehrer, 1977; Diwan et al., 1979), and they more strongly inhibited Na^+-dependent Ca^{2+} release from rat liver mitochondria compared to the effects of Li^+, Cs^+, and Rb^+ (Nalyvajko et al., 2006).

Univalent thallium increased mitochondrial respiration in sucrose-containing media (Barrera and Gomez-Puyou, 1975; Melnick et al., 1976; Bragadin et al., 2003; Korotkov et al., 2008a). Cyclic movement of Tl^+ ions across the IMM can explain this increased respiration of energized mitochondria (Melnick et al., 1976; Bragadin et al., 2003; Korotkov et al., 2007). Inorganic phosphate and nonactin forming a solid complex with Tl^+ ions (Cornelius et al., 1974) accelerated electrophoretic transport of $^{201}Tl^+$ or $^{204}Tl^+$ into the matrix and increased respiration and swelling of energized mitochondria in media containing thallium acetate and thallium nitrate (Skul'skiĭ et al., 1978a, 1979a,b, 1980a, 1981; Saris et al., 1981; Skul'skiĭ and Glazunov, 1981, 1982; Korotkov et al., 2005, 2007, 2008a). The latter is consistent with a marked increase in respiration of energized mitochondria in a medium containing thallium acetate and nonactin (Korotkov et al., 2005, 2007). At the same time, 25–75 mM Tl^+ has no noticeable effect on mitochondrial respiration in 3 or $3U_{DNP}$ states (Barrera and Gomez-Puyou, 1975; Melnick et al., 1976; Korotkov et al., 2007, 2008a). The use of isolated mitochondria to assess mechanisms underlying mitochondrial toxicity of various chemicals and heavy metals is quite justified instead of complex and expensive experiments with animals or all kinds of multivariate methods to evaluate the toxicity of different substances to humans (Ogata et al., 1983; Ekwall et al., 1990).

Some heavy metals (Cd^{2+}, Hg^{2+}, Pb^{2+}, Zn^{2+}, Cu^{2+}) were previously shown to have induced the MPTP opening in the inner membrane in calcium-free experiments, ADP, and CsA can inhibit this pore (Miyahara and Utsumi, 1975; Chávez and Holguín, 1988; Zoratti and Szabó, 1995; Wudarczyk et al., 1999; Belyaeva et al., 2002; Dorta et al., 2003; Lee et al., 2005). At the same time, the ability of Tl^+ to induce this pore opening turned out to be practically unexplored. We tried to answer the question in the proposed study "Can Tl^+ ions themselves, like the action of above bivalent heavy metals, induce the opening of the MPTP with an increase of the inner membrane permeability?" We studied the Tl^+-induced MPTP opening in calcium-loaded mitochondria, energized by glutamate and malate or succinate alone, respectively, respiratory substrates of first or second respiratory complexes. In addition, we research the effects of the pore inhibitors (ADP, bongkrekic acid [BKA]). Since the MPTP opening depends on the conformation of the ANT being the pore regulatory component (Halestrap and Davidson, 1990), effects of reagents fixing this exchanger in "m" conformation (ADP, BKA, NEM) or "c" (PAO, tBHP, diamide; Halestrap and Brenner, 2003). On the other hand, the phosphate cotransporter (PiC) belongs to the main components of the pore (Halestrap, 2010). Therefore, the Tl^+-induced MPTP opening did investigate in the presence of the PiC inhibitor mersalyl (MSL).

4.2 Effect of thallium on the inner membrane potential and ion permeability and the mitochondrial respiration

4.2.1 Thallium effect on ion permeability and potential

The inner mitochondrial membrane is known to be easily permeable to the NO_3^- anion. Therefore minor swelling of nonenergized mitochondria in media containing 125 mM nitrates (KNO_3, NH_4NO_3, $NaNO_3$) concluded that the membrane possesses low passive permeability for K^+, H^+, and Na^+ cations, respectively (Mitchell and Moyle, 1969; Brierley et al., 1970, 1977). On the contrary, the marked swelling of mitochondria in a medium containing 20–80 mM $TlNO_3$ made it possible to conclude that this membrane has a high passive ionic permeability to Tl^+ (Saris et al., 1981; Skul'skiĭ et al., 1984). Taking into account the results of the above experiments, we correspondingly studied the passive permeability to cations (K^+, Na^+, Li^+, H^+) in the swelling experiments with nonenergized mitochondria in a 400 mOsm medium containing 0–75 mM $TlNO_3$ and 125 mM nitrate (KNO_3, $NaNO_3$, $LiNO_3$, NH_4NO_3). Since the IMM is highly permeable to Tl^+, the results of these experiments might compare with ones of similar experiments in the medium with 125 mM nitrate replaced by 250 mM sucrose, for which the membrane is impermeable. Swelling of nonenergized rat liver and rat heart mitochondria was noticeably higher in the above medium with $TlNO_3$ and nitrates (KNO_3, $NaNO_3$, $LiNO_3$, NH_4NO_3) in comparison with the swelling of these organelles in a similar medium with $TlNO_3$ and sucrose (Figs. 4.1 and 4.2 and Table 4.1; Korotkov, 2009; Korotkov et al., 2013, 2020).

Experimental conditions and additions are the same as in Fig. 4.2. Mitochondria (1.5 mg/mL of protein) were added into medium containing 75 mM $TlNO_3$, along with 250 mM sucrose (Sucrose) or 125 mM nitrates

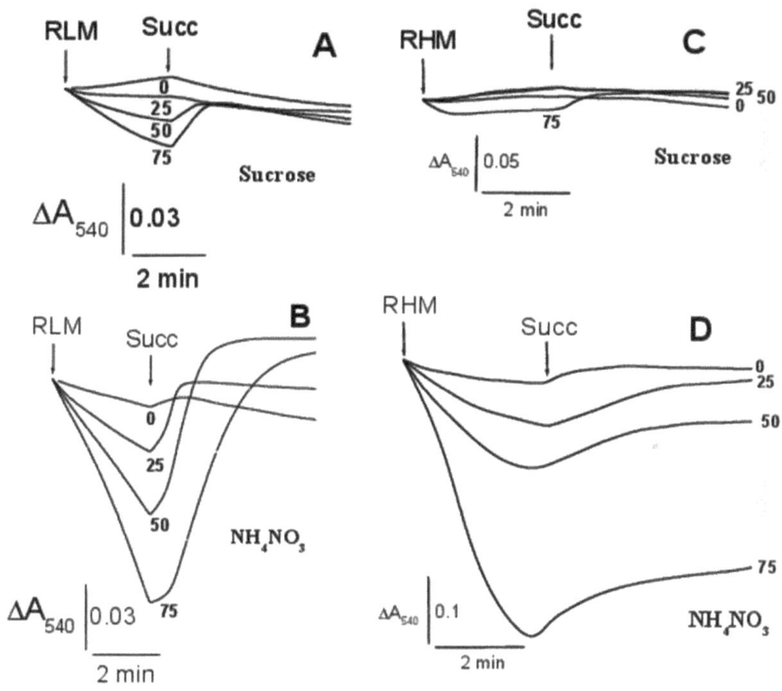

Figure 4.1 Effects of Tl^+ on swelling of nonenergized rat liver and rat heart mitochondria. Mitochondria of liver (1.5 mg/mL of protein, panels A and B) and rat heart (1.0 mg/mL of protein, panels C and D) were added into the medium containing 0–75 mM $TlNO_3$, 5 mM Tris-NO_3 (pH 7.3), 4 μM rotenone, and 3 μg/mL oligomycin, as well as 250 mM sucrose (A, C) or 125 mM NH_4NO_3 (B, D). All media osmolarity was maintained by sucrose in this and subsequent figures in demonstrating results of our experiments. Numbers near traces show concentration of $TlNO_3$ (mM) in the medium. Arrows indicate addition of rat liver mitochondria (RLM), rat heart mitochondria (RHM), and succinate till 5 mM concentration (Succ). The figure shows traces representing results in three independent experiments.

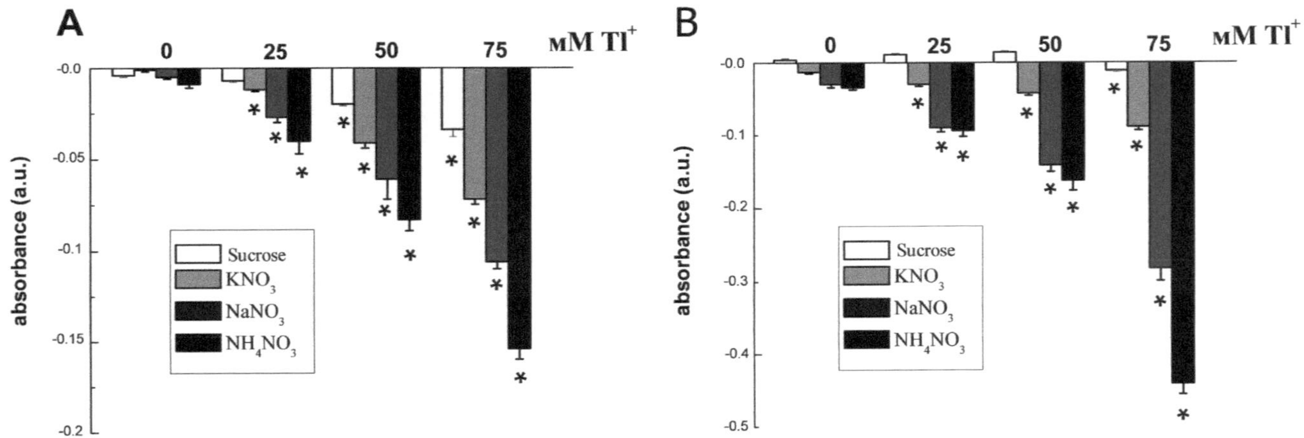

Figure 4.2 Effects of Tl^+ on passive swelling of rat liver (A) and rat heart (B) mitochondria. Mitochondria were injected into the 400 mOsm medium containing 0–75 mM $TlNO_3$ along with 250 mM sucrose (Sucrose) or 125 mM nitrates (KNO_3, $NaNO_3$, NH_4NO_3). The medium other components and experimental conditions were as in Fig. 4.1. Ordinate shows change in absorbance (ΔA_{540}) after 3 min after adding mitochondria and before adding succinate into the medium (see Fig. 4.1). X-axis is $TlNO_3$ concentration (mM). Asterisks indicate significant differences ($P < .05$) from the control (experiments free $TlNO_3$).

Table 4.1 Change in absorbance (ΔA_{540}) in suspension of nonenergized rat liver mitochondria.

Concentration of $TlNO_3$ (mM) in the medium	Media				
	Sucrose	KNO_3	$NaNO_3$	NH_4NO_3	$LiNO_3$
0 (Control)	−0.004 ± 0.001 (3)	−0.001 ± 0.001 (3)	−0.005 ± 0.001 (3)	−0.009 ± 0.005 (3)	−0.003 ± 0.001 (4)
25	−0.007 ± 0.001 (3)	−0.012 ± 0.001 (3)*#	−0.027 ± 0.002 (3)*#	−0.040 ± 0.007 (3)*#	
50	−0.020 ± 0.001 (3)*	−0.041 ± 0.003 (3)*#	−0.061 ± 0.006 (3)*#	−0.083 ± 0.006 (3)*#	
75	−0.034 ± 0.004 (3)*	−0.072 ± 0.003 (3)*#	−0.106 ± 0.004 (3)*#	−0.154 ± 0.006 (3)*#	−0.125 ± 0.006 (4)*#

Asterisk (*) indicate significant differences from the control, experiments free of $TlNO_3$ ($P < .05$).
The hash signs (#) significant differences between experiments in the medium with nitrates from those that were carried out in the medium with sucrose for the same concentration of $TlNO_3$ ($P < .05$).

(KNO_3, $NaNO_3$, NH_4NO_3, $LiNO_3$), as well as 4 μM rotenone, and 1 μg/mL oligomycin. Number of experiments indicated in brackets. The absorbance changes are shown in 3 min after addition of mitochondria into the medium (see Fig. 4.2A). Asterisks indicate significant differences from the control, experiments free of $TlNO_3$ ($P < .05$). The hash signs mark significant differences between experiments in the medium with nitrates from those that were carried out in the medium with sucrose for the same concentration of $TlNO_3$ ($P < .05$).

The Cd^{2+}-induced swelling of mitochondria in nitrate or chloride media was shown earlier to have increased in the order of $Li^+ < Na^+ < K^+ < NH_4^+$ (Skul'skiĭ et al., 1988; Korotkov et al., 1998; Lee et al., 2005). Experiments presented here (Figs. 4.1 and 4.2 and Table 4.1) show that swelling of nonenergized rat liver and rat heart mitochondria in a medium with $TlNO_3$ increased in a series of sucrose < KNO_3 < $NaNO_3$ < $LiNO_3$ < NH_4NO_3 (Korotkov and Brailovskaya, 2001). These results can conclude that weak effect of Tl^+ ions on swelling of nonenergized mitochondria in the medium with KNO_3 and $TlNO_3$, in contrast to a similar effect of Cd^{2+} in a medium with KNO_3 (Skul'skiĭ et al., 1988; Korotkov et al., 1998; Lee et al., 2005), may indicate a weak effect of Tl^+ on the potassium permeability of the inner membrane. Perhaps this is due to an inhibition by Tl^+ ions of passive transport of K^+ into the matrix (Barrera and Gomez-Puyou, 1975; Diwan and Lehrer, 1977; Korotkov and Brailovskaya, 2001), in contrast to Cd^{2+}, having been stimulated this transport (Skul'skiĭ et al., 1988; Korotkov et al., 1998; Lee et al., 2005).

Thus, the presented data suggest that Tl^+ (like Cd^{2+}) increases the passive ionic permeability of the inner mitochondrial membrane (Korotkov and Brailovskaya, 2001; Korotkov, 2009; Korotkov et al., 2013).

Another possible reason for the increased nonenergized mitochondria swelling in a medium containing $TlNO_3$ and monovalent cation nitrate could be increased transport of Tl^+ into the matrix by a concentration gradient. To verify this hypothesis, we researched the effects of quinine (blocker of the mitochondrial K^+/H^+ exchanger) which inhibited a contraction of energized mitochondria preswollen in media containing KNO_3, $NaNO_3$, or NH_4NO_3 (Bernardi, 1999; Nakashima and Garlid, 1982; Jung et al., 1984; Diwan, 1986; Garlid and Paucek, 2003). The energized mitochondria pre-swollen in saline media was found to have contracted due to extrusion of univalent cations (K^+, NH_4^+, Na^+, Li^+, Rb^+, Cs^+) from the matrix by the participation of the inner membrane K^+/H^+ exchanger (Mitchell and Moyle, 1969; Brierley et al., 1977;

Diwan, 1986; Garlid and Paucek, 2003). In parallel, NO_3^- anion freely leaves mitochondria since the IMM is freely permeable for this anion, and protons that appeared in the matrix, when exchanged for monovalent matrix cations, are removed with the participation of the respiratory chain complexes. We found that the contraction of succinate-energized RLM preswollen in the medium with $TlNO_3$ and nitrate (KNO_3, $NaNO_3$, NH_4NO_3) inhibited the presence of quinine (Fig. 4.3) (Korotkov, 2009).

Noticeable inhibition did observe for the contraction of rat liver mitochondria preswollen in the medium containing KNO_3, $TlNO_3$, and quinine after the addition of succinate (Fig. 4.3). Complete contraction inhibition showed in a similar medium with $TlNO_3$ and KNO_3 replaced by $NaNO_3$ or NH_4NO_3 (Korotkov, 2009).

The inner membrane passive ion permeability is known to increase by inorganic phosphate (Pi), but it significantly decreased in the presence of

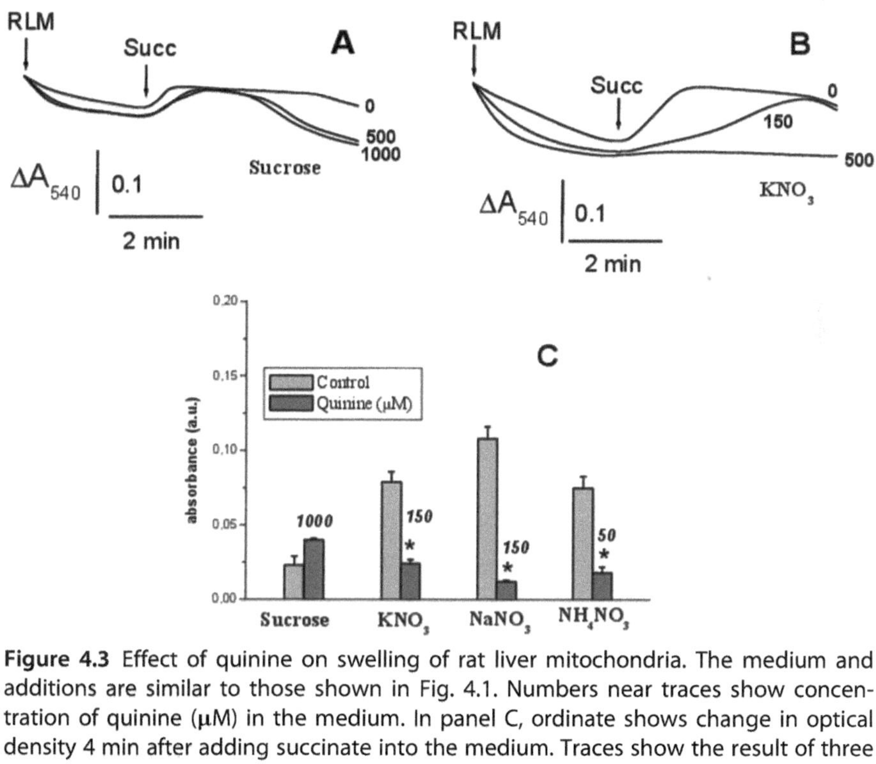

Figure 4.3 Effect of quinine on swelling of rat liver mitochondria. The medium and additions are similar to those shown in Fig. 4.1. Numbers near traces show concentration of quinine (μM) in the medium. In panel C, ordinate shows change in optical density 4 min after adding succinate into the medium. Traces show the result of three independent experiments carried out. Asterisks indicate significant differences from the control, experiments free of quinine ($P < .05$).

Mg^{2+} and ADP (Zoratti and Szabo, 1995; Ichas and Mazat, 1998). At the same time, Pi inhibited contraction of energized mitochondria preswollen in media with KNO_3 or $NaNO_3$ (Brierley et al., 1977; Devin et al., 1996, 1997a,b) or with KI and valinomycin (Azzone et al., 1976b). On the other hand, Pi increased the transport of K^+ ions through the IMM (Barrera and Gomez-Puyou, 1975; Bernardi, 1999). Cyclic ionophore nonactin facilitated the transport of Tl^+ ions into the mitochondrial matrix (Saris et al., 1981; Skul'skiĭ et al., 1984). Fig. 4.4 shows that swelling of nonenergized mitochondria markedly reduced in the presence of 2 mM ADP.

At the same time, mitochondrial contraction after succinate application was maximal in the medium with $TlNO_3$, KNO_3, and ADP (traces one and 2), but it inhibited in experiments with nonactin (trace 3). Similar results did obtain in a medium with $TlNO_3$, ADP, and KNO_3 substituted by $NaNO_3$ or NH_4NO_3 (Korotkov, 2009).

After succinate injection (Fig. 4.5), potential ($\Delta\Psi_{mito}$) appeared on the RLM inner membrane, clearly showing the sharp decrease in the safranin O fluorescence. In experiments with quinine and $TlNO_3$ (trace 5), there was a noticeable decrease in the IMM potential compared to experiments using only $TlNO_3$ (trace 2). Replacing $TlNO_3$ with sucrose eliminated this effect (trace 1). The potential decline in media with $TlNO_3$ was not evident in experiments with 2 mM ADP (trace 4) or with Mg^{2+} in sucrose medium with $TlNO_3$ (Fig. 4.5A, trace 3). Thus, we found the quinine inhibition contraction of energized mitochondria preswollen in the medium with

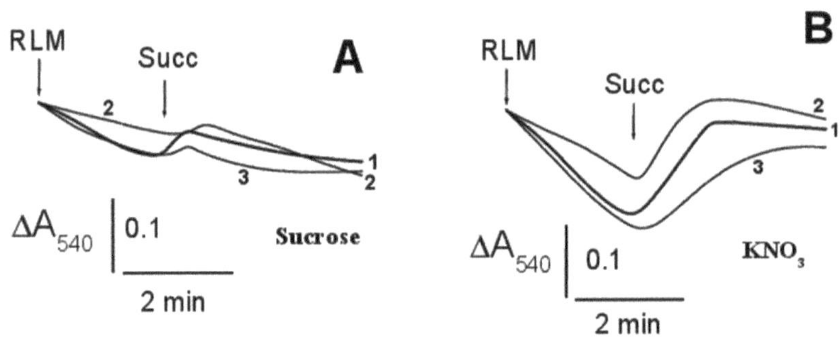

Figure 4.4 Influence of ADP and nonactin on swelling of rat liver mitochondria. Mitochondria (1.5 mg/mL of protein) were added to a medium containing 75 mM $TlNO_3$, 4 μM rotenone, and 3 μg/mL oligomycin along with 250 mM sucrose (A) or 125 mM KNO_3 (B). Numerals to the right of the traces show additions into the medium before mitochondria: 1, free of additions - bold curves (control), 2—2 mM ADP, 3—10^{-8} M nonactin. The arrows indicate addition of RLM (RLM) and 5 mM succinate (Succ). The traces show data obtained from three independent experiments.

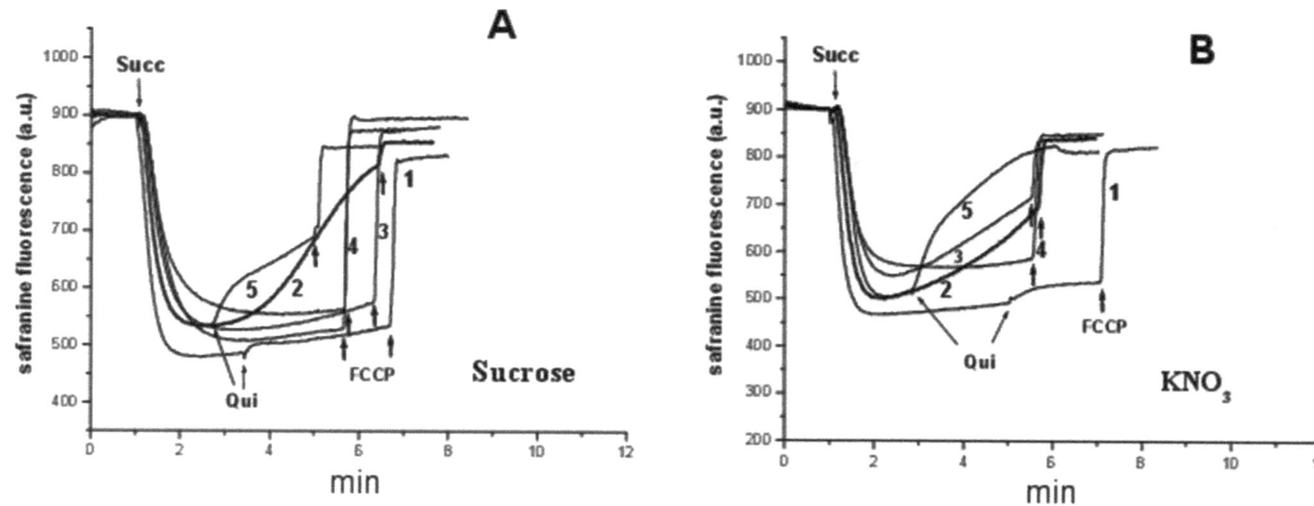

Figure 4.5 The Change in safranin O fluorescence (a.u.) in the mitochondrial suspension in the presence of ADP, Mg^{2+}, Pi, and quinine. Mitochondria (0.5 mg/mL of protein) were added into a medium containing 60 mM sucrose (1) or 30 mM $TlNO_3$ (2–5), 5 mM Tris-NO_3 (pH 7.3), 1 mM Tris -PO_4, 3 μM safranin O, 5 μM rotenone, and 3 μg/mL oligomycin along with 250 mM sucrose (A) or 125 mM KNO_3 (B). Succinate additions to a concentration of 5 mM succinate (Succ) and FCCP to a concentration of 1 μM (FCCP), as well as quinine (Qui) are indicated by arrows. The next additions were supplemented before mitochondria: free of additions (1), 2-medium with $TlNO_3$ free of additions, bold trace (2). In the remaining experiments (3–5), substances were added into the medium with $TlNO_3$ before mitochondria: 5 mM $Mg(NO_3)_2$ (3), and 2 mM ADP (4), as well as quinine (5) up to concentration of 1000 μM (A) or 500 μM (B). This figure shows the data obtained from three independent experiments.

$TlNO_3$ and nitrate, on the one hand, and the potential decline in experiments with quinine (Fig. 4.5), on the other hand. So, these findings combined with the Tl^+ weak effect on respiratory complexes (Melnick et al., 1976; Woods and Fowler, 1986; Korotkov et al., 2008a) can suggest that such mitochondria contraction (Figs. 4.1–4.4) occurs due to the removal of monovalent cations (K^+, Na^+, NH_4^+) from the matrix in participating the mitochondrial K^+/H^+ exchanger (Korotkov, 2009).

On the other hand, mitochondria preswollen in a medium containing $TlNO_3$ and sucrose (Fig. 4.3A) contracted despite the presence of 1 mM quinine. Perhaps this is because extrusion of Tl^+ from the matrix in this medium was carried out without the participation of the mitochondrial K^+/H^+ exchanger and proceeded according to a different mechanism of Tl^+/H^+ exchange postulated earlier by Saris and Skul'skiĭ (Saris et al., 1981). Our data comparison suggests the nonenergized mitochondria swelling increased in the nitrate medium (Figs. 4.1 and 4.2) can associate with univalent cations' large influx into the matrix due to the thallium-induced increase of the inner membrane ionic permeability. We previously believed that this increased swelling in nitrate media is associated only with Tl^+ transport into the matrix (Korotkov and Brailovskaya, 2001).

Thus, we have the following results. The research of nonenergized mitochondria swelling in nitrate media containing (NH_4NO_3, KNO_3, $NaNO_3$, $TlNO_3$) revealed low permeability of the inner mitochondrial membrane to H^+, K^+, and Na^+, and high one to Tl^+ (Mitchell and Moyle, 1969; Brierley et al., 1977; Saris et al., 1981). On the other hand, we showed that the effect of Cd^{2+} on mitochondria in similar experiments with nitrate media increased this membrane permeability to H^+, K^+ and Na^+ (Skul'skiĭ et al., 1988; Korotkov et al., 1998). Taking these results into account, our studies of nonenergized mitochondria swelling a medium with $TlNO_3$ and nitrates (NH_4NO_3, KNO_3, $NaNO_3$) made it possible to conclude that Tl^+ (like Cd^{2+} and other heavy metals) increases the inner membrane passive permeability to H^+, K^+, and Na^+ (Korotkov and Brailovskaya, 2001; Korotkov, 2009). So, the inhibition of contraction of succinate-energized rat liver mitochondria preswollen in a medium with $TlNO_3$ and nitrates (KNO_3, NH_4NO_3, $NaNO_3$) in the presence of quinine (K^+/H^+ exchanger inhibitor) allowed us to conclude that this exchanger takes part in the processes of removing thallium-induced excess of monovalent cations (K^+, NH^{4+}, and Na^+) from the matrix, respectively.

4.2.2 Thallium effects on mitochondrial respiration

According to some researchers (Devin et al., 1996, 1997a,b; Bernardi, 1999), mitochondrial respiration rate in state 4 depends on mitochondrial volume. It may increase due to an increase in passive ionic permeability of the IMM the acceleration in the active transport of cations into the matrix. The need to keep the matrix volume constant triggers exchange of excess matrix cations by the medium protons in participating K^+/H^+ exchanger (Garlid and Paucek, 2003). After, proton pumps of the mitochondrial respiratory chain remove the protons from the matrix. The oxygen uptake increase in states 4_0 and 4_1 accompany this process. We found previously that the RLM state 4 respiration markedly increased with 5–10 μM Cd^{2+} due to raised IMM ionic permeability (Skul'skiĭ et al., 1988; Korotkov et al., 1998; Belyaeva and Korotkov, 2003). Toxic effects of organic hydrophilic complex Cd^{2+} with anabasine and, especially, the hydrophobic complex of Cd^{2+} with chimcoccide on isolated rat liver mitochondria was significantly higher than the effect of Cd^{2+} alone (Korotkov et al., 1999). At higher concentrations of Cd^{2+} (15–40 μM), RLM respiration in state 4 in a medium with KCl or NH_4NO_3 decreased by about 20%–30% due to inhibition of the mitochondrial respiratory chain. This Cd^{2+} effect did eliminate in the presence of thiol reagent (dithiothreitol), and there was a noticeable increase in this respiration (Korotkov et al., 1998, 2008b; Belyaeva and Korotkov, 2003). Similar respiration increase showed earlier in potassium media supplemented to valinomycin ionophore (Azzi and Azzone, 1966; Azzone et al., 1976b).

As mentioned earlier, Tl^+, unlike divalent heavy metals (Cd^{2+}, Hg^{2+}, Pb^{2+}, Zn^{2+}, Cu^{2+}), did not significantly affect respiration in state 3, but it significantly increased respiration of these organelles in state 4. Considering that removing univalent cations from the matrix in participating K^+/H^+ exchanger depends on the subsequent pumping of protons by the mitochondrial respiratory complexes, we studied mitochondrial respiration in states: 4_0 or $3U_{DNP}$ (DNP-uncoupled).

Mitochondria (Fig. 4.6) did add into the medium with 25–75 mM $TlNO_3$ along with 125 mM nitrate (KNO_3, NH_4NO_3, $NaNO_3$) or 250 mM sucrose. Before mitochondria, substrates of first complex (glutamate with malate) or second complex (succinate in the presence of rotenone) did inject into the medium (Fig. 4.1, accordingly, panels A, B, and C, D). Respiration of rat liver and rat heart mitochondria did accelerate in states 4_0 and 4_1 (Figs. 4.6–4.9 and Tables 4.2 and 4.3) in increasing the concentration of Tl^+ ions from 25 to 75 mM. At the same time, the

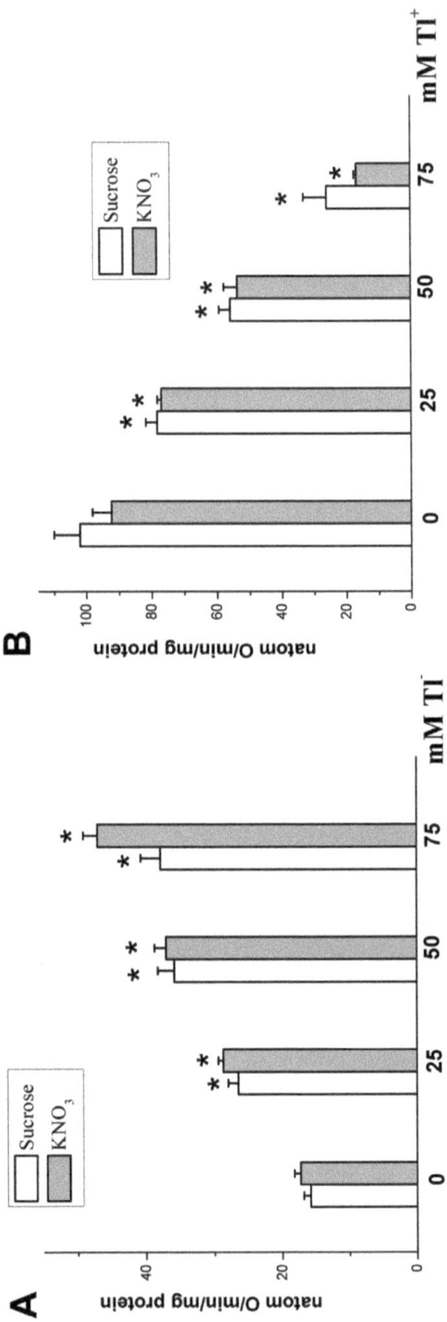
A
natom O/min/mg protein
0
20
40
Sucrose
KNO3
0
25
50
75
mM Tl+
B
natom O/min/mg protein
0
20
40
60
80
100
Sucrose
KNO3
0
25
50
75
mM Tl+

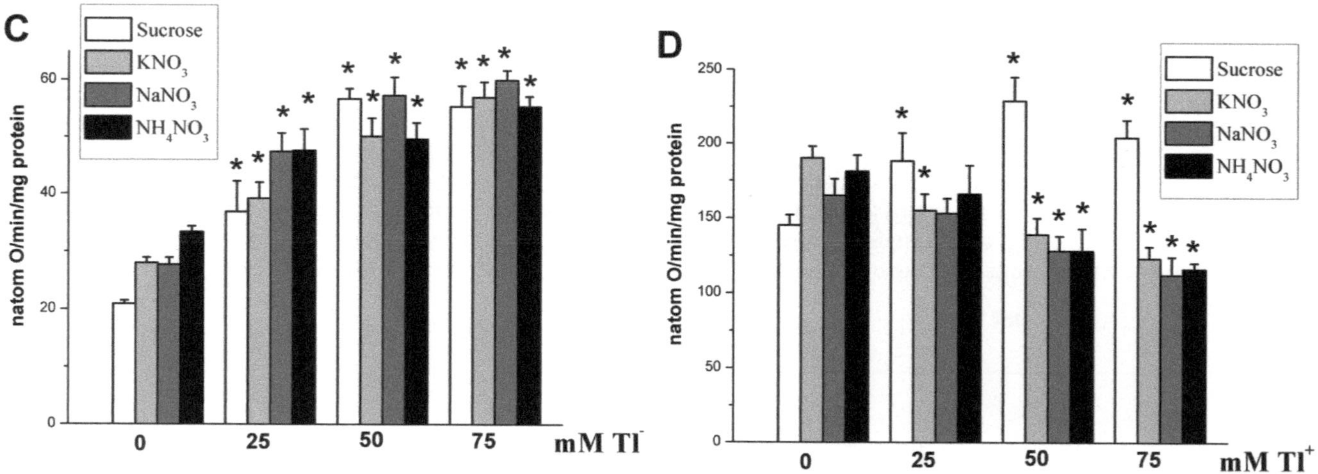

Figure 4.6 Effect of Tl^+ on the oxygen consumption rate in rat liver mitochondria. Mitochondria were injected into medium containing 0–75 mM $TlNO_3$ and 5 mM Tris-NO_3 (pH 7.3) along with 250 mM sucrose or 125 mM nitrates (KNO_3, $NaNO_3$, NH_4NO_3), as well as 5 mM glutamate and 5 mM malate (A, B), or 5 mM succinate with 2 μM rotenone (C, D). Two minutes after the injection of mitochondria and registration of respiration at state 4_0, 30 μM 2,4-dinitrophenol (B, D) was added into the medium. The rates (n atom O/min per mg of protein) is displayed on the ordinate axis, and concentration of $TlNO_3$ (mM) is displayed on the abscissa axis. Significant differences ($P < .05$) from the control (experiments free of $TlNO_3$) are shown with asterisks.

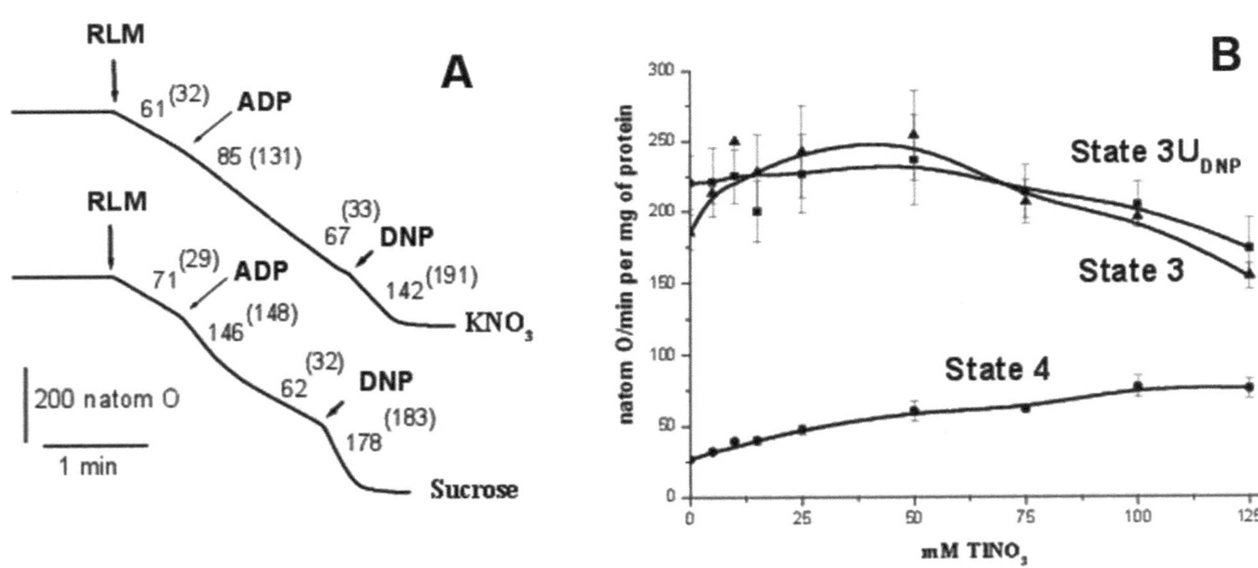

Figure 4.7 Effects of Tl^+ on oxygen consumption rates in succinate-energized rat liver mitochondria. Mitochondria (1.5 mg/mL of protein) were added into 400 mOsm medium (panel A) containing 75 mM $TlNO_3$ along with 250 mM sucrose (Sucrose) or 125 mM KNO_3 (KNO_3). Arrows indicate mitochondrial supplementation (RLM), 130 μM ADP (ADP) and 30 μM DNP (DNP). The oxygen uptake data (n atom O/min per mg of protein) are presented by numbers above the traces. In parentheses (panel A) are indicated the oxygen consumption rates of RLM added into a medium where 75 mM $TlNO_3$ was replaced by 150 mM sucrose to compare changes in the rates under influence of Tl^+ ions. Panel A shows experimental traces from three independent experiments. In other experiments (panel B), the medium contained 0–125 mM $TlNO_3$ and 0–250 mM sucrose to maintain the required osmolality in this medium (290 mOsm). In addition, all media (A, B) contained 3 mM $Mg(NO_3)_2$, 3 mM Tris-NO_3 (pH 7.3), 3 mM Tris-PO_4, 5 mM succinate, and 4 μM rotenone. In panel B, ordinate shows the rates, and the abscissa shows the $TlNO_3$ concentration (mM). Traces: 1—state 4, 2—state 3, and 3—state $3U_{DNP}$.

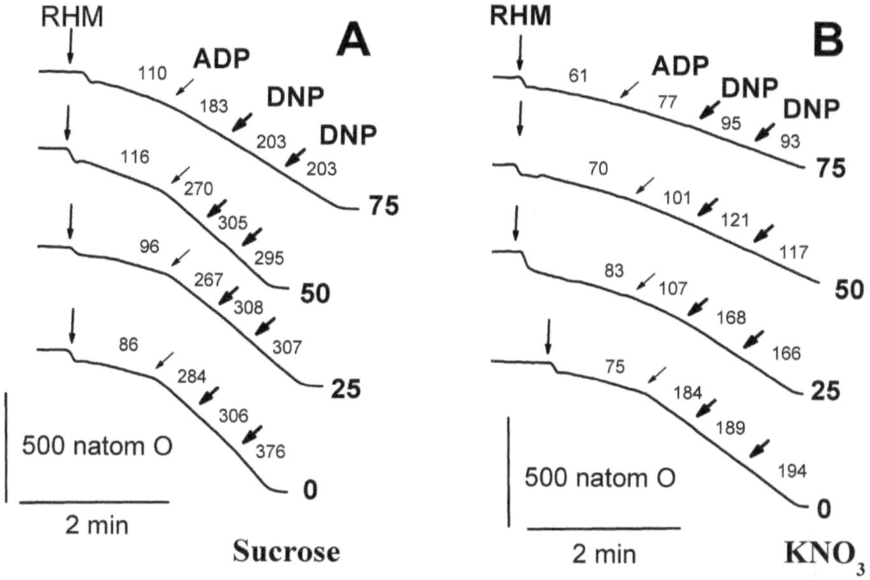

Figure 4.8 Effects of Tl^+ on oxygen consumption rates in succinate-energized rat heart mitochondria. Mitochondria were injected into 400 mOsm medium containing 0–75 mM $TlNO_3$ as well as 250 mM sucrose (A) or 125 mM KNO_3 (B). All media contained 3 mM Tris-PO_4, 5 mM succinate, and 2 μM rotenone. Numbers above the traces show the rates (n atom O/min per mg of protein). Arrows indicate mitochondrial supplementation (RHM), 130 μM ADP (ADP), and 10 μM DNP (DNP). The numbers near traces show concentration (mM) of $TlNO_3$ in the medium. Typical traces for three independent experiments are presented.

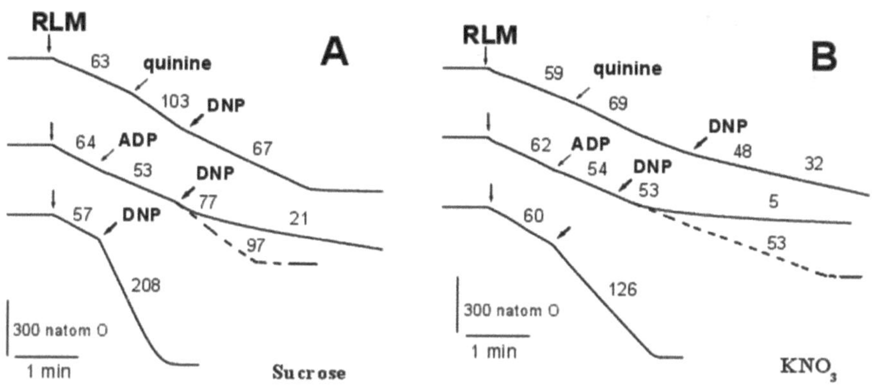

Figure 4.9 Influence of ADP and quinine on oxygen consumption rates in succinate-energized rat heart mitochondria. Mitochondria (1.5 mg/mL of protein) were added into 400 mOsm medium containing 75 mM $TlNO_3$, 5 mM Tris-NO_3 (pH 7.3), 5 mM succinate, 4 μM rotenone, and 3 μg/mL oligomycin along with 250 mM sucrose (A) or 125 mM KNO_3 (B). The *arrows* show addition of mitochondria (RLM), 2 mM ADP (ADP), 30 μM DNP (DNP), and quinine (quinine) to concentrations of 1 mM (A) or 0.5 mM (B). The rates (n atom O min/mg of protein) are indicated numbers above the traces. The figure shows typical traces obtained from three independent experiments.

Table 4.2 Effects of thallium nitrate on oxygen consumption rates (n atomO/mg of protein) by succinate-energized rat liver mitochondria.

Respiratory states	Media			
	Sucrose	KNO_3	NH_4NO_3	$NaNO_3$
State 4_0	60 ± 1 (3)	59 ± 2 (3)	67 ± 3 (3)	63 ± 1 (3)
State 3	126 ± 5 (3)	86 ± 3 (3)	80 ± 6 (3)	79 ± 5 (3)
State 4_1	61 ± 6 (3)	69 ± 3 (3)	73 ± 8 3)	67 ± 3 (3)
RCR	2.10 ± 0.19 (3)	1.25 ± 0.04 (3)	1.11 ± 0.04 (3)	1.18 ± 0.02 (3)
State $3U_{DNP}$	223 ± 18 (3)	132 ± 14 (3)	109 ± 8 (3)	104 ± 10 (3)
RCR_{DNP}	3.69 ± 0.13 (3)	1.90 ± 0.15 (3)	1.51 ± 0.06 (3)	1.56 ± 0.10 (3)

Asterisk (*) indicate significant differences from the control, experiments free of $TlNO_3$ ($P < .05$).

decrease of respiration in state 3 and state $3U_{DNP}$ in experiments with succinate-energized RLM was more noticeable in the medium containing $TlNO_3$ and nitrates than $TlNO_3$ and sucrose (Figs. 4.6D and 4.7A and Table 4.2, correspondingly). However, there was no noticeable decrease in states 3 or $3U_{DNP}$ respiration of the RLM injected into 290 mOsm medium containing 5−75 mM $TlNO_3$ and sucrose (Fig. 4.7B). However, there was no noticeable decrease in states 3 or 3UDNP respiration of the RLM injected into 290 mOsm medium containing 5−75 mM $TlNO_3$ and sucrose (Fig. 4.7B). The decrease did find only in experiments with 100−125 mM $TlNO_3$ (Fig. 4.7B). A decrease in state 3UDNP respiration in a medium with $TlNO_3$ and nitrates did see in similar experiments with RHM (Fig. 4.8 and Table 4.3).

Mitochondria (1.5 mg/mL of protein) were added into 400 mOsm medium containing 75 mM $TlNO_3$ along with 250 mM sucrose (sucrose) or 125 mM nitrates (KNO_3, $NaNO_3$, NH_4NO_3). The experimental procedure as well as the medium other components and additions of ADP and DNP are similar to those shown in Fig. 4.7A.

Medium, additions, and experimental conditions are identical to those presented in Fig. 4.8. Asterisks show significant differences from control (experiments without $TlNO_3$) ($P < .05$).

Like earlier experiments (Melnick et al., 1976; Diwan and Lehrer, 1977; Bragadin et al., 2003), thallium increased respiration in states 4_0 and 4_1 of RLM or RHM, added in sucrose medium, and energized by glutamate with malate (Fig. 4.6A) or succinate (Figs. 4.6C−4.8 and Table 4.3). In nitrate media, Tl^+ decreased respiration in four and $3U_{DNP}$

Table 4.3 Effects of thallium nitrate on oxygen consumption rates (ng atomO/mg of protein) by succinate-energized rat heart mitochondria.

	Media					
	Sucrose	KNO_3	Sucrose	KNO_3	Sucrose	KNO_3
$TlNO_3$ (mM)	State 4_0		State 3		State $3U_{DNP}$	
0 (Control)	82 ± 5 (3)	66 ± 4 (4)	253 ± 19 (3)	112 ± 5 (4)	379 ± 3 (3)	123 ± 5 (4)
25	97 ± 2 (3)*	62 ± 2 (3)	265 ± 5 (3)	88 ± 8 (3)*	303 ± 7 (3)*	96 ± 10 (3)*
50	114 ± 2 (3)*	54 ± 4 (4)*	256 ± 8 (3)	61 ± 6 (4)*	289 ± 9 (3)*	65 ± 4 (4)*
75	107 ± 2 (3)*	42 ± 3 (4)*	176 ± 5 (3)*	50 ± 2 (3)*	190 ± 8 (3)*	51 ± 5 (3)*

Asterisk (*) indicate significant differences from the control, experiments free of $TlNO_3$ ($P < .05$).

states (Fig. 4.6C) and increased swelling (Fig. 4.1) in the series sucrose < KNO_3 < $NaNO_3$ ≤ NH_4NO_3 in experiments with energized RLM (Korotkov, 2009). Similarly, Cd^{2+} decreased respiration in 4, 3, and $3U_{DNP}$ states and increased swelling of RLM in adding the metal into salt media (Korotkov et al., 1998). The effect of Cd^{2+} increased in the following row: sucrose < KCl < NaCl. These data above allow us to conclude that Tl^+, like Cd^{2+} (Sanadi et al., 1981; Rasheed et al., 1984; Korotkov et al., 1998; Belyaeva and Korotkov, 2003; Lee et al., 2005), Pb^{2+} (Scott et al., 1971; Miyahara and Utsumi, 1975) and Hg^{2+} (Belyaeva et al., 2004a), can increase the ionic permeability of the inner membrane of energized mitochondria.

State 4_0 respiration of RLM decreased in experiments with all media supplemented 2 mM ADP, and it increased by phosphate and quinine (Fig. 4.9 and Table 4.4). At the same time, mitochondrial respiration in $3U_{DNP}$ state did not change after injection of Mg^{2+}, Pi, or 0.5 mM ADP into a medium with $TlNO_3$ and nitrates (Table 4.4). However, this respiration significantly decreased after adding quinine or 2 mM ADP into the medium (Fig. 4.9 and Table 4.4). This ADP effect did eliminate in the medium supplemented 3 mM Mg^{2+} and 3 mM phosphate (Fig. 4.9, dotted traces).

The experimental procedure is similar to that shown in Fig. 4.9. Mitochondria were introduced into a medium containing 75 mM $TlNO_3$ along with 250 mM sucrose or 125 mM nitrates (KNO_3, $NaNO_3$, NH_4NO_3). The media other components are identical to those shown in Fig. 4.8. After mitochondria, 0.5 mM ADP (ADP), 2 mM ADP (ADP (2)), 1 μM CsA (CsA), 5 mM $Mg(NO_3)_2$, 3 mM Tris-PO_4 (Pi), and 7.5 μM PK (PK) were added. The quinine supplement (μM) into the medium with sucrose, KNO_3, $NaNO_3$, or NH_4NO_3 was 1000, 500, 150, and 75, respectively. The number of experiments is indicated in parentheses. The state $3U_{DNP}$ respiration value (after its short-term activation) is given in square brackets (Fig. 4.9). Notation in the table: "4_0"—state 4 respiration before first ADP addition, "foa"—free of additions, "+ reagt"—a reagent of 1 min after adding mitochondria into the medium, "$3U_{DNP}$"—the DNP-uncoupled (state after injection of 30 μM DNP after 1 min after additions of the reagent), "RCR_{DNP}"—state $3U_{DNP}$/state 4_0. Asterisks indicate a difference from the control, experiments free of additions ($P < .05$).

Increased state 4 respiration (Fig. 4.9 and Table 4.4) and increased swelling in succinate-energized mitochondria (Fig. 4.4) were markedly

Table 4.4 Effects of ADP, CsA, Mg^{2+}, ruthenium red, inorganic phosphate, and quinine on oxygen consumption rates (ng atomO/mg of protein) by succinate-energized rat heart mitochondria.

Supplements after mitochondria	Media							
	75 mM $TlNO_3$ and 250 mM sucrose				75 mM $TlNO_3$ and 125 mM KNO_3			
	Foa	+ Reagt	$3U_{DNP}$	RCR_{DNP}	Foa	+ Reagt	$3U_{DNP}$	RCR_{DNP}
A control	58 ± 2 (3)		197 ± 9 (3)	3.42 ± 0.22 (3)	59 ± 2 (3)		122 ± 8 (3)	2.08 ± 0.12 (3)
Mg^{2+}	58 ± 3 (3)	56 ± 1 (3)	173 ± 10 (3)*	3.00 ± 0.18 (3)	62 ± 2 (3)	60 ± 2 (3)	131 ± 6 (3)	2.13 ± 0.09 (3)
CsA	58 ± 7 (3)	57 ± 6 (3)	177 ± 10 (3)	3.12 ± 0.25 (3)	56 ± 4 (3)	53 ± 4 (3)	81 ± 5 (3)*	1.48 * ± 0.12 (3)
ADP	62 ± 4 (3)	61 ± 5 (3)	181 ± 18 (3)	2.95 ± 0.36 (3)	61 ± 3 (3)	61 ± 3 (3)	124 ± 12 (3)	2.06 ± 0.18 (3)
ADP + Mg^{2+}	48 ± 5 (3)	47 ± 4 (3)*	174 ± 15 (3)	3.66 ± 0.35 (3)	52 ± 5 (3)	46 ± 4 (3)*	108 ± 18 (3)	2.14 ± 0.17 (3)
ADP + CsA + Mg^{2+}	54 ± 5 (3)	44 ± 2 (3)*	168 ± 7 (3)*	3.13 ± 0.15 (3)	53 ± 5 (3)	43 ± 3 (3)*	95 ± 14 (3)	1.76 ± 0.10 (3)
ADP (2)	63 ± 1 (3)	53 ± 1 (3)*	84 ± 7 [22 ± 2] (3)*	1.35 ± 0.11 (3)*	57 ± 2 (3)	51 ± 2 (3)*	68 ± 13 [8 ± 2] (3)*	1.21 ± 0.27 (3)*
Pi	64 ± 2 (3)	81 ± 3 (3)*	184 ± 9 (3)	2.89 ± 0.11 (3)*	63 ± 1 (3)	77 ± 3 (3)*	132 ± 11 (3)	2.07 ± 0.16 (3)
RR	61 ± 3 (3)	61 ± 3 (3)	199 ± 14 (3)	3.25 ± 0.26 (3)	59 ± 2 (3)	59 ± 2 (3)	128 ± 11 (3)	2.17 ± 0.14 (3)
Quinine	61 ± 2 (3)	95 ± 5 (3)*	67 ± 1 (3)*	1.10 ± 0.02 (3)*	58 ± 1 (3)	69 ± 2 (3)*	35 ± 8 [27 ± 4] (3)*	0.60 ± 0.13 (3)*

Supplements after mitochondria	Media							
	75 mM $TlNO_3$ and 125 mM $NaNO_3$				75 mM $TlNO_3$ and 125 mM NH_4NO_3			
	4_0	+ Reagt	$3U_{DNP}$	RCR_{DNP}	4_0	+ Reagt	$3U_{DNP}$	RCR_{DNP}
A control	60 ± 2 (3)		112 ± 4 (3)	1.63 ± 0.04 (3)	59 ± 2 (3)		117 ± 7 (3)	2.01 ± 0.11 (8)
Mg^{2+}	64 ± 1 (3)	56 ± 1 (3)	129 ± 14 (3)*	2.01 * ± 0.20 (3)	61 ± 1 (3)	56 ± 1 (3)	106 ± 4 (3)	1.75 * ± 0.08 (3)
CsA	60 ± 3 (3)	59 ± 4 (3)	89 ± 7 (3)	1.46 ± 0.04 (3)	54 ± 2 (3)	54 ± 2 (3)	88 ± 7 (3)*	1.64 * ± 0.07 (3)
ADP	62 ± 1 (3)	62 ± 1 (3)	119 ± 16 (3)*	1.89 ± 0.23 (3)	63 ± 1 (3)	68 ± 2 (3)*	126 ± 7 (3)	2.01 ± 0.14 (3)
ADP + Mg^{2+}	58 ± 3 (3)	46 ± 2 (3)*	87 ± 11 (3)	1.50 ± 0.11 (3)	50 ± 4 (3)	45 ± 3 (3)*	96 ± 6 (3)*	1.90 ± 0.10 (3)
ADP + CsA + Mg^{2+}	61 ± 2 (3)	42 ± 2 (3)*	91 ± 4 (3)	1.52 ± 0.23 (3)	60 ± 2 (3)	47 ± 3 (3)*	99 ± 3 (3)*	1.64 * ± 0.08 (3)
ADP (2)	63 ± 2 (3)	52 ± 3 (3)*	54 ± 6 [8 ± 3] (3)*	0.86 ± 0.06 (3)*	58 ± 1 (3)	46 ± 1 (3)*	56 ± 2 [8 ± 3] (3)*	0.96 ± 0.09 (3)*
Pi	62 ± 1 (3)	76 ± 4 (3)*	134 ± 16 (3)*	2.16 ± 0.25 (3)	64 ± 1 (3)	73 ± 5 (3)*	120 ± 6 (3)	1.88 ± 0.08 (3)
RR	60 ± 1 (3)	60 ± 1 (3)	147 ± 22 (3)*	2.43 * ± 0.33 (3)	59 ± 1 (3)	59 ± 1 (3)	128 ± 13 (3)*	2.17 ± 0.20 (3)
Quinine	65 ± 2 (3)	68 ± 3 (3)*	59 ± 7 [13 ± 2] (3)*	0.92 ± 0.14 (3)*	58 ± 2 (3)	58 ± 2 (3)	68 ± 3 [63 ± 2] (3)*	1.17 ± 0.06 (3)*

Asterisk (*) indicate significant differences from the control, experiments free of $TlNO_3$ ($P < .05$).

decreased in the medium containing $TlNO_3$ and nitrate (KNO_3, $NaNO_3$, NH_4NO_3) in the presence of 2 mM ADP. Still, they increased even more in the medium with Pi. Thus, such an unequal change in these parameters in experiments with MPTP inhibitor (ADP) and inorganic phosphate promoting the accumulation of calcium in the matrix may be associated with the thallium-induced MPTP opening in the inner membrane. This pore opening can lead to an increase in both the matrix Tl^+ concentration and swelling of mitochondria.

On the other hand, a Tl^+ concentration increase resulted in a parallel rise of respiration in 4_0 and 4_1 states (Figs. 4.6–4.8 and Table 4.3). It increased markedly in the presence of inorganic phosphate and, especially, ionophore nonactin (Fig. 4.10; Korotkov et al., 2005, 2007). The increase in state 4 may be due to acceleration of Tl^+ transport through the IMM in the presence of nonactin, which forms a potent complex with Tl^+ ions (Skul'skiĭ, 1977, 1978a, 1979b, 1981). On the other hand, mitochondrial respiration in 3 and $3U_{DNP}$ states did not change in the presence of inorganic phosphate and nonactin (Figs. 4.10 and 4.11).

As mentioned earlier, the presence of Pi accelerated energy-dependent K^+ uptake catalyzed by the potassium uniport mechanism (Azzi and Azzone, 1966; Barrera and Gomez-Puyou, 1975; Tedeschi, 1981; Garlid and Paucek, 2003). Inorganic phosphate is known to contribute to an even more significant increase in already stimulated state 4 respiration and swelling of mitochondria in experiments with Cd^{2+} or K^+ and valinomycin (Azzi and Azzone, 1966; Koike et al., 1991; Belyaeva et al., 2002, 2004b). The Tl^+ stimulated state 4_0 respiration of succinate-energized RLM increased in the medium with Pi (Fig. 4.10A). Noticeable increase both state 4 respiration (Fig. 4.10B) and swelling of succinate-energized RLM in media containing Tl-acetate and nonactin (Melnick et al., 1976; Korotkov et al., 2005, 2007) found. This fact can explain that nonactin forms a potent complex with ions Tl^+ that significantly facilitates this cation entry into the mitochondrial matrix (Skul'skiĭ et al., 1978a, 1981, 1984; Saris et al., 1981; Skul'skiĭ and Glazunov, 1981, 1982). Thus, an increase in state 4_0 respiration of RLM added into media containing $TlNO_3$ and P_i (Fig. 4.10A) or with Tl acetate and nonactin (Fig. 4.10B) may be the result of increased cyclic movement of Tl^+ and H^+ ions through the IMM (Korotkov et al., 2005, 2007, 2008b) due to acceleration of cyclic Tl^+ transport into energized mitochondria with subsequent extrusion of Tl^+ ions from the matrix via Tl^+/H^+ exchange (Saris et al., 1981; Diwan et al., 1990). However, the medium presence of P_i (Fig. 4.10C) and nonactin (Fig. 4.11) did not affect

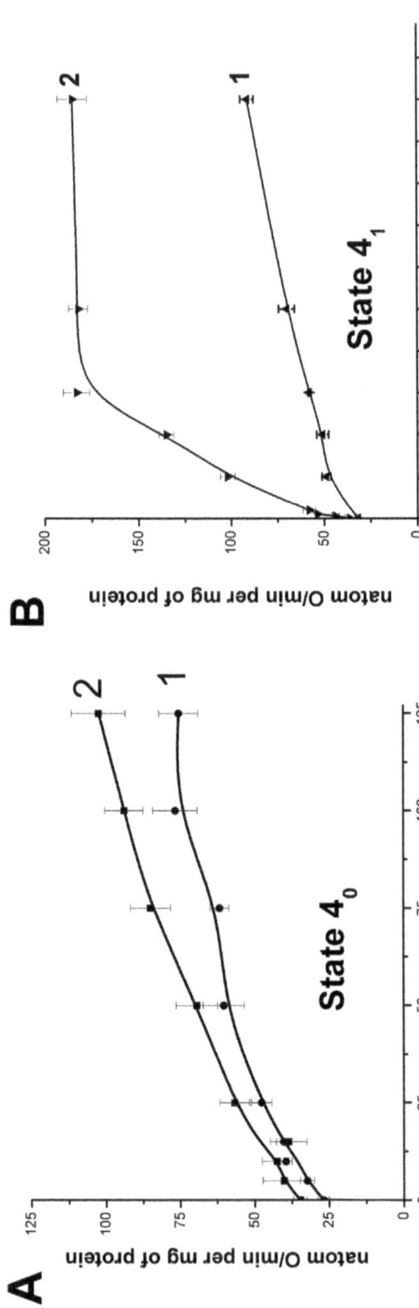
A
natom O/min per mg of protein
State 4$_0$
TlNO$_3$ (mM)
B
natom O/min per mg of protein
State 4$_1$
Tl acetate (mM)
1
2

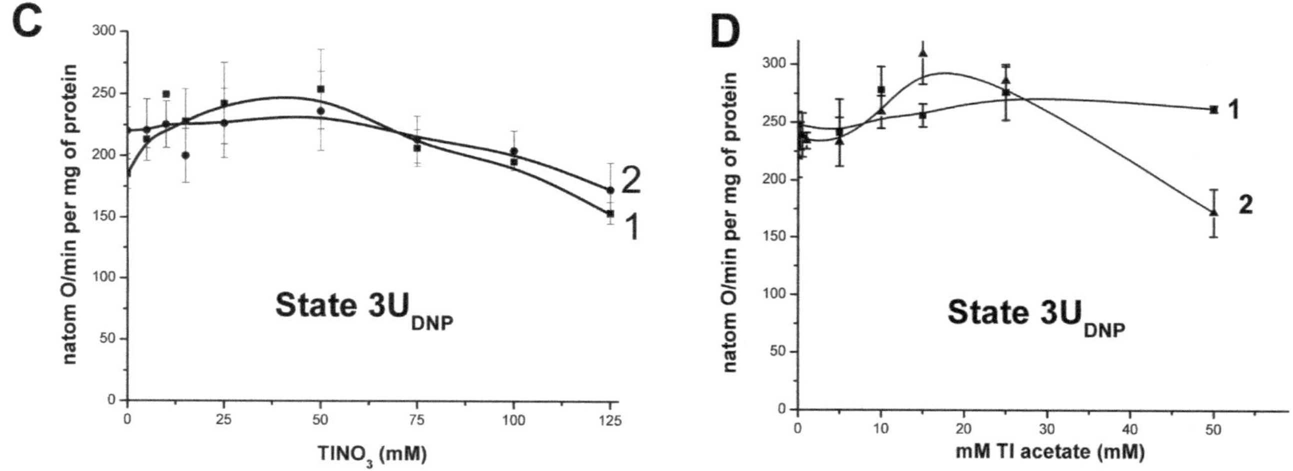

Figure 4.10 Influence of inorganic phosphate and nonactin on oxygen consumption rates in succinate-energized rat liver mitochondria. Mitochondria (1.5 mg/mL of protein) were added into 290 mOsm medium containing 0–125 mM $TlNO_3$ and 5 mM Tris-NO_3 (A, C) or 0–50 mM Tl acetate and 5 mM Tris acetate along with 5 mM succinate and 4 μM rotenone (B, D). Before mitochondria (traces 2) 3 mM tris-PO_4 (A, C) or 10^{-8} M nonactin (B, D) were supplemented into the medium. DNP of 30 μM was added 2 min after mitochondria. The ordinate shows the rates (n atom/min per mg of protein), and the abscissa shows the concentration (mM): $TlNO_3$ (A, C) or Tl-acetate (B, D).

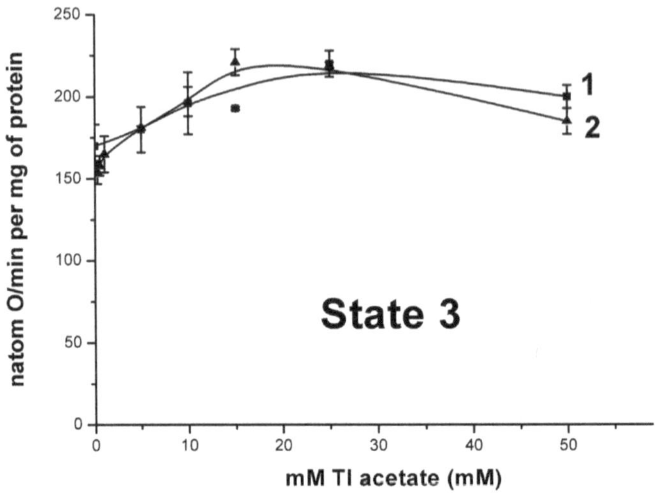

Figure 4.11 State 3 respiration of succinate-energized rat liver mitochondria in the presence of nonactin. Nonactin of 10^{-8} M was added before mitochondria (trace 2), and 100 μM ADP was added 2 min after mitochondria to initiate state 3. The medium is as in Fig. 4.10B. Before mitochondria, 3 mM Tris-PO_4 and 5 mM $Mg(NO_3)_2$ were added. The ordinate shows oxygen consumption rates (ng atom/min per mg of protein), and the abscissa shows concentration of $TlNO_3$ (mM).

the oxygen consumption rates in succinate-energized rat liver mitochondria in states 3 and $3U_{DNP}$. These data, accordingly, allow us to assume that neither phosphate nor nonactin affected the interaction of Tl^+ ions with mitochondrial enzymes involved in electron transport along the respiratory chain and the implementation of oxidative phosphorylation processes. At the same time, the DNP-induced swelling of energized by succinate or glutamate with malate RLM increased more strongly in nitrate media than swelling in sucrose media (Fig. 4.12). This result may be associated with the MPTP opening in the inner membrane and a high univalent cations' uptake into the matrix.

Uncoupled respiration (state 3U) is known to depend on the activity of the respiratory chain complexes. State 3 also depends on enzymes participating in the oxidative phosphorylation processes (Devin et al., 1996, 1997a,b). The interaction of heavy metals (Cd^{2+}, Pb^{2+}, Hg^{2+}, Ag^+) with free thiol groups of mitochondria inhibits respiratory enzymes and increases the inner membrane ionic permeability. This interaction was accompanied by a noticeable decrease the respiration in 3 and 3U states, as well as a slowdown in the contraction of energized mitochondria preswollen in nitrate or chloride media (Riley and Lehninger, 1964; Scott et al., 1971;

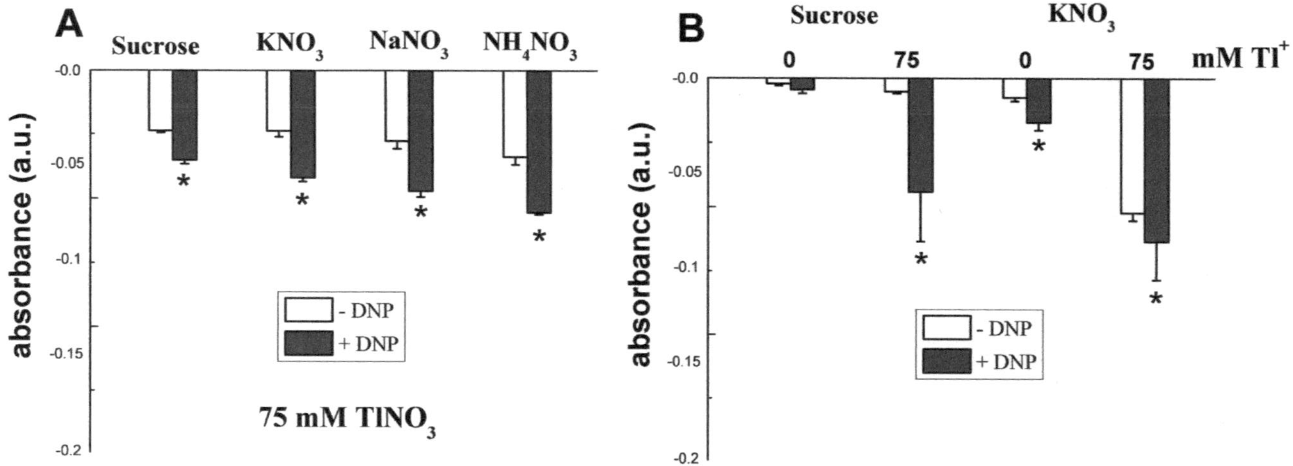

Figure 4.12 Effects of DNP on swelling of energized rat liver mitochondria. Mitochondria (1.5 mg/mL of protein) were added into a medium containing 75 mM $TlNO_3$ (A) or 0–75 mM $TlNO_3$ (B), 3 μg/mL oligomycin, as well as 250 mM sucrose (Sucrose) or 125 mM nitrates (KNO_3, $NaNO_3$, NH_4NO_3) along with 5 mM succinate and 2 μM rotenone (A) or 5 mM glutamate with 5 mM malate (B). The ordinate axis shows the absorbance change (ΔA_{540}) 3 min after adding mitochondria into the medium and 30 μM DNP, respectively. The abscissa (B) shows concentration of $TlNO_3$ (mM). Asterisks indicate significant differences from control, experiments free of DNP.

Sanadi et al., 1981; Weinberg et al., 1982; Rasheed et al., 1984; Miccadei and Floridi, 1993; Korotkov and Skul'skiĭ, 1996; Korotkov et al., 1998; Rikans and Yamano, 2000; Belyaeva et al., 2002; Belyaeva and Korotkov, 2003; Belyaeva et al., 2004a,b). Cd^{2+}-induced decrease in respiration of succinate-energized RLM and Cd^{2+}-inhibition in the contraction of these organelles preswollen in saline media were noticeably eliminated by dithiothreitol, which reduces blocked thiol groups (Belyaeva and Korotkov, 2003; Belyaeva et al., 2004a,b). On the other hand, Tl^{+} weakly interacts with the mitochondrial respiratory enzymes (Melnick et al., 1976; Woods and Fowler, 1986; Mulkey and Oehme, 1993). It forms lesser stable complexes with thiol groups of organic compounds, and the stability constants of Tl complexes were two orders of magnitude less than the above heavy metals (Perrin, 1979). Probably, for this reason, even 75 mM $TlNO_3$ did not inhibit the RLM respiration in 3 or 3U states, and the Tl^{+} affect up to 125 mM was insignificant (Fig. 4.7B) (Barrera and Gomez-Puyou, 1975; Melnick et al., 1976; Diwan and Lehrer, 1977; Korotkov et al., 2007, 2008a).

Oxygen consumption rates of succinate-energized RLM in states 4, 3, and $3U_{DNP}$ rose under increasing $TlNO_3$ concentration in sucrose medium (Figs. 4.5 and 4.6), which can explain the effects of $TlNO_3$, DNP, and ADP. On the other hand, a noticeable decrease in respiration in 3 and $3U_{DNP}$ states observes after increasing $TlNO_3$ concentration in nitrate media (Figs. 4.5 and 4.6). This respiration decrease finds in experiments with energized glutamate with malate RLM added into the $TlNO_3$ media containing either sucrose or KNO_3 (Fig. 4.5). The swelling of DNP-uncoupled RLM energized by glutamate with malate (Fig. 4.12B) or succinate (Fig. 4.12A) was more potent in nitrate media. The activity of the respiratory complexes turned out to be a limiting factor for mitochondrial respiration in 3 or $3U_{DNP}$ states in media containing KCl or NaCl but not sucrose (Devin et al., 1996, 1997a,b). Since most respiratory enzymes are part of complexes that cross the inner membrane, their enzymatic activity largely depends on the structure of this membrane. Tl^{+} was found to have affected only enzymes closely related to the IMM (Woods and Fowler, 1986). One of the possible reasons for the respiration decrease in rat liver and rat heart mitochondria being in 3 or 3UDNP states (Figs. 4.6–4.8) can assume to be a change in the inner membrane structure due to increased mitochondria volume caused by swelling of these organelles (Fig. 4.12) (Korotkov and Brailovskaya, 2001; Korotkov, 2009, 2015b). This assumption is consistent with our found correlation between a decrease in

$3U_{DNP}$ state respiration of succinate-energized RLMs and their swelling in media containing $TlNO_3$, nitrates, and quinine (Korotkov, 2009). Other possible reasons for such a decrease in mitochondrial respiration in 3 and $3U_{DNP}$ states (Figs. 4.6B, D, 4.7A, 4.8) may be a decrease in the matrix K^+ concentration due to the Tl^+-induced increase in the IMM ionic permeability, as well as due to the ability of Tl^+ to form coordination complexes with matrix proteins having active groups ($-SH$, $-COOH$, $-NH_2$) (Perrin, 1979; Woods and Fowler, 1986; Mulkey and Oehme, 1993; Huang et al., 2012; Korotkov et al., 2014a). Another reason for this decrease is possibly the ability of Tl^+ to react with thiol groups in the inner membrane of RLM injected into $TlNO_3$ media. At the same time, as mentioned earlier, succinate-energized rat liver and rat heart mitochondria, preswollen in nitrate media with $TlNO_3$ (Fig. 4.1), retained their ability to contract in experiments with Tl^+ but not with Cd^{2+} (Belyaeva and Korotkov, 2003; Belyaeva et al., 2004b; Korotkov et al., 2008b). Interestingly, energized RLM could be contracted even in the presence of DNP in a medium with $TlNO_3$ and NH_4NO_3 (not published). The data obtained suggest that thallium-induced swelling of mitochondria can lead to disruption of the inner membrane structure. This disruption may be possibly due to the interaction of Tl^+ with this membrane thiol groups. That may be the leading cause of a decrease in the activity of respiratory enzymes, which results in a reduction of mitochondrial respiration in states 3 and $3U_{DNP}$ (see next section).

So, we found that Tl^+ up to 75 mM did not affect mitochondrial respiration in states 3 and $3U_{DNP}$ in a sucrose medium with $TlNO_3$. These results are in contrast to ones found in experiments with bivalent heavy metals (Hg^{2+}, Cd^{2+}, Pb^{2+}) since the letters' micromolar concentrations inhibited respiration of mitochondria uncoupled by DNP, FCCP, or CCP. This result could be associated with a weak reaction of Tl^+ with respiratory enzymes of mitochondria (Melnick et al., 1976; Woods and Fowler, 1986; Mulkey and Oehme, 1993; Korotkov, 2009). However, experiments in a medium containing 125 mM nitrates (KNO_3, $NaNO_3$, NH_4NO_3) revealed a noticeable decrease in the respiration of rat liver and rat heart mitochondria in states 3 and $3U_{DNP}$, which did not observe in a similar medium with $TlNO_3$ and 250 mM sucrose (Korotkov and Brailovskaya, 2001; Korotkov, 2009). Nevertheless, the decrease in respiration in experiments with Tl^+ was not so pronounced compared to the complete inhibition of this respiration by bivalent heavy metals and Ag^{+}, which may indicate differences in mechanisms of the toxic effect of Tl^{+} and these heavy metals on mitochondria.

4.3 Thallium effects on the concentration of free thiol groups in mitochondrial proteins

Thallium was found to deplete the reduced glutathione pool, to increase ROS production in cells, and to decline the body's antioxidants defenses (Herman and Bensch, 1967; Hasan and Ali, 1981; Aoyama et al., 1988; Mulkey and Oehme, 1993; Galván-Arzate et al., 2005; Verstraeten, 2006; Puga Molina and Verstraeten, 2008; Hanzel and Verstraeten, 2009; Cvjetko et al., 2010; Kiliç and Kutlu, 2010; Méndez-Armenta et al., 2011; Osorio-Rico et al., 2015). This Tl^{+}-induced decrease in reduced glutathione intracellular concentration, the inhibition of enzymes containing active thiol groups, and the ROS production increase probably relates to Tl^{+} reaction with various cell structures and mitochondrial structures, in particular (Zierold, 2000; Hanzel and Verstraeten, 2009). According to several researchers, such effects of thallium being in similarity to those of heavy metals (Ag^{+}, Pb^{2+}, Hg^{2+}, Cd^{2+}, In^{3+}, AsO_2^{-}) can explain the toxic impact of this metal on living organisms (Herman and Bensch, 1967; Woods and Fowler, 1986; Hanzel and Verstraeten, 2009). In this regard, one should expect a decrease in the content of free thiol groups in the inner membrane after the action of Tl^{+} with mitochondrial thiol groups might be one of the causes of thallium toxicity (Korotkov et al., 2008a, 2009, 2013; Korotkov and Saris, 2011). In this regard, one should expect a decrease in the content of free thiol groups in the inner membrane after the action of Tl^{+} on mitochondria, by analogy with the effects of other thiol reagents (Riley and Lehninger, 1964; Kowaltowski and Castilho, 1997; Kowaltowski et al., 1997).

Currently, there are no data on the interaction of Tl^{+} with thiol groups of proteins of the inner membrane and the mitochondrial matrix. We studied the reaction of Tl^{+} with free thiol groups of these proteins in mitochondria, energized by succinate (Fig. 4.13A and C) or glutamate with malate (Fig. 4.13B and D) (Korotkov et al., 2014a). For this purpose, mitochondria were incubated in 400 mOsm medium containing 25–75 mM $TlNO_3$, as well as 125 mM KNO_3 (Fig. 4.13A and B) or 250 mM sucrose (Fig. 4.13C and D). Determination of free thiol groups did carry out according to Ellman's method (1959) in the modification of

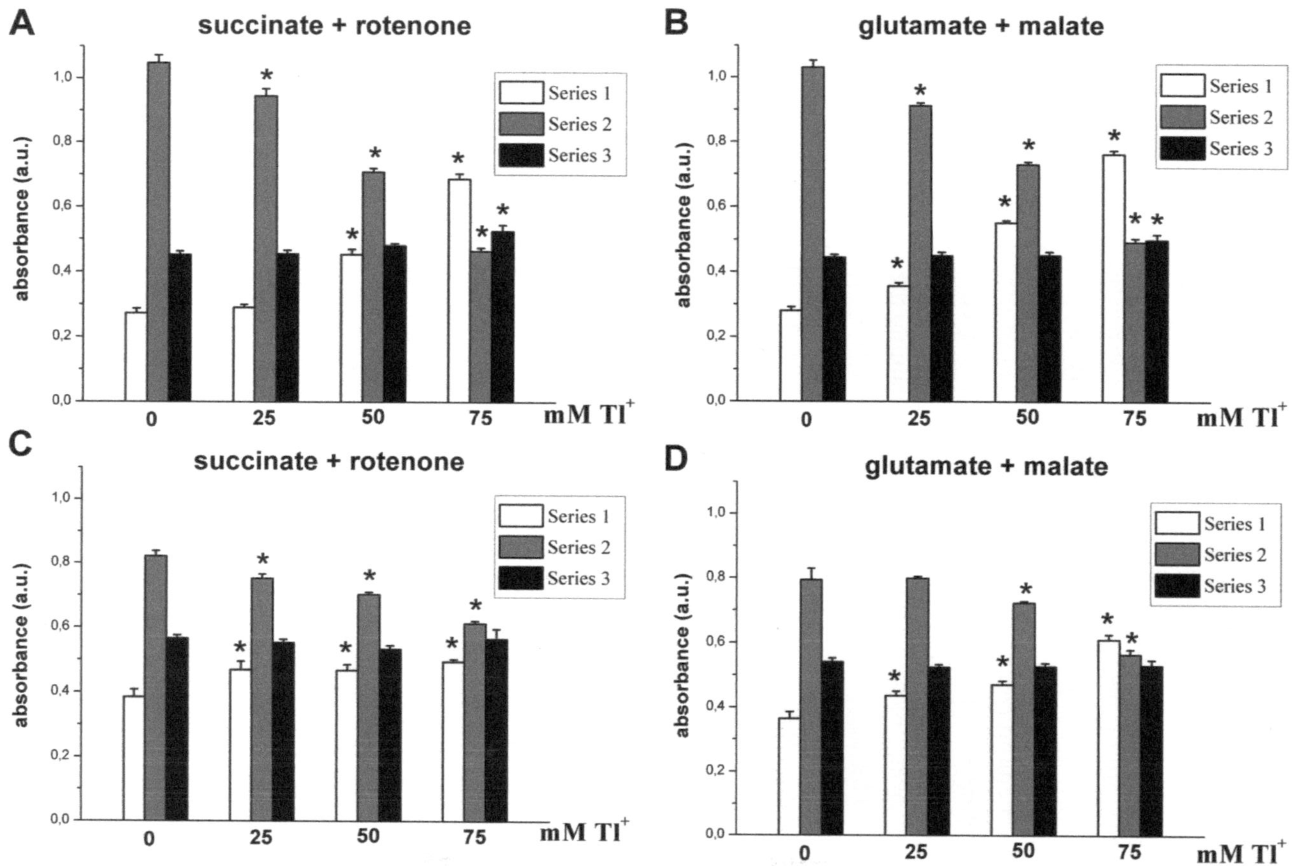

Figure 4.13 Effects of Tl^+ on content of free thiol groups in energized rat liver mitochondria incubated in a medium with $TlNO_3$ and KNO_3 (A–B) or with $TlNO_3$ and sucrose (C–D). The ordinate shows the absorbance changes (ΔA_{412}), and the abscissa shows concentration of $TlNO_3$ (mM). Asterisks indicate significant differences from control, experiments free of $TlNO_3$ ($P < .05$).

Kowaltowski and Castilho (Kowaltowski and Castilho, 1997; Kowaltowski et al., 1997). We used threefold freezing/thawing of mitochondrial samples in the above medium to maximize the separation of the inner membrane proteins from soluble matrix proteins (Fig. 4.13). One can see that with an increase in this medium $TlNO_3$ concentration, the concentration of free SH groups precipitated in the IMM fraction has increased noticeably 3(series 1). Still, it decreased accordingly in the supernatant containing the fraction of soluble matrix proteins (series 2). If RLM were incubated in the $TlNO_3$ medium, but after being washed out and frozen/thawed in free thallium medium containing KNO_3, the content of detected SH groups in the membrane fraction did not depend on the concentration of Tl^+ in the initial medium with $TlNO_3$ (series 3).

Thiol groups play an essential role in maintaining the spatial structure of proteins. The activity of many mitochondrial enzymes is known to be determined by the presence of free SH groups in their active centers. The concentration of free thiol groups in the IMM is 80−95 nmol per mg of protein (Riley and Lehninger, 1964; Bindoli et al., 1997). These groups participate in oxidative phosphorylation processes, adenine nucleotide translocation, succinate dehydrogenation, and inorganic phosphate transport into mitochondria (Le Quoc et al., 1979; Wehrle and Pedersen, 1981; McStay et al., 2002). Oxidation of mitochondrial SH groups promotes the MPTP opening in the inner membrane (Kowaltowski and Castilho, 1997). Heavy metals (Ag^+, Cd^{2+}, Hg^{2+}, Pb^{2+}, Zn^{2+}) showed high chemical affinity to free SH groups that determine the activity of many cellular and mitochondrial enzymes, which explains the high toxicity of these metals (Perrin, 1979). For example, Ag^+, Hg^{2+}, Cu^{2+}, and Pb^{2+} increased ionic permeability of cell membranes (Kone et al., 1988, 1990; Tang et al., 1996). Strong interaction with SH groups of the IMM found Ag^+ (Chappell and Greville, 1954; Riley and Lehninger, 1964), Cu^{2+} (Leblondel and Allain, 1984), AsO_2^- (Kowaltowski and Castilho, 1997), Cd^{2+} (Miccadei and Floridi, 1993; Stohs et al., 2001; Dorta et al., 2003; Korotkov et al., 2008b), Pb^{2+} (Parr and Harris, 1976; Ma et al., 2017), and Hg^{2+} (Brawer et al., 1998; Houston, 2011). This interaction resulted in inhibition of mitochondrial respiratory enzymes manifested by decreased oxygen consumption rates in these organelles in 4_0, 4_1, 3, and 3U (uncoupled) states. In addition, increased ROS production and high-amplitude mitochondrial swelling did observe due to an increase in the ionic permeability of the inner membrane.

Studies have shown that respiration, swelling, and energy-dependent contraction of mitochondria, as well as the IMM ionic permeability, depended on essentially the number of active mitochondrial thiol groups in respiratory enzymes and the inner membrane proteins (Riley and Lehninger, 1964; Lê-Quôc and Lê-Quôc, 1985; Bindoli et al., 1997). Intense mitochondrial swelling showed after 10%–20% of SH groups in mitochondria reacted with a thiol reagent (Riley and Lehninger, 1964; Bindoli et al., 1997). The use of chelators with free thiol groups (captopril, 2,3-dimercaptopropane-1-sulfonate, dithiothreitol) prevented poisoning of animals by heavy metal compounds (Cd^{2+}, Hg^{2+}, Pb^{2+}, As^{3+}, cisplatin) in experiments in vivo (Chávez et al., 1991; Kelley, 1999; Flora et al., 2008; Pachauri et al., 2009). At the same time, thiol chelators that had a similar effect (2,3-dimercaptopropanol, dimercaptopropanesulfonic acid, methionine, N-acetylcysteine) did not show a protective effect against thallotoxicosis (Taber, 1964; Zook et al., 1968; Hakala, 1984; Hoffman, 2003; Ramsden, 2007; Rusyniak et al., 2010). Compounds containing SH groups showed inefficiency (dimercaptol, monothioglycerol, 3-mercaptopropane) or weak protective effect (aminoethylisothiuronium, aminopropylisothiuronium, dithizone) against thallium poisoning (Oehme, 1978). On the other hand, Prussian blue, which does not have SH groups and forms a potent complex with Tl^{+}, resulted in a decrease in thallium concentrations in blood and in various organs of rats in experiments in vivo (Barroso-Moguel et al., 1994; Ibrahim et al., 2006; Montes et al., 2011). Thus, the decrease of heavy metals' toxicity in experiments with thiol-containing chelators, on the one hand, the lack of such effect in similar experiments with thallium, on the other hand, and the weakening of thallotoxicosis by Prussian blue, on the third hand, allows us to draw a preliminary conclusion, the toxic effect of thallium salts on living organisms can much less relate with the interaction of Tl^{+} with thiol groups in cellular and mitochondrial enzymes.

Indeed, some authors claim a significant interaction of Tl^{+} with SH groups (Truhaut, 1960; Galván-Arzate and Santamaría, 1998; Hasan and Haider, 1989). The authors refer further to Gross et al. (1948), who showed a decrease in thallium toxicity with the introduction of cystine into the diet of white rats. Other researchers, based on the above analogy with the action of Tl^{+} and bivalent heavy metals (Cd^{2+}, Pb^{2+}, Hg^{2+}), the damaging effect of thallium compounds on cells and mitochondria was associated with the interaction of Tl^{+} with thiol groups of molecules that make up living organisms (Woods and Fowler, 1986; Mulkey and Oehme, 1993; Kiliç and Kutlu, 2010). The latter three authors refer to Herman and Bensch (1967), who substantiated such action of Tl^+ by unproven analogies in effects of Tl^+ compared to Cu^{2+}, Ag^+, Hg^{2+}, Cd^{2+}, and Fe^{3+}. Many authors limit themselves to stating this fact without references confirming such toxic action by Tl^+ (Taber, 1964; Innis and Moses, 1978; Cavanagh, 1991; Goyer and Clarkson, 2001; Hoffman, 2003). The authors of other publications refer to manuscripts that are ultimately irrelevant to this problem (Chandler and Scott, 1986; Mulkey and Oehme, 1993). Tl^+ was shown previously to have interacted weakly with enzymes with active thiol groups in their structure (Oehme, 1978; Douglas et al., 1990).

An increase in the inner membrane concentration of free SH groups was observed parallel rise from 25 to 75 mM $TlNO_3$ in the medium (Fig. 4.13, series 1), instead of our expected decrease in this concentration in the case of solid binding of Tl^+ to the inner membrane SH groups. The above in the SH group concentration increase was eliminated in similar experiments with RLM incubated with $TlNO_3$ but only if the experiments' pellet was then washed and subjected to three cycles of freezing/thawing in a medium containing 125 mM KNO_3 instead of 25–75 mM $TlNO_3$ (Fig. 4.13, series 3). In this case, the concentration of SH groups in the membrane fraction no longer depends on the concentration of $TlNO_3$ in the medium. Comparison of these results (series 1 and 3) suggested that this increase could be associated with co-precipitation of specific Tl^+ complexes with soluble matrix proteins along with the inner membrane fraction (Korotkov et al., 2014a). This assumption did confirm by analyzing supernatant obtained from experiments in the $TlNO_3$ medium (series 2). Indeed, we found a decrease in the concentration of free SH groups in the supernatant fraction, increasing $TlNO_3$ concentration in the medium from 25 to 75 mM (Fig. 4.13, series 2). This result confirms the previously stated assumption that the toxicity of thallium compounds for living organisms can base on the ability of Tl^+ to form ligand complexes with cytoplasm and mitochondrial enzymes (Flesch and Goldstone, 1950; Truter, 1975; Melnick et al., 1976; Prick, 1979; Chandler and Scott, 1986; Woods and Fowler, 1986; Mulkey and Oehme, 1993; Huang et al., 2012; Kiliç and Kutlu, 2010). However, the total concentration of thiol groups was constant when adding concentrations found in the inner membrane fraction (series 1) and the non-membrane protein fraction (series 3) summarized. In this regard, it is impossible to exclude the formation of Tl^+ ligand complexes with the matrix proteins and (in particular) with protruding parts of I and II multienzyme complexes of the respiratory chain (Korotkov et al., 2014a).

Another explanation for the increase in the concentration of thiol groups in the membrane fraction of mitochondria with 25−75 mM $TlNO_3$ may be protein denaturation of mitochondrial proteins. However, the latter assumption does not agree with our results found under the experimental conditions where succinate-energized mitochondria showed state 4, 3, and $3U_{DNP}$ respiration (Figs. 4.6−4.9) and marked contraction (Figs. 4.1−4.4, and 4.10), which can not be if the structure of proteins could disturb during their denaturation. Proteins' denaturation with a violation of their tertiary structure should completely lose their catalytic activity. So, a decrease but not a complete inhibition of mitochondrial respiration in various energy states at $TlNO_3$ concentrations used, in our opinion, is more consistent with the distortion of the structure of these proteins due to swelling of organelles. The concentration of thiol groups in the inner membrane fraction of mitochondria incubated in the medium with $TlNO_3$ and 125 mM KNO_3 (Fig. 4.13A and B, series 1) was slightly higher than that found in the medium with $TlNO_3$ and 250 mM sucrose (Fig. 4.13C and D, series 1). This result may be due to the greater availability of amino acid R-groups in the matrix proteins to Tl^+ in the medium with $TlNO_3$ and KNO_3, which could contribute to a more intensive formation of complexes of these proteins with Tl^+ ions in comparison with a similar medium containing $TlNO_3$ and sucrose. According to the results of our studies, we also found that the concentration of free thiol groups in the IMM fraction did not depend on the presence of Ca^{2+} in this medium (Korotkov et al., 2014a) or the type of substrate used (Fig. 4.13). This finding allows us to conclude that calcium load and method of the mitochondria energizing have a weak effect on the interaction of Tl^+ with the inner membrane thiol groups. Our data are consistent with the minor impact of Tl^+ both on mitochondrial respiration in 3 or 3U states and on mitochondrial respiratory enzymes (Fig. 4.6B) (Melnick et al., 1976; Woods and Fowler, 1986; Korotkov et al., 2008a).

Thiol reagents (PAO, Cd^{2+}, Ag^+) significantly reduced free thiol groups' content in the inner mitochondrial membrane (Riley and Lehninger, 1964; Kowaltowski et al., 1997; Zazueta et al., 2000). The decrease in cystine's toxic effects of thallium in experiments in vivo (Gross et al., 1948) made it possible to assume that the interaction of thallium with the thiol groups of proteins, similar to other heavy metals, is the main reason for this metal toxicity (Herman and Bensch, 1967). Thallium in experiments in vivo and in vitro depleted intracellular pool of reduced glutathione, cut the body's antioxidants defense and promoted ROS production increase (Hasan and Ali, 1981;

Aoyama et al., 1988; Mulkey and Oehme, 1993; Galván-Arzate et al., 2005; Hanzel and Verstraeten, 2009; Kiliç and Kutlu, 2010; Osorio-Rico et al., 2015). However, our experiments with 25–75 mM $TlNO_3$ (Korotkov et al., 2014a) did not reveal any decrease in the number of free thiol groups in the inner membrane of rat liver mitochondria. This fact is in complete agreement that fivefold excess of Tl^+ acetate did not affect the content of free SH groups determined in the use of Elman reagent in 0.1 M glutathione solution (Bunni et al., 1987). However, the respiration decrease in the medium with $TlNO_3$ and nitrates that we found may be due to changes in the inner membrane structure due to thallium-induced mitochondrial swelling and interaction of Tl^+ with soluble matrix proteins (Korotkov et al., 2014a). The latter assumption is consistent with scientific literature data on the ability of thallium(I) to form ligand complexes with intracellular enzymes and enzymes of the inner mitochondrial membrane (Flesch and Goldstone, 1950; Truter, 1975; Prick, 1979; Chandler and Scott, 1986; Woods and Fowler, 1986; Mulkey and Oehme, 1993; Huang et al., 2012).

Comparing these results with our data and literature data allowed us to conclude about the insignificant effect of Tl^+ on the concentration of free SH groups in the inner membrane of the RLM (Korotkov et al., 2014a). At the same time, the decrease in mitochondrial respiration in nitrate media with $TlNO_3$ can associate with the Tl^+-induced increase in ionic permeability of the IMM and subsequent swelling of these organelles with distorting their spatial and structural integrity. This assumption, along with the formation of Tl^+ ligand complexes with mitochondrial respiratory complexes, can reduce the latter's activity (Korotkov et al., 2014a).

4.4 Tl^+ induces the opening of the mitochondrial permeability transition pore in the inner membrane of calcium-loaded mitochondria: The effects of the pore inhibitors

4.4.1 Tl^+ induces the opening of the mitochondrial permeability transition pore

Mitochondria calcium overload accompanies the saturation of matrix-facing specific Ca^{2+} sites near the adenine nucleotide translocase (ANT). Subsequent $\Delta\Psi_{mito}$ decline induces the opening of the mitochondrial permeability transition pore (MPTP) in the inner membrane in a high-conductance state (Bernardi et al., 1992; Zoratti and Szabó, 1995; Ichas and Mazat, 1998; Bernardi, 1999; Halestrap and Brenner, 2003; Halestrap,

2009). In this case, the inner membrane shows permeability for molecules up to 300–1500 Da (particularly for sucrose molecules). And also, intensive mitochondria swelling, the outflow of Ca^{2+} and ATP from the matrix, $\Delta\Psi_{mito}$ decline, and release of cytochrome C into the intermembrane space take place (Bernardi and Pietrobon, 1982; Nicholls and Akerman, 1982; Ichas and Mazat, 1998; Gunter et al., 2004; Gunter and Sheu, 2009). If the calcium-binding sites are insufficiently saturated, then the MPTP opens in a low-conductance state. The IMM becomes permeable to small molecules (up to 300 Da) or ions (H^+, K^+, Na^+, Li^+, Ca^{2+}), and the IMM potential is partially retained (Ichas and Mazat, 1998; Bernardi, 1999).

Early research believed that ANT, cyclophilin D (CyP-D), and voltage-dependent anion channel (VDAC) belong to the main components of this MPTP (Halestrap and Davidson, 1990; Zoratti and Szabó, 1995; Bernardi, 1999; Halestrap and Brenner, 2003). However, the main components of MPTP underwent a particular revision after studies on the reconstruction of mitochondria deficient in CyP-D and ANT (Kokoszka et al., 2004; Basso et al., 2008; Leung et al., 2008; Baines, 2009; Halestrap, 2009; Zorov et al., 2009; Beutner et al., 2017; Morciano et al., 2021). So, phosphate symporter (PiC) and cyclophilin D (CyP-D) became the main components of MPTP. The adenine nucleotide translocase (ANT) is the MPTP regulatory part. Voltage-dependent anion channel (VDAC) excluded entirely from this pore community (Leung et al., 2008; Baines, 2009; Zorov et al., 2009; Halestrap, 2010; Giorgio et al., 2009; Elrod and Molkentin, 2013; Halestrap, 2014; Gutiérrez-Aguilar and Baines, 2015). Recently, there have been results that have led to another revision of the MPTP main components. The MPTP main parts are the phosphate symporter (PiC), cyclophilin D (CyP-D), and other ATP synthase subunits. The pore regulatory part is still the ANT (Bernardi, 2013; Chinopoulos and Szabadkai, 2013; Giorgio et al., 2013; Elrod and Molkentin, 2013; Gutiérrez-Aguilar and Baines, 2015; Halestrap and Richardson, 2015; Bonora et al., 2017). According to the most recent data, the MPTP main components MPTP can be considered the complex of the Ca^{2+}-modified ANT and the Ca^{2+}-modified ATP synthase (Halestrap, 2014; Beutner et al., 2017; Morciano et al., 2021; Carraro et al., 2019; Brustovetsky, 2020; Bernardi et al., 2021).

We studied the possibility of the MPTP induction by Tl^+ ions in the inner membrane of calcium-loaded mitochondria of rat liver and heart. In experiments with rotenone, swelling of nonenergized RLM became more in increasing Tl^+ concentration in the medium, regardless of the presence of Ca^{2+} in it (Fig. 4.14A and B). On the other hand, the nonenergized

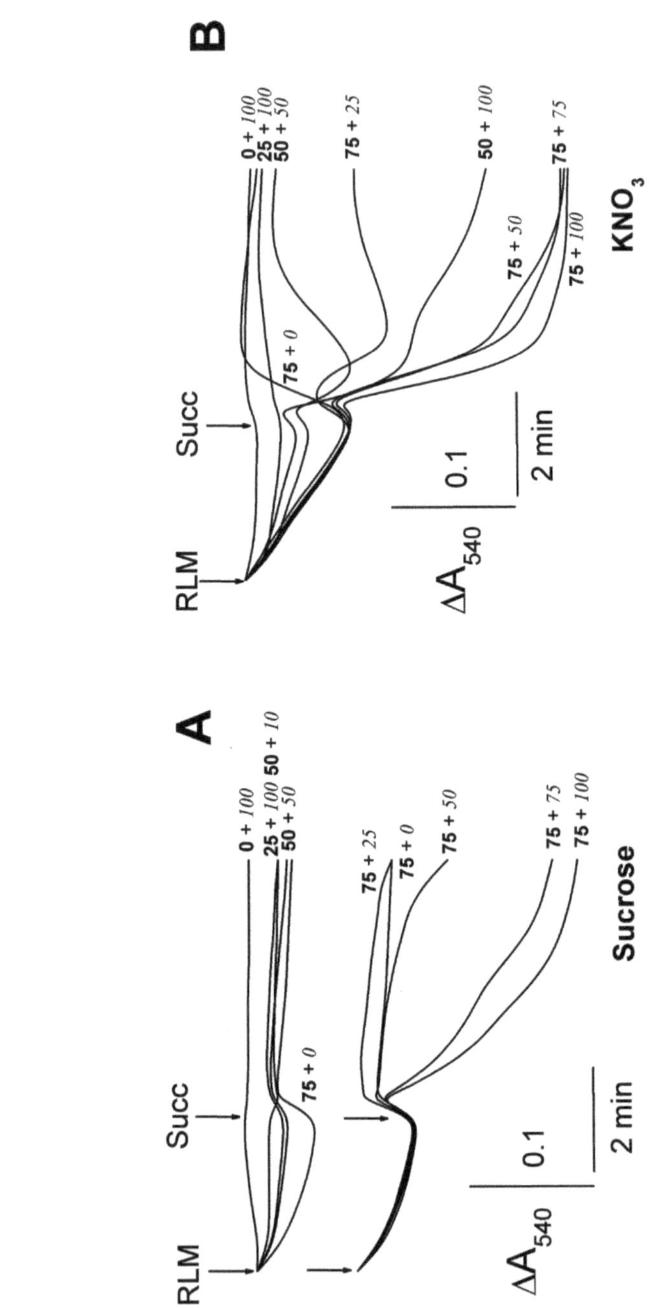
A
RLM
Succ
0 + 100
25 + 100
50 + 50
50 + 10
75 + 0
75 + 25
75 + 50
75 + 75
75 + 100
ΔA540
0.1
2 min
Sucrose
B
RLM
Succ
0 + 100
25 + 100
50 + 50
75 + 25
75 + 0
50 + 100
75 + 50
75 + 75
75 + 100
ΔA540
0.1
2 min
KNO3

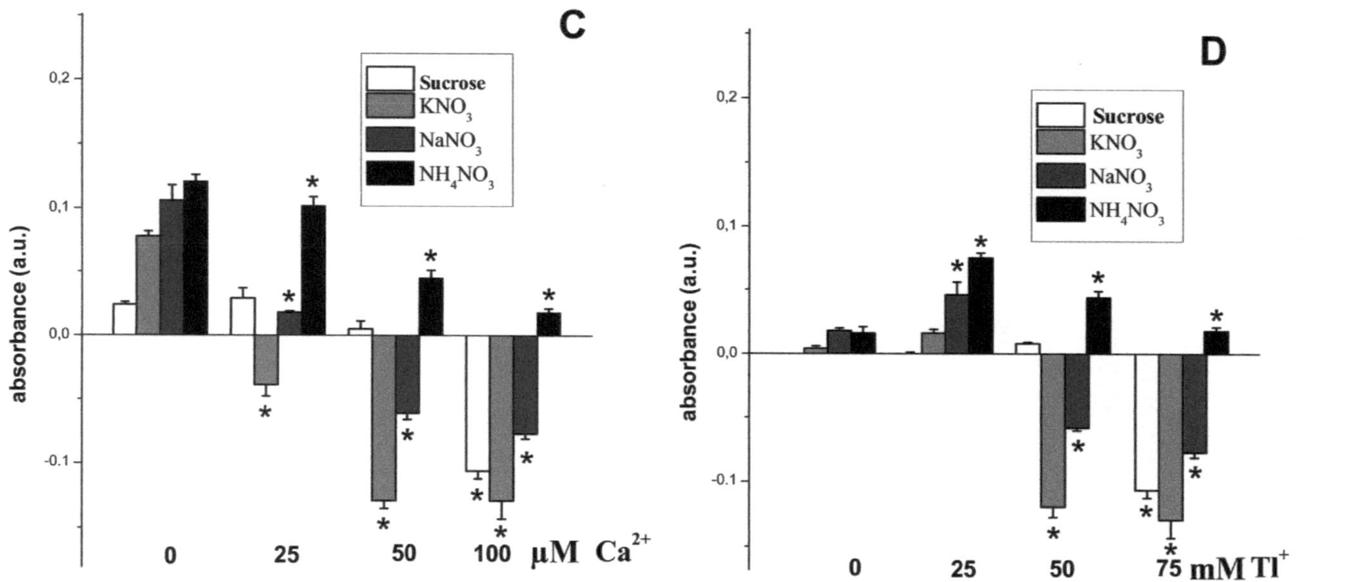

Figure 4.14 Influence of Tl^+ and Ca^{2+} on swelling of rat liver mitochondria. Mitochondria (1.5 mg/mL of protein) were injected into medium containing 0–75 mM $TlNO_3$ (A, B, D) or 75 mM $TlNO_3$ (C), as well as 5 mM Tris-NO_3 (pH 7.3), 100 μM Ca^{2+} (D), 2 μM rotenone and 1 μg/mL oligomycin together (where indicated) with 250 mM sucrose (Sucrose) or 125 mM nitrates (KNO_3, $NaNO_3$, NH_4NO_3). Arrows (A, B) show addition of mitochondria (RLM) and 5 mM succinate (Succ). The presence of $TlNO_3$ (mM) and Ca^{2+} (μM) in the medium before mitochondria is accordingly noted to the right of traces in bold and italics numbers. The ordinate (C–D) shows the change in absorbance (ΔA_{540}) in 3 min after addition of succinate into the medium, as in panels A and B. Concentrations of Ca^{2+} (μM) and Tl^+ (mM) are accordingly showed on the abscissa of panel C and panel D. The figure shows typical traces for each of three independent experiments. Asterisks show significant differences ($P < .05$) from the control, experiments free of Ca^{2+} (C) or free of $TlNO_3$ (D).

mitochondria swelling in a rotenone-free medium depended on the concentrations of Tl^+ and Ca^{2+} (Figs. 4.15 and 4.16). This difference may be due to the uptake of Ca^{2+} by mitochondria with the participation of endogenous substrates, followed by thallium-induced MPTP opening (Korotkov et al., 2015b). This result is consistent with slightly changed mitochondria swelling after adding glutamate and malate into the medium (Fig. 4.15 and Tables 4.7 and 4.8). The increased swelling of calcium-loaded and succinate-energized RLM in the presence of rotenone (Fig. 4.14 and Tables 4.5 and 4.6) may also be associated with the thallium-induced MPTP opening. The swelling of succinate-energized RLM at the same $TlNO_3$ concentration depended on Ca^{2+} concentration in the

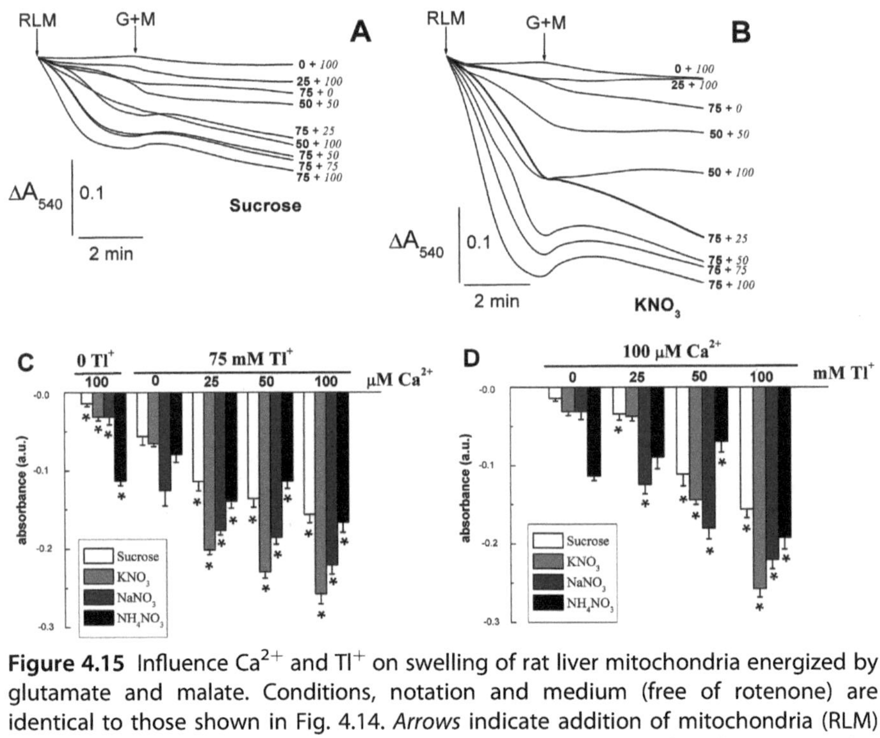

Figure 4.15 Influence Ca^{2+} and Tl^+ on swelling of rat liver mitochondria energized by glutamate and malate. Conditions, notation and medium (free of rotenone) are identical to those shown in Fig. 4.14. *Arrows* indicate addition of mitochondria (RLM) and 5 mM glutamate with 5 mM malate (G + M). Panels A and B show traces displaying results in for each of three independent experiments. The ordinate axis (C–D) shows change in absorbance (ΔA_{540}) 6 minutes after injection of mitochondria into the medium, as in panels A and B. Concentrations of Ca^{2+} (μM) and Tl^+ (mM) are respectively represented by numbers on the abscissa of panel C and panel D. Significant differences from control (experiments free of Ca^{2+} (C) or free of $TlNO_3$ (D) are noted by asterisks ($P < .05$).

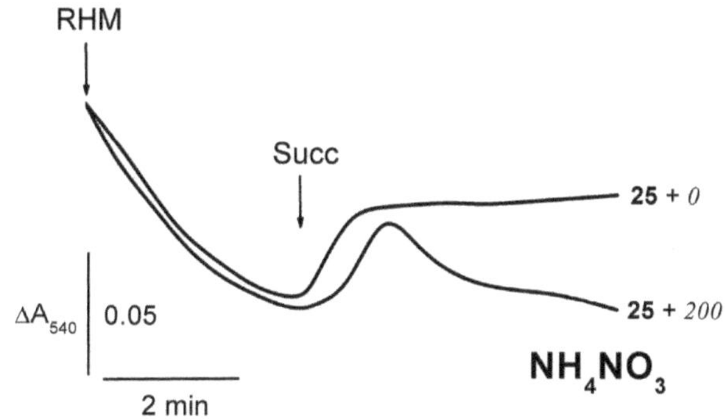

Figure 4.16 The effect of Ca^{2+} on the swelling of rat heart mitochondria. Mitochondria (1 mg/mL of protein) were introduced into the medium with 75 mM $TlNO_3$ and 125 mM NH_4NO_3. Other conditions and designations are similar to Fig. 4.14. Arrows show adding mitochondria (RHM) and succinate (Succ) into the medium. The traces shown results in for each of three independent experiments.

medium (Fig. 4.14 and Table 4.5). In contrast, this swelling depends on the medium's Ca^{2+} and Tl^+ concentrations, like experiments with glutamate and malate (Fig. 4.15 and Tables 4.7 and 4.8). The swelling of energized mitochondria in media containing nitrate (KNO_3, NH_4NO_3, $NaNO_3$, $LiNO_3$) and $TlNO_3$ was more than that found in media with $TlNO_3$ and sucrose (Figs. 4.14 and 4.15, Tables 4.5, 4.6, 4.7, 4.8, Korotkov and Saris, 2011). These results allow us to conclude that the effect of Tl^+ on mitochondria was an increase in the IMM's permeability for monovalent cations forming above nitrates due to the MPTP opening in the membrane in a low-conductance state (Mitchell and Moyle, 1969; Brierley et al., 1977; Ichas and Mazat, 1998; Bernardi, 1999). The swelling of calcium-loaded mitochondria occurred only in a medium containing 75 mM $TlNO_3$ and sucrose (Fig. 4.14A), probably due to the MPTP opening in the inner membrane in a high-conductance state.

Mitochondria (1.5 mg/mL of protein) were added into 400 mOsm medium containing 75 mM $TlNO_3$, 5 mM Tris-NO_3 (pH 7.3), 2 μM rotenone, and 1 μg/mL oligomycin along with 250 mM sucrose (Sucrose), or 125 mM nitrates (KNO_3, NH_4NO_3, $NaNO_3$, $LiNO_3$). Succinate was added into the medium to 5 mM concentration as in Fig. 4.14A. Before mitochondria, 0−100 μM Ca^{2+} (Ca^{2+}) was added into the medium. The number of experiments is indicated in brackets. The table shows the change in absorbance (ΔA_{540}) in the mitochondrial suspension 3 min after additions

Table 4.5 Change in absorbance (ΔA_{540}) of rat liver mitochondria suspension after adding succinate into the medium.

Addition of Ca^{2+} (μM) into the medium before mitochondria	Media				
	Sucrose	**KNO_3**	**NH_4NO_3**	**$NaNO_3$**	**$LiNO_3$**
0 (Control)	0.024 ± 0.002 (4)	0.078 ± 0.004 (4)	0.121 ± 0.005 (6)	0. 106 ± 0.012 (4)	0.083 ± 0.006 (4)
25	0.029 ± 0.008 (3)	−0.039 ± 0.009 (4)*	0.102 ± 0.007 (6)*	0.018 ± 0.001 (3)*	0.051 ± 0.016 (4)*
50	0.005 ± 0.006 (4)	−0.129 ± 0.006 (4)*	0.045 ± 0.006 (4)*	−0.061 ± 0.005 (4)*	−0.055 ± 0.008 (4)*
100	−0.106 ± 0.006 (4)*	−0.129 ± 0.014 (4)*	0.018 ± 0.003 (6)*	−0.077 ± 0.004 (5)*	−0.092 ± 0.003 (4)*

Asterisk (*) indicate significant differences from the control, experiments free of $TlNO_3$ ($P < .05$).

Table 4.6 Change in absorbance (ΔA_{540}) in succinate-energized rat liver mitochondria suspension in the medium containing 100 μM Ca^{2+}.

	Media				
$TlNO_3$ (mM)	Sucrose	KNO_3	NH_4NO_3	$NaNO_3$	$LiNO_3$
0 (Control)	−0.002 ± 0.001 (3)	0.004 ± 0.002 (3)	0.016 ± 0.005 (3)	0.018 ± 0.002 (3)	0.007 ± 0.002 (4)
25	0 ± 0.001 (3)	0.016 ± 0.003 (3)*	0.075 ± 0.004 (3)	0.046 ± 0.010 (3)	0.025 ± 0.004 (4)*
50	0.008 ± 0.001 (3)	−0.119 ± 0.008 (3)*	0.044 ± 0.005 (3)*	−0.058 ± 0.002 (3)*	−0.038 ± 0.008 (4)*
75	−0.106 ± 0.006 (4)*	−0.129 ± 0.014 (4)*	0.018 ± 0.003 (6)*	−0.077 ± 0.004 (5)*	−0.092 ± 0.003 (4)*

Asterisk (*) indicate significant differences from the control, experiments free of $TlNO_3$ ($P < .05$).

Table 4.7 Change in absorbance (ΔA_{540}) in suspension of rat liver mitochondria energized by glutamate and malate in the medium containing 75 mM $TlNO_3$.

Addition of Ca^{2+} (μM) into the medium before mitochondria	Media				
	Sucrose	KNO_3	NH_4NO_3	$NaNO_3$	$LiNO_3$
Free of Ca^{2+} and free of $TlNO_3$	−0.014 ± 0.004 (3)*	−0.031 ± 0.005 (3)*	−0.113 ± 0.006 (3)*	−0.031 ± 0.001 (3)*	−0.025 ± 0.005 (3)*
0 (Control)	−0.056 ± 0.011 (3)	−0.065 ± 0.004 (3)	−0.079 ± 0.010 (3)	−0.125 ± 0.020 (3)	−0.062 ± 0.015 (3)
25	−0.114 ± 0.012 (3)*	−0.201 ± 0.006 (3)*	−0.139 ± 0.009 (3)*	−0.176 ± 0.006 (3)*	−0.148 ± 0.018 (3)*
50	−0.136 ± 0.011 (3)*	−0.229 ± 0.008 (3)*	−0.113 ± 0.010 (3)*	−0.185 ± 0.009 (3)*	−0.154 ± 0.006 (3)*
100	−0.156 ± 0.011 (3)*	−0.257 ± 0.013 (3)*	−0.166 ± 0.013 (3)*	−0.220 ± 0.012 (3)*	−0.184 ± 0.006 (3)*

Asterisk (*) indicate significant differences from the control, experiments free of $TlNO_3$ ($P < .05$).

Table 4.8 Change in absorbance (ΔA_{540}) in suspension of rat liver mitochondria energized by glutamate and malate in the medium containing 100 μM Ca^{2+}.

$TlNO_3$ (mM)	Media				
	Sucrose	KNO_3	NH_4NO_3	$NaNO_3$	$LiNO_3$
0 (Control)	−0.014 ± 0.004 (3)	−0.031 ± 0.005 (3)	−0.113 ± 0.006 (3)	−0.031 ± 0.010 (3)	−0.029 ± 0.005 (3)
25	−0.034 ± 0.008 (3)*	−0.037 ± 0.006 (3)	−0.089 ± 0.015 (3)	−0.124 ± 0.012 (3)*	−0.038 ± 0.008 (3)
50	−0.111 ± 0.015 (3)*	−0.144 ± 0.011 (3)*	−0.069 ± 0.014 (3)*	−0.180 ± 0.014 (3)*	−0.113 ± 0.013 (3)*
75	−0.156 ± 0.011 (3)*	−0.257 ± 0.011 (3)*	−0.192 ± 0.015 (3)*	−0.220 ± 0.012 (3)*	−0.184 ± 0.006 (3)*

Asterisk (*) indicate significant differences from the control, experiments free of $TlNO_3$ ($P < .05$).

of succinate into medium (Fig. 4.14A and B). Asterisks mark significant differences ($P < .05$) from the control, experiments in the medium containing 75 mM $TlNO_3$ and free of Ca^{2+}.

Mitochondria (1.5 mg/mL of protein) were injected into 400 mOsm medium containing 0–75 mM $TlNO_3$ and 100 μM Ca^{2+}. Other components of the medium and designations are similar to those presented in Table 4.5. Asterisks indicate significant differences from the control in experiments with 100 μM Ca^{2+} free of $TlNO_3$ ($P <.05$).

The medium, additions and designations are similar to those presented in Table 4.5, but the medium contained 5 mM glutamate and 5 mM malate instead of succinate and rotenone. The change in absorbance (ΔA_{540}) 6 min after the injection of mitochondria into the medium is shown as in Fig. 4.15A and B. Asterisks mark significant differences from the control (experiments in Ca^{2+} free medium containing 75 mM $TlNO_3$) ($P < .05$).

The medium, additions, and designations are similar to those presented in Table 4.6, except that 5 mM glutamate and 5 mM malate instead of succinate and rotenone were present in the medium. The change in absorbance (ΔA540) 6 min after the injection of mitochondria into the medium is shown similarly to Fig. 4.15A and B. Asterisks mark significant differences from the control (experiments in $TlNO_3$ free medium containing 100 μM Ca^{2+}) ($P <.05$).

A noticeable swelling of succinate-energized rat heart mitochondria (RHM) did observe in a medium containing 25 mM $TlNO_3$ and 200 μM Ca^{2+} (Fig. 4.16). Mitochondria swell noticeably in a rotenone-free medium containing $TlNO_3$ and nitrate (KNO_3, NH_4NO_3, $NaNO_3$) even before the substrate addition into the medium (Fig. 4.17B; Korotkov and Saris, 2011). The subsequent energization of these organelles with any substrate stimulated their weak contraction. At the same time, the mitochondrial swelling blocked even before substrate addition in the rotenone medium.

Rotenone is known to block complex I of the mitochondrial respiratory chain (Enriquez, 2016). In this case, pyridine nucleotides are in a more reduced state, which reduces the probability of the MPTP opening in the inner mitochondrial membrane (Rigobello et al., 1995; Fontaine et al., 1998; Waldmeier et al., 2002; Baines, 2009). The MPTP opening requires more high calcium concentrations in a medium with rotenone than ones found in rotenone-free experiments with mitochondria, energized by the first complex substrates or succinate alone (Fontaine et al., 1998; Belyaeva et al., 2004b; Baines, 2009). This phenomenon is called substrate specificity, and it may be associated with the participation of complex I in the MPTP

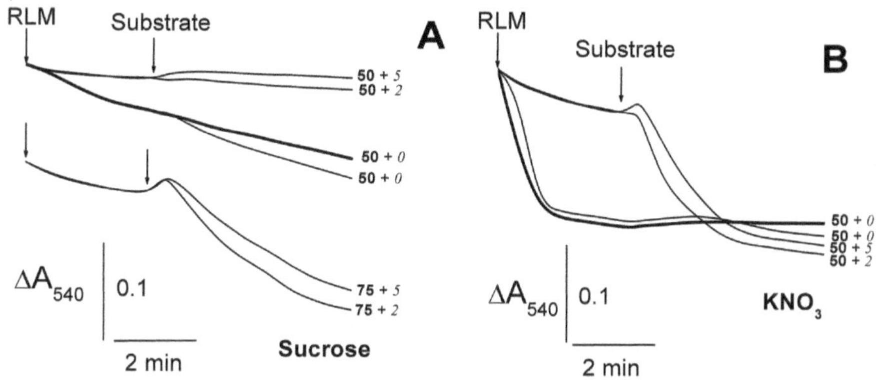

Figure 4.17 Effects of Ca^{2+} and rotenone on swelling of rat liver mitochondria. The medium and experimental conditions are similar to those shown in Fig. 4.15. Arrows mark additions into the medium of mitochondria (RLM) and 5 mM glutamate with 5 mM malate (Substrate). Bold trace shows similar experiments with 5 mM succinate, added into this medium instead of glutamate and malate. To the right of the traces, $TlNO_3$ concentration (mM) and rotenone (μM) concentration are shown in bold. Typical traces are shown for three independent experiments.

formation in the inner membrane (Fontaine et al., 1998; Ichas and Mazat, 1998; Fontaine and Bernardi, 1999; Leverve and Fontaine, 2001; Halestrap and Brenner, 2003). Oxidation of mitochondrial pyridine nucleotides and NAD(P)H, as well as an increase in the concentration of reactive oxygen species in mitochondria, stimulated the MPTP opening by increasing the number of sites available to CyP-D from the matrix side (Lê-Quôc and Lê-Quôc, 1989; Bindoli et al., 1997; Fontaine and Bernardi, 1999; García et al., 2005; Li et al., 2012; Aldakkak et al., 2013).

Noticeably lower concentrations of heavy metals (Cd^{2+}, Zn^{2+}, Pb^{2+}, Hg^{2+}, Cu^{2+}) than Ca^{2+} can induce the MPTP opening in the inner membrane. The interaction of these metals with the inner membrane leads to an increase in its potassium and proton permeability, to an extensive mitochondrial swelling, to a decrease in oxygen consumption rates of mitochondria being different energy states, in a reduction in $\Delta\Psi_{mito}$, and to a decline in the concentration of pyridine nucleotides, ATP and Ca^{2+} in the mitochondrial matrix (Miyahara and Utsumi, 1975; Chavez and Holguin, 1988; Zoratti and Szabó, 1995; Korotkov and Skul'skiĭ, 1996; Korotkov et al., 1998; Bernardi, 1999; Wudarczyk et al., 1999; Zazueta et al., 2000; Belyaeva et al., 2001, 2002, 2004a,b; Belyaeva and Korotkov, 2003; Dorta et al., 2003; Lee et al., 2005; Ma et al., 2017). The opening of the MPTP, induced by heavy metals (Cd^{2+}, Hg^{2+}, Zn^{2+}, Pb^{2+}, Cu^{2+}, As^{3+}), showed the substrate specificity. This conclusion made taking into account the following data: the inhibition of respiration in states 3 and 3U, the oxidation of pyridine nucleotides, and mitochondrial swelling were more substantial in experiments with mitochondria, energized by glutamate and malate but not succinate in the presence of rotenone (Cameron et al., 1986; Chávez et al., 1987; Skul'skiĭ et al., 1988; Chávez and Holguín, 1988; Lund et al., 1993; Miccadei and Floridi, 1993; Bates et al., 1994; Zoratti and Szabò, 1995; Brown et al., 2000; Belyaeva et al., 2002; Belyaeva and Korotkov, 2003; Shilo et al., 2003; Bustamante et al., 2005; Dineley et al., 2005; Belyaeva et al., 2011). The Cd^{2+}-induced MPTP opening in our experiments was due to the interaction of this metal both with calcium sites on the matrix side and with SH groups of the I-th and III mitochondrial respiratory complexes (Belyaeva et al., 2004b).

The induction of MPTP by heavy metals in the inner membrane is due to their interaction with vicinal SH groups, which are part of active centers of mitochondrial respiratory complexes and ANT (Leung et al., 2008; Baines, 2009; Halestrap, 2009; Zorov et al., 2009). At the same time, bivalent heavy metals induced the MPTP opening even in calcium-free media (Wudarczyk et al., 1999; Zazueta et al., 2000; Belyaeva et al., 2002, 2004a,b; Belyaeva and Korotkov, 2003; Lee et al., 2005). Heavy metals (Cd^{2+}, Pb^{2+}, As^{3+}) and selenite in the presence of Ca^{2+} induced even more significant mitochondrial swelling, and also they more reduced $\Delta\Psi_{mito}$, respiration, and the Ca^{2+} matrix retention in experiments with cells and mitochondria (Chávez et al., 1987; Miccadei and Floridi, 1993; Belyaeva et al., 2001; Shilo et al., 2003; Bustamante et al., 2005; Adiele et al., 2012). Tl^{+}, like those mentioned above bivalent heavy metals, has an uncoupling effect on mitochondria (it stimulates state 4 and reduces $\Delta\Psi_{mito}$) (Melnick et al., 1976; Saris et al., 1981; Korotkov et al., 2007, 2008a; Korotkov, 2009). Our experiments show (Figs. 4.14–4.16 and Tables 4.5–4.8) that Tl^{+} in contrast with Ag^{+} (Inoue et al., 2009) and Cd^{2+} (Korotkov et al., 1998; Belyaeva et al., 2001, 2002, 2004b; Belyaeva and Korotkov, 2003) inducts high-amplitude swelling of energized mitochondria only in the incubation medium with Ca^{2+}, regardless of what complex the substrate used.

Swelling of mitochondria can affect the IMM structure and, as a result, alter the activity of the respiratory complexes embedded into this membrane (Devin et al., 1996, 1997a,b). Since calcium-loaded mitochondria swelled intensively in 400 mOsm media containing $TlNO_3$, sucrose, and nitrates (Figs. 4.14–4.16 and Tables 4.5–4.8), it was interesting to study these organelles' respiration rates in different energy states under these conditions. Respiration of calcium-loaded RLMs in states 3 or $3U_{DNP}$ decreased more strongly in media with nitrate and $TlNO_3$, compared with respiration in similar experiments in media with sucrose and $TlNO_3$ (Fig. 4.18 and Table 4.9; Korotkov and Saris, 2011). Similar results we found in similar experiments with succinate-energized RHM (Fig. 4.19; Korotkov et al., 2013, 2020).

Mitochondria (1.5 mg/mL of protein) were added into 400 mOsm medium containing 75 mM $TlNO_3$, as well as 250 mM sucrose (Sucrose) or 125 mM nitrates (KNO_3, $NaNO_3$, NH_4NO_3). Experimental procedure and other additions of Ca^{2+}, ADP and DNP as ones shown on Fig. 4.18A.

The appearance of $\Delta\Psi_{mito}$ on the inner membrane after succinate injection into the medium led to a noticeable decrease in concentration and, accordingly, the fluorescence of safranin O due to voltage-dependent uptake of this dye by energized mitochondria of rat liver and rat heart (Fig. 4.20, trace 1 and Table 4.10; Korotkov et al., 2013, 2015a,b). Short-term increase in safranin O fluorescence after injection of Ca^{2+} into the medium (Fig. 4.20, trace two and Table 4.10) is due to a decrease in potential-dependent uptake of Ca^{2+} by mitochondria with the simultaneous release of safranin O into the medium. Further, after some potential recovery, safranin O rerelease was observed in the medium due to the MPTP opening and the associated decrease in $\Delta\Psi_{mito}$ in the inner membrane (Zoratti and Szabó, 1995; Ichas and Mazat, 1998; Bernardi, 1999; Skulachev, 2001; Panov, 2015).

The succinate medium contained: 20 mM $TlNO_3$, 110 mM sucrose, 125 mM KNO_3, 2 μM rotenone, and 5 mM succinate (succinate) followed by addition of 75 μM Ca^{2+} (Ca^{2+}). Medium with 5 mM glutamate and 5 mM malate (G + M) included 15 mM $TlNO_3$, 120 mM sucrose, 125 mM KNO_3 followed by addition of 40 μM Ca^{2+} (Ca^{2+}). Other additions, conditions, and order of the experiment are as in Fig. 4.20. Asterisks show significant differences from the control, experiments free of Ca^{2+} ($P < .05$).

Experiments with calcium-loaded mitochondria revealed the effect of substrate specificity. This effect was that mitochondria energized by the complex I substrates were less resistant to calcium action in comparison with organelles energized by succinate (the complex II substrate) (Fontaine et al., 1998; Fontaine and Bernardi, 1999; Leverve and Fontaine, 2001; Belyaeva et al., 2011). This effect found with calcium-loaded mitochondria

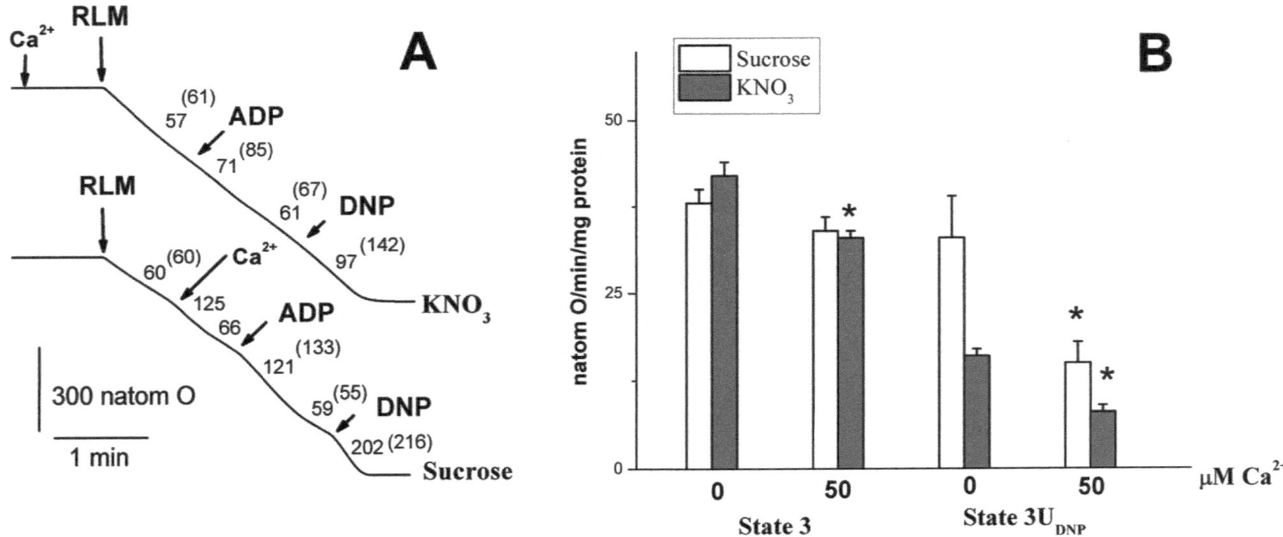

Figure 4.18 Effect of Tl^+ and Ca^{2+} on oxygen consumption rates in rat liver mitochondria energized by 5 mM succinate (A) or 5 mM glutamate and 5 mM malate (B). Mitochondria (1.5 mg/mL of protein) were introduced into the medium containing 75 mM $TlNO_3$. The rest designations, media composition and additions of ADP and DNP were the same as shown in Fig. 4.16A. Ca^{2+} addition on Panel A was 100 μM. Respiratory rates found in similar experiments performed free of Ca^{2+} are indicated in parentheses (Fig. 4.16A). Numbers above traces show oxygen consumption rates (n atom O/min per mg of protein). Typical traces are shown for three independent experiments. To induce 3 and $3U_{DNP}$ states, 200 μM ADP and 30 μM DNP were sequentially added into rotenone-free medium after mitochondria (panel B). The ordinate shows the rates, and the abscissa shows concentration of Ca^{2+} (μM). Asterisk indicate significant differences from control in experiments free of Ca^{2+} (B) ($P < .05$).

Table 4.9 Effect of thallium nitrate on oxygen consumption rates (n atom O/mg of protein) in succinate-energized calcium-loaded rat liver mitochondria.

	Media			
	Sucrose	**KNO_3**	**NH_4NO_3**	**$NaNO_3$**
State 4_0	60 ± 2 (3)	55 ± 3 (3)	57 ± 1 (3)	60 ± 2 (3)
State 4_0 + Ca^{2+}	71 ± 6 (3)	60 ± 3 (3)	64 ± 2 (3)	67 ± 3 (3)
State 3	116 ± 5 (3)	68 ± 2 (3)	69 ± 1 (3)	81 ± 2 (3)
State 4	72 ± 13 (3)	57 ± 2 (3)	63 ± 1 (3)	67 ± 3 (3)
RCR	1.68 ± 0.24 (3)	1.19 ± 0.02 (3)	1.09 ± 0.01 (3)	1.20 ± 0.05 (3)
State $3U_{DNP}$	202 ± 18 (3)	80 ± 9 (3)	73 ± 11 (3)	96 ± 4 (3)
RCR_{DNP}	2.88 ± 0.30 (3)	1.40 ± 0.11 (3)	1.15 ± 0.18 (3)	1.42 ± 0.06 (3)

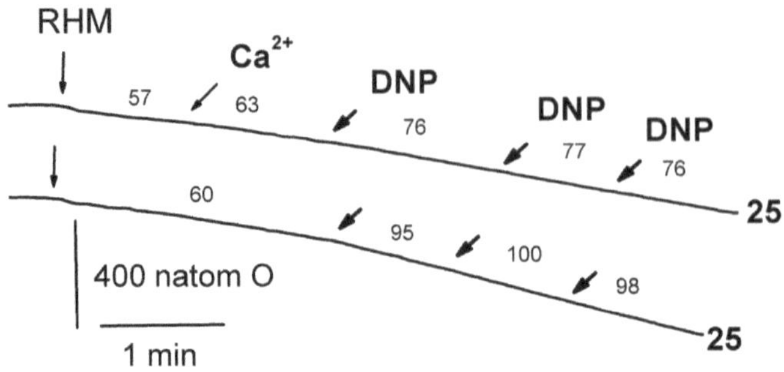

Figure 4.19 Effect of Tl^+ and Ca^{2+} on oxygen consumption rates in succinate-energized rat heart mitochondria. Mitochondria (1 mg/mL) were added into 400 mOsm medium containing 25 mM $TlNO_3$, 100 mM sucrose, 125 mM KNO_3, KNO_3, 5 mM succinate, 2 μM rotenone, and 1 μg/mL oligomycin. Numbers above traces show oxygen consumption rates (n atom O/min per mg of protein). Mitochondrial additions (RHM), 75 μM Ca^{2+} (Ca^{2+}) and 10 μM DNP (DNP) are indicated by arrows. Typical traces are shown for three independent experiments.

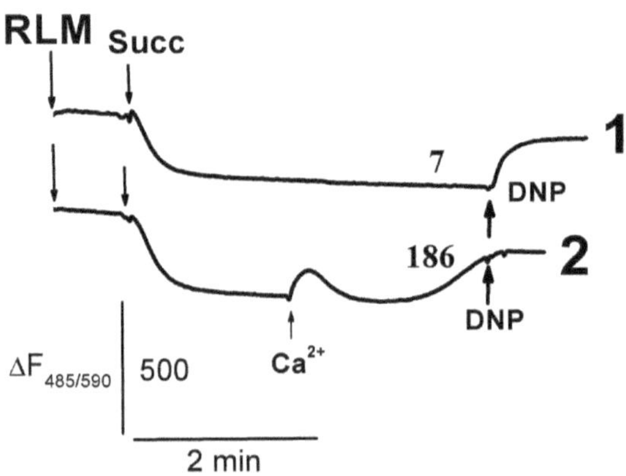

Figure 4.20 Effects of Ca^{2+} on change in safranin O fluorescence (a.u.) in rat liver mitochondria suspension. Mitochondria (0.5 mg/mL of protein) were placed in a medium containing 20 mM $TlNO_3$, 110 mM sucrose, 125 mM KNO_3, 5 mM Tris-NO_3 (pH 7.3), 1 mM Tris-PO_4, 2 μM rotenone, 3 μM safranin O, and 1 μg/mL oligomycin. Arrows show addition of rat liver mitochondria (RLM), 5 mM succinate (Succ), 75 μM Ca^{2+} (Ca^{2+}), and 20 μM DNP (DNP) into the medium. The ordinate shows change in fluorescence ($\Delta F_{485/590}$), the abscissa shows time (min). Control are experiments free of Ca^{2+} (trace 1). Numbers above traces indicate the rate of potential decrease (arb.units/min). Typical traces are shown for three independent experiments.

was that the $\Delta\Psi_{mito}$ decline, states 4, 3, or 3U (DNP- or FCCP-uncoupled) respiration decrease, and the swelling increase was more substantial in the presence of the first complex substrates than succinate. We previously found that concentrations of Cd^{2+} required to inhibit

Table 4.10 Influence of Ca^{2+} on the rate of potential decrease (arb.units/min) of the inner membrane of energized rat liver and heart mitochondria.

Additions after mitochondria (substrate)	RLM		RHM
	Media		Media
	Sucrose	KNO_3	KNO_3
Free of Ca^{2+} (succinate) control		7 ± 2 (6)	27 ± 7 (4)
+ Ca^{2+} (succinate)		186 ± 13 (6)*	97 ± 9 (4)*
Free of Ca^{2+} (G + M) control	13 ± 5 (3)	5 ± 3 (3)	
+ Ca^{2+} (G + M)	165 ± 12 (3)*	121 ± 11 (3)*	

Asterisk (*) indicate significant differences from the control, experiments free of $TlNO_3$ ($P < .05$).

mitochondrial respiration of RLM energized by glutamate and malate in 3 or $3U_{DNP}$ states were lower than those found in similar experiments with succinate (Skul'skiĭ et al., 1988; Korotkov et al., 1998; Belyaeva and Korotkov, 2003).

Similar to Cd^{2+} (Belyaeva and Korotkov, 2003) and Ca^{2+} (Fontaine et al., 1998; Belyaeva et al., 2004b; Baines, 2009), the substrate specificity effect with Tl^+ manifested in that Ca^{2+} induced swelling of RLM, energized by glutamate and malate, was more (Fig. 4.15 and Tables 4.7 and 4.8) than swelling of RLM energized by succinate (Figs. 4.14 and 4.17 and Tables 4.5 and 4.6). It was previously assumed that the respiratory chain complex I play an essential role in the regulation of the activity of the MPTP Ca^{2+}-binding sites (Fontaine et al., 1998; Ichas and Mazat, 1998; Fontaine, Bernardi, 1999; Leverve and Fontaine, 2001; Halestrap and Brenner, 2003; Baines, 2009). A comparison of scientific literature data and our results (Figs. 4.14, 4.15 and 4.17 and Tables 4.5–4.8) suggested that sites located near the respiratory complex I calcium can be involved in the MPTP opening in calcium-loaded RLM in the medium containing $TlNO_3$ and nitrates (KNO_3, NH_4NO_3, $NaNO_3$) (Korotkov and Saris, 2011). This result is in good agreement with our earlier data on the role of these sites in the study of cadmium-induced swelling of rat liver mitochondria (Belyaeva et al., 2004b).

Experiments with isolated mitochondria showed that Tl^+ decreased transport of K^+ ions into mitochondria and their release from the matrix (Barrera and Gomez-Puyou, 1975; Diwan and Lehrer, 1977). However, thallium-induced swelling of Ca^{2+}-loaded succinate-energized RLM was maximal in media containing $TlNO_3$ and KNO_3 compared to the same swelling in media with $TlNO_3$ and NH_4NO_3, or $TlNO_3$ and $NaNO_3$

(Figs. 4.14 and 4.15B). It is possible that much more intense swelling of Ca^{2+} loaded RLM injected into the medium with $TlNO_3$ and KNO_3 can be associated with thallium stimulation of potassium permeability of the inner membrane in energized mitochondria. It should be noted that swelling of energized RLM stimulated by Tl^+ (Figs. 4.14 and 4.15B), Cd^{2+} and Ca^{2+} were always below in sucrose media than swelling in electrolyte media (Korotkov et al., 1998; Belyaeva et al., 2002, 2004b; Belyaeva and Korotkov, 2003).

More substantial, both RLM swelling increase and decreased respiration in states 4, 3, and $3U_{DNP}$ was previously found more in a medium containing Cd^{2+} and Ca^{2+} (Skul'skiĭ et al., 1988; Korotkov and Skul'skiĭ, 1996; Belyaeva et al., 2001, 2004b). The joint action of Cd^{2+} and Ca^{2+} did eliminate after adding into a medium of dithiothreitol or MPTP inhibitors (ADP, CsA, and Mg^{2+}) (Belyaeva and Korotkov, 2003; Belyaeva et al., 2004b). Ions of Tl^+ did not affect noticeably on respiration of mitochondria in 3 or $3U_{DNP}$ states in the sucrose media (Figs. 4.5–4.7 and Table 4.2) due to the lack of noticeable inhibition of mitochondrial respiratory enzymes by this metal (Herman and Bensch, 1967; Barrera and Gomez-Puyou, 1975; Melnick et al., 1976; Woods and Fowler, 1986; Bragadin et al., 2003; Korotkov et al., 2007, 2008a; Korotkov, 2009). However, the respiration decrease did find in calcium-free nitrate media containing $TlNO_3$, 125 mM nitrates (KNO_3, $NaNO_3$, NH_4NO_3), 5 mM succinate, and 2 μM rotenone (Figs. 4.5D and 4.6A and Table 4.2) (Korotkov, 2009, 2013). We hypothesized that such a decrease in respiration of RLM and RHM in 3 or $3U_{DNP}$ states might be due to swelling of these organelles (Korotkov, 2009, 2013). After calcium loading rat liver and rat heart mitochondria, their respiration in states 4, 3, or $3U_{DNP}$ decreased even more in nitrate media with $TlNO_3$ (Figs. 4.18 and 4.19) with noticeable swelling of these organelles (Figs. 4.14 and 4.16) (Korotkov and Saris, 2011; Korotkov et al., 2013).

Correlation between state 4 mitochondrial respiration and the matrix volume found previously (Devin et al., 1996; 1997a,b). At the same time, state 3 respiration of mitochondria injected into hyperosmotic medium containing sucrose or electrolytes was dependent on the functional activity of ANT and respiratory enzymes. So, we can assume that the decrease in states 4, 3, or $3U_{DNP}$ respiration of calcium-loaded RLM added in the medium with Ca^{2+} and nitrates (Figs. 4.18 and 4.19 and Table 4.9) might be related to the respiratory enzymes' activity decrease due to high-amplitude mitochondrial swelling which induces the IMM structural changes (Figs. 4.14 and 4.16 and Tables 4.5—4.8) (Korotkov and Saris, 2011, 2015b) (Korotkov and Saris, 2011). The observed increase in swelling and decrease in both $\Delta\Psi_{mito}$ and mitochondrial respiration in experiments with Tl^{+} and Ca^{2+} suggest that these phenomena are associated with thallium-induced MPTP opening in the inner membrane of these organelles. The next chapter of the study in experiments with Ca^{2+} loaded mitochondria added in media containing $TlNO_3$ and nitrates (KNO_3, NH_4NO_3, $NaNO_3$, $LiNO_3$) presents this assumption evidence.

Summarizing the data described above and analyzing the results of experiments with energized mitochondria, we can conclude that Tl^{+} (like Cd^{2+} and Ca^{2+}), regardless of the type of substrate used, induces the MPTP opening in a low-conductance state (Ichas and Mazat, 1998; Bernardi, 1999). Since increased swelling of these organelles, reduced respiration rates, and the inner membrane potential were observed significantly more in the medium with $TlNO_3$ and nitrates than in the medium with $TlNO_3$ and sucrose (Figs. 4.4—4.10 and Tables 4.5—4.10). However, Tl^{+}-stimulated MPTP opening was possible only in the presence of Ca^{2+} in the medium, in contrast to the action on mitochondria of bivalent heavy metals (Cd^{2+}, Zn^{2+}, Hg^{2+}, Pb^{2+}) and Ag^{+}, where the presence of Ca^{2+} was not necessary. This result may be the fact that single-charge Tl^{+}, in contrast to these bivalent heavy metals' cations and Ag^{+}, is not actively reacting with the IMM calcium sites and thiol groups involved in the MPTP induction due to the weak affinity of Tl^{+} ions to molecular SH groups (Perrin, 1979; Korotkov et al., 2014a). The effects of MPTP inhibitors (ADP, CsA, Mg^{2+}) and MPTP activator (P_i) on swelling, the inner membrane potential, and mitochondrial respiration studied in media containing $TlNO_3$, nitrates (KNO_3, NH_4NO_3, $NaNO_3$, $LiNO_3$) or sucrose. To prove that the Tl^{+}-induced MPTP opening in the inner membrane of calcium-loaded energized mitochondria occurs.

Thus, we have the following summary of this chapter. Both calcium and heavy metals induced the MPT pore opening in the inner mitochondrial membrane. As a result, potent mitochondrial swelling, the release of ions and metabolites from the matrix, and $\Delta\Psi_{mito}$ decline occurred (Zoratti and Szabó, 1995; Ichas and Mazat, 1998). Our study of rat liver and heart mitochondria in a medium with $TlNO_3$ and nitrates (KNO_3, NH_4NO_3, $NaNO_3$, $LiNO_3$) showed that the thallium-induced MPTP opening in the inner membrane was only in experiments with calcium-loaded mitochondria. And this moment markedly distinguishes Tl^{+} from other heavy metals, where the presence of Ca^{2+} was not necessary (Korotkov and

Lapin, 2003; Korotkov and Saris, 2011; Korotkov et al., 2013). Significant swelling of calcium-loaded mitochondria, as well as a decrease in their respiration and the inner membrane potential, resulted in the MPTP opening. The appearance of such a pore in medium with $TlNO_3$ proves a noticeable weakening of these effects when mitochondria are affected by inhibitors of this pore (CsA, ADP, BKA, NEM) (Korotkov and Lapin, 2003; Korotkov et al., 2013, 2015b,c, 2016a; Korotkov and Saris, 2011). These results are in good agreement with Pourahmad et al. (2010). These results are in good agreement with Pourahmad et al. (2010). Experiments with isolated hepatocytes found that MPTP inhibitors (CsA and carnitine) decreased cytotoxic effects of Tl^+, including $\Delta\Psi_{mito}$ decline.

4.4.2 Effects of MPTP inhibitors

Famously, MPTP inhibitors (ADP, CsA, and Mg^{2+}) reduce this pore opening in the inner membrane, but this opening increases in the presence of MPTP modulator (P_i; Zoratti and Szabó, 1995; Ichas and Mazat, 1998; Bernardi, 1999). In our experiments (Fig. 4.21 and Table 4.11) (Korotkov et al., 2008a; Korotkov and Saris, 2011), swelling of calcium-loaded succinate-energized RLM in medium containing $TlNO_3$ and nitrates (KNO_3, NH_4NO_3, $NaNO_3$), or $TlNO_3$ and sucrose decreased markedly in the presence of these inhibitors and increased in the medium with P_i. Interestingly, this swelling stopped after introducing ADP, CsA, Mg^{2+}, RR (mitochondrial calcium uniporter blocker), and EGTA into the medium (Fig. 4.21D and Table 4.11).

The medium, additions and designations are similar to those shown in Fig. 4.21. Absorbance change (ΔA_{540}) is shown 3 min after adding succinate into the medium (Fig. 4.21A and B). Asterisks mark significant differences from control (experiments with Ca^{2+} alone) ($P < .05$).

The MPTP inhibitors (ADP, CsA, Mg^{2+}) did reduce markedly swelling of calcium-loaded RLM energized by glutamate and malate in medium containing $TlNO_3$ along with nitrate (KNO_3, NH_4NO_3, $NaNO_3$, $LiNO_3$) or sucrose (Fig. 4.22 and Table 4.12) (Korotkov and Saris, 2011). This effect persisted after DNP additions (Fig. 4.23).

The medium, additions and notation are similar to those of Fig. 4.22. The table shows the absorption changes (ΔA_{540}) 6 min after additions of mitochondria into the medium (Fig. 4.22). The number of experiments is indicated in brackets. Asterisks show significant differences from control in experiments with Ca^{2+} alone ($P < .05$).

Maximal decrease in swelling of energized and calcium-loaded RLM and RHM occurred in the simultaneous presence of ADP and CsA in the

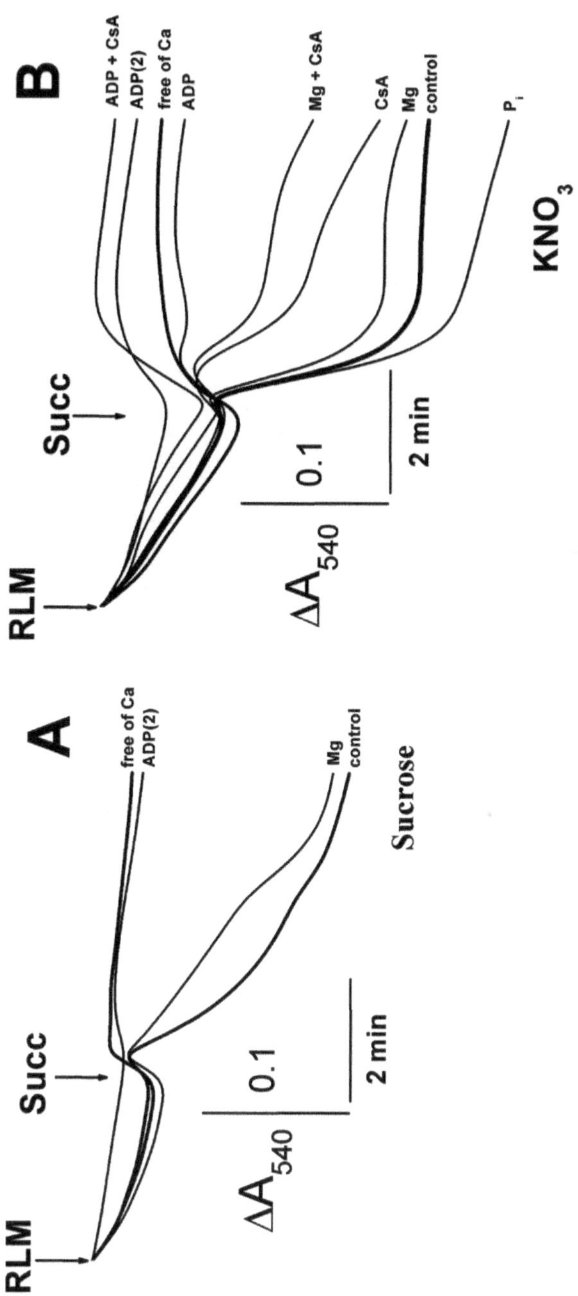
A
RLM
Succ
free of Ca
ADP(2)
Mg
control
Sucrose
ΔA_{540}
0.1
2 min
B
RLM
Succ
ADP + CsA
ADP(2)
free of Ca
ADP
Mg + CsA
CsA
Mg
control
P_i
KNO_3
ΔA_{540}
0.1
2 min

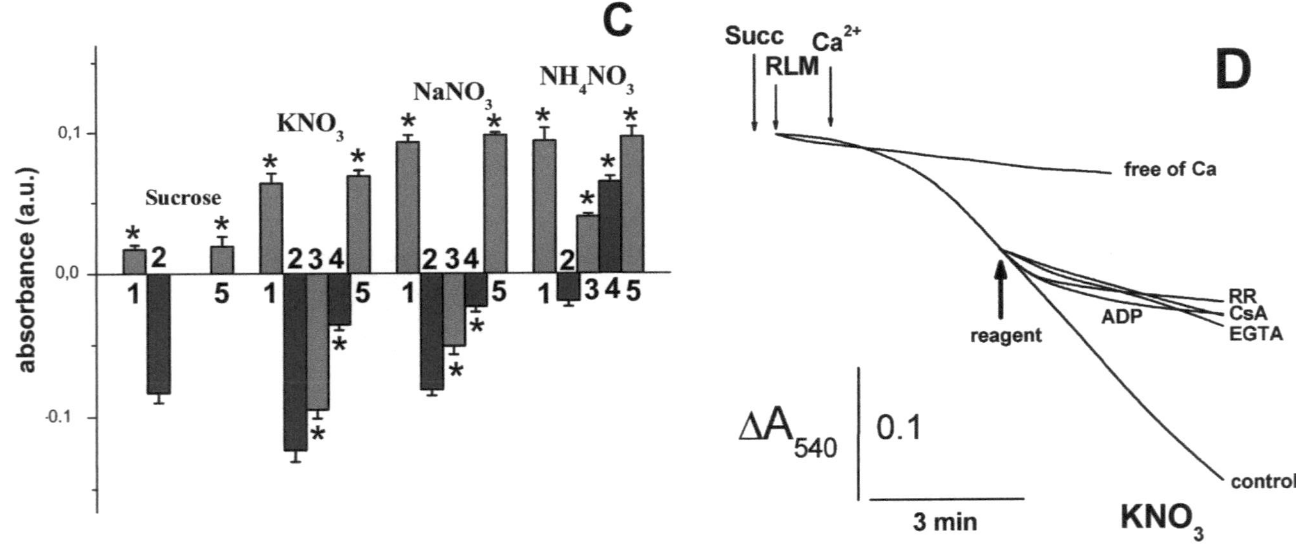

Figure 4.21 Effects of ADP, RR, CsA, EGTA, Mg^{2+} and Pi on swelling of rat liver mitochondria. Mitochondria (1.5 mg/mL of protein) were added into 400 mOsm medium containing 75 mM $TlNO_3$, 100 μM Ca^{2+}, 5 mM Tris-NO_3 (pH 7.3), 2 μM rotenone, and 1 μg/mL oligomycin along with 250 mM sucrose (Sucrose) or 125 mM nitrates (KNO_3, $NaNO_3$, NH_4NO_3). To the right of traces, the next substances were added into the medium before mitochondria: 500 μM ADP (ADP), 2 mM ADP (ADP (2)), 5 mM Mg^{2+} (Mg^{2+}), 7.5 μM ruthenium red (RR), 3 mM Tris-PO_4 (Pi), and 1 μM CsA. The arrows show addition of mitochondria (RLM), 5 mM succinate (Succ), and 100 μM Ca^{2+} (Ca^{2+}). Bold traces show results of experiments with Ca^{2+} alone (control) and no additions (free of Ca). The figure shows typical traces for three independent experiments. Y-axis (C): change in absorbance (ΔA_{540}) 3 min after adding succinate into the medium. Abscissa (C): 1—free of Ca^{2+}, 2—100 μM Ca^{2+} (control). In other cases, along with Ca^{2+} (3–5), 1 μM CsA (3), 500 μM ADP (4), ADP + CsA (5) were added into the medium before mitochondria. Significant differences ($P < .05$) from the control (experiments with Ca^{2+} alone) designated asteriks. The *bold arrow* (panel D) shows the next additions: 500 μM ADP (ADP), 1 μM CsA (CsA), 7 μM ruthenium red (RR), and 1 mM EGTA (EGTA). Each of the traces lines corresponds to one of three independent experiments.

Table 4.11 Change in absorbance (ΔA_{540}) in suspension of succinate-energized calcium-loaded rat liver mitochondria in the presence of MPTP inhibitors.

	Media			
Additions before mitochondria	**Sucrose**	**KNO_3**	**NH_4NO_3**	**$NaNO_3$**
Free of Ca^{2+}	0.017 ± 0.003 (5)*	0.064 ± 0.007 (7)*	0.094 ± 0.009 (5)*	0.093 ± 0.005 (5)*
Control	−0.083 ± 0.007 (5)	−0.123 ± 0.008 (7)	−0.019 ± 0.004 (5)	−0.081 ± 0.009 (5)
Ca^{2+} + RR	0.010 ± 0.004 (3)*	0.071 ± 0.006 (4)*	0.099 ± 0.009 (4)*	0.085 ± 0.006 (4)*
Ca^{2+} + Mg^{2+}	−0.065 ± 0.007 (5)	−0.111 ± 0.008 (5)	0.020 ± 0.004 (3)*	−0.071 ± 0.005 (3)*
Ca^{2+} + CsA	0.020 ± 0.002 (3)*	−0.095 ± 0.006 (5)*	0.040 ± 0.002 (3)*	−0.051 ± 0.006 (3)*
Ca^{2+} + CsA + Mg^{2+}	0.022 ± 0.005 (3)*	−0.034 ± 0.006 (5)*	0.060 ± 0.004 (3)*	−0.033 ± 0.002 (3)*
Ca^{2+} + ADP	0.021 ± 0.004 (3)*	0.022 ± 0.005 (3)*	0.065 ± 0.004 (3)*	−0.023 ± 0.004 (3)*
Ca^{2+} + ADP(2)	0.002 ± 0.004 (3)*	0.034 ± 0.005 (3)*	0.037 ± 0.003 (3)*	0.039 ± 0.004 (3)*
Ca^{2+} + ADP + Mg^{2+}		0.055 ± 0.006 (3)*	0.065 ± 0.004 (3)*	0.051 ± 0.002 (3)*
Ca^{2+} + ADP + CsA	0.019 ± 0.007 (3)*	0.069 ± 0.004 (3)*	0.097 ± 0.007 (3)*	0.098 ± 0.002 (3)*
Ca^{2+} + ADP + CsA + Mg^{2+}	0.021 ± 0.005 (3)*	0.059 ± 0.002 (3)*	0.126 ± 0.008 (3)*	0.110 ± 0.009 (3)*
Ca^{2+} + P_i		−0.184 ± 0.023 (4)*	−0.031 ± 0.002 (4)*	−0.108 ± 0.006 (4)*

Asterisk (*) indicate significant differences from the control, experiments free of $TlNO_3$ ($P < .05$).

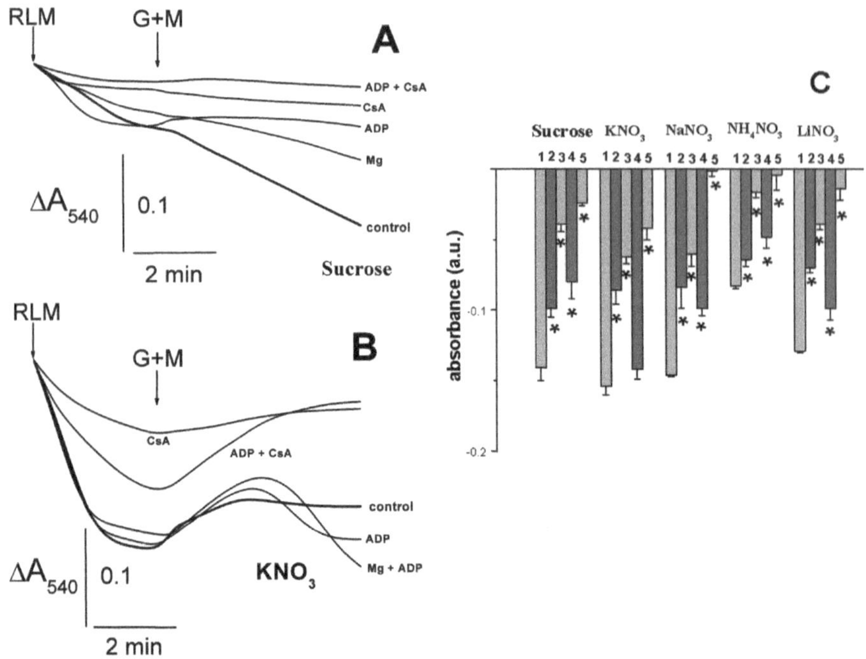

Figure 4.22 Influence of ADP, CsA and Mg^{2+} on swelling of rat liver mitochondria. Mitochondria (1.5 mg/mL of protein) were introduced into 400 mOsm medium containing 50 mM $TlNO_3$, 100 μM Ca^{2+}, 5 mM Tris-NO_3 (pH 7.3) and 1 μg/mL oligomycin along with 250 mM sucrose (Sucrose), or 125 mM nitrate (KNO_3, $NaNO_3$, $LiNO_3$ and NH_4NO_3). Additions and designations are identical to Fig. 4.21. Typical traces for three independent experiments are shown. Y-axis (C): change in absorbance (ΔA_{540}) 6 min after the injection of mitochondria into the medium. On the abscissa: 1—100 μM Ca^{2+} (control). In other cases (3—5), following substances were added into the medium before mitochondria along with Ca^{2+}: 7.5 μM RR (2), 1 μM CsA (3), 500 μM ADP (4), ADP + CsA (5). Asterisks show significant differences from control in experiments with Ca^{2+} alone ($P < .05$).

medium (Figs. 4.21—4.24 and Tables 4.11—4.13) (Korotkov et al., 2015b). In all cases, mitochondrial swelling decreased in the following order: control > CsA > ADP > ADP + CsA.

Media, additions and notation are shown similarly to Fig. 4.24. The change in absorbance (ΔA_{540}) is shown 3 min after succinate additions into the medium (Fig. 4.24). The number of experiments is indicated in brackets. Asterisks show significant differences from the control, experiments with Ca^{2+} alone ($P < .05$).

Decrease in state $3U_{DNP}$ respiration of calcium-loaded RLM energized by succinate (Fig. 4.25 and Table 4.14) or glutamate and malate (Fig. 4.26)

Table 4.12 Change in absorbance (ΔA_{540}) in suspension of calcium-loaded rat liver mitochondria energized by glutamate with malate in the presence of MPTP inhibitors.

	Media				
Additions before mitochondria	**Sucrose**	**KNO_3**	**NH_4NO_3**	**$NaNO_3$**	**$LiNO_3$**
Control	−0.141 ± 0.009 (3)	−0.154 ± 0.006 (3)	−0.083 ± 0.002 (3)	−0.146 ± 0.001 (3)	−0.129 ± 0.001 (3)
Ca^{2+} + RR	−0.099 ± 0.006 (3)*	−0.086 ± 0.010 (3)*	−0.064 ± 0.005 (3)*	−0.084 ± 0.015 (3)*	−0.070 ± 0.003 (3)*
Mg^{2+}	−0.115 ± 0.012 (3)*	−0.123 ± 0.007 (3)*	−0.053 ± 0.005 (3)*	−0.103 ± 0.007 (3)*	−0.103 ± 0.004 (3)*
ADP	−0.080 ± 0.012 (3)*	−0.142 ± 0.007 (3)	−0.048 ± 0.008 (3)*	−0.099 ± 0.005 (3)*	−0.099 ± 0.008 (3)*
ADP + Mg^{2+}	−0.069 ± 0.010 (3)*	−0.132 ± 0.008 (3)	−0.041 ± 0.009 (3)*	−0.089 ± 0.008 (3)*	−0.096 ± 0.006 (3)*
CsA	−0.039 ± 0.005 (3)*	−0.062 ± 0.005 (3)*	−0.016 ± 0.004 (3)*	−0.060 ± 0.009 (3)*	−0.039 ± 0.004 (3)*
CsA + Mg^{2+}	−0.036 ± 0.001 (3)*	−0.059 ± 0.009 (3)*	−0.021 ± 0.004 (3)*	−0.043 ± 0.007 (3)*	−0.044 ± 0.003 (3)*
ADP + CsA	−0.024 ± 0.002 (3)*	−0.042 ± 0.008 (3)*	−0.004 ± 0.011 (3)*	0.001 ± 0.004 (3)*	−0.014 ± 0.008 (3)*
ADP + CsA + Mg^{2+}	−0.017 ± 0.001 (3)*	−0.042 ± 0.006 (3)*	−0.000 ± 0.008 (3)*	0.009 ± 0.005 (3)*	−0.013 ± 0.003 (3)*

Asterisk (*) indicate significant differences from the control, experiments free of $TlNO_3$ ($P < .05$).

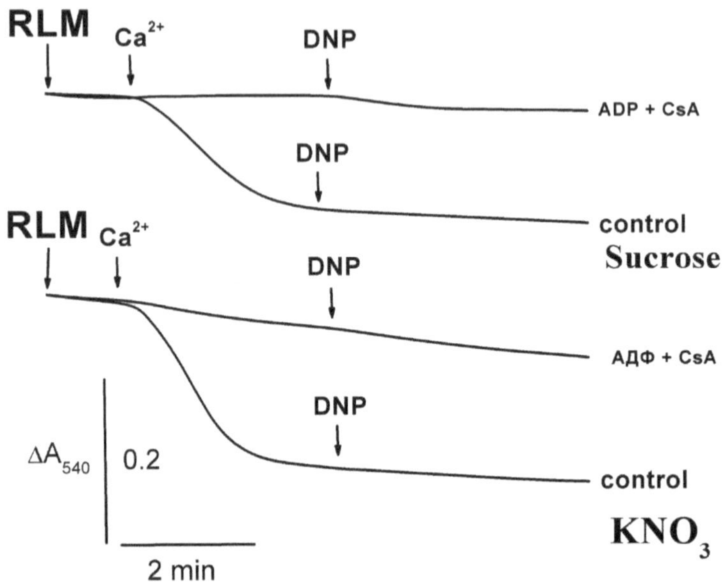

Figure 4.23 Effects of ADP, CsA and DNP on swelling of rat liver mitochondria energized by glutamate and malate. Medium, additions and notation are similar to those in Fig. 4.22. DNP addition was 30 μM. Shown are typical traces from three independent experiments. Similar to experiments with RLM (Fig. 4.21 and Table 4.11), MPTP inhibitors (ADP and CsA) under similar conditions reduced swelling of succinate-energized rat heart mitochondria (RHM) in medium containing $TlNO_3$ and nitrates (Fig. 4.24 and Table 4.13).

was even more noticeable in experiments with P_i (Korotkov and Saris, 2011; Korotkov et al., 2015b). In contrast, the decrease in this respiration was markedly less in the presence of ADP, CsA, Mg^{2+}, and RR compared to the control (experiments free of additions). The maximal protective effect against the calcium-induced decrease in $3U_{DNP}$ respiration in experiments with RLM (Fig. 4.25 and Table 4.14) and RHM (Fig. 4.26) did observe in the presence of ADP and CsA.

The medium, additions, and experimental procedure are similar to those shown in Fig. 4.25. The number of tests carried out is indicated in parentheses. Designations in the table: "4_0"—state 4 respiration before DNP injection into the medium, "+ Ca^{2+}"—respiration after Ca^{2+} adding into the medium, "$3U_{DNP}$"—state $3U_{DNP}$ uncoupled respiration after injection of 30 μM DNP followed the Ca^{2+} addition, "RCR_{DNP}"—the ratio of state $3U_{DNP}$/state 4_0. Asterisks show significant differences ($P < .05$) from the control, experiments with Ca^{2+} alone.

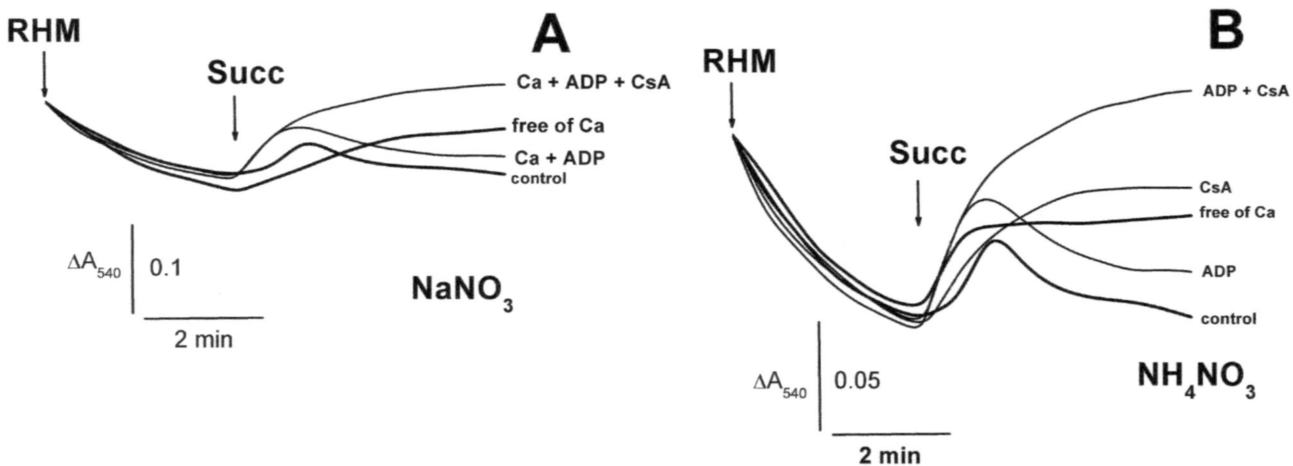

Figure 4.24 Influence of ADP and CsA on swelling of rat heart mitochondria. Mitochondria (1 mg/mL of protein) were injected into 400 mOsm medium containing 25 mM $TlNO_3$, 100 mM sucrose, 200 μM Ca^{2+} (where indicated), 1 mM Tris-PO_4, 2 μM rotenone, and 1 μg/mL oligomycin along with 125 mM nitrates ($NaNO_3$, NH_4NO_3). Arrows indicate addition of mitochondria (RHM) and 5 mM succinate (Succ). To the right of the traces are injected into the medium before mitochondria: 500 μM ADP (ADP), 1 μM CsA (CsA), Ca^{2+} alone (control), and experiments free of additions and calcium (free of Ca). Typical traces are shown for three independent experiments.

Table 4.13 The absorbance change (ΔA_{540}) in suspension of calcium-loaded rat heart mitochondria in the presence of MPTP inhibitors after adding succinate into the medium.

	Media	
75 mM $TlNO_3$	**$NaNO_3$**	**NH_4NO_3**
Free of Ca^{2+}	0.049 ± 0.005 (3)*	0.048 ± 0.004 (5)*
Control	0.002 ± 0.006 (3)	0.004 ± 0.004 (6)
CsA	0.072 ± 0.005 (3)*	0.063 ± 0.003 (3)*
ADP	0.094 ± 0.003 (3)*	0.035 ± 0.003 (3)*
ADP + CsA	0.106 ± 0.006 (3)*	0.094 ± 0.002 (3)*

Asterisk (*) indicate significant differences from the control, experiments free of $TlNO_3$ ($P < .05$).

Calcium-induced $\Delta\Psi_{mito}$ decline noticeably inhibited for rat liver and heart mitochondria energized by succinate or glutamate and malate in the presence of ADP and CsA (Fig. 4.27 and Table 4.15) (Korotkov et al., 2013, 2015b, 2016a,b).

Mitochondria (0.5 mg/mL of protein) were added into 400 mOsm medium containing 20 mM $TlNO_3$ (RLM) or 15 mM $TlNO_3$ (RHM), as well as 5 mM Tris-NO_3 (pH 7.3), 1 mM Tris-PO_4, 2 μM rotenone (where indicated), 3 μM safranin O and 1 μg/mL oligomycin along with 250 mM sucrose (Sucrose) or 125 mM KNO_3 (KNO_3). Before mitochondria (where indicated) 0.5 mM ADP (ADP) and 1 μM CsA (CsA) were added into the medium. After mitochondria injection (as in Fig. 4.27), 5 mM glutamate with 5 mM malate and 40 μM Ca^{2+} (Ca^{2+}) or 5 mM succinate and 75 μM Ca^{2+} (Ca^{2+}) were successively added into the medium. Asterisks show significant differences from control, experiments with Ca^{2+} alone ($P < .05$).

According to results of earlier studies, ADP reduces the affinity of Ca^{2+} ions to calcium sites located on the matrix side, CsA (interacting with CyP-D) prevents its reaction with the inner membrane, and Mg^{2+} and Li^+, acting on this membrane from the outside, inhibit Ca^{2+} transport into the matrix. This finding allowed considering all these substances as MPTP inhibitors (Halestrap and Davidson, 1990; Novgorodov et al., 1994; Zoratti and Szabó, 1995; Ichas and Mazat, 1998; Waldmeier et al., 2002; Halestrap and Brenner, 2003; Shalbuyeva et al., 2006, 2007; Leung et al., 2008; Halestrap, 2009; Zorov et al., 2009).

Cd^{2+} and Ca^{2+} effect on mitochondria was shown to have caused potent of these organelles due to the MPTP opening in their inner membrane (Novgorodov et al., 1994; Bravo et al., 1997; Belyaeva et al.,

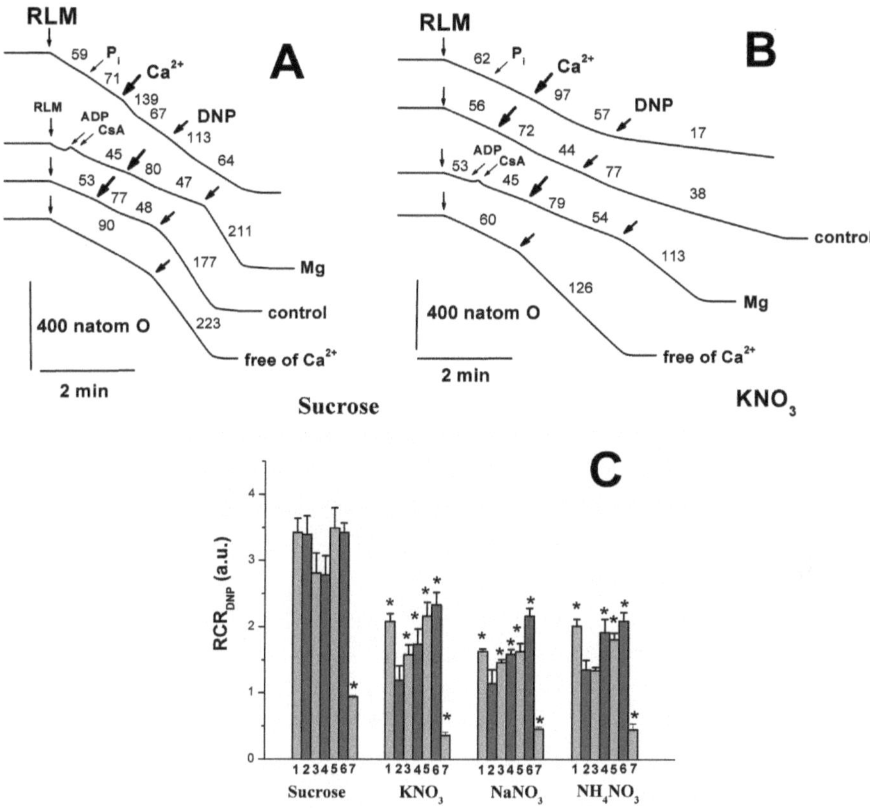

Figure 4.25 Effects of Ca^{2+}, ADP, CsA, Mg^{2+}, RR, and P_i on oxygen consumption rates and RCR_{DNP} value in experiments with succinate-energized rat liver mitochondria. Mitochondria (1.5 mg/mL of protein) were added into 400 mOsm medium containing 75 mM $TlNO_3$, 5 mM succinate, 2 μM rotenone, 3 μg/mL oligomycin along with 250 mM sucrose (sucrose) or 125 mM nitrates (KNO_3, $NaNO_3$, NH_4NO_3). Addition of 5 mM $Mg(NO_3)_2$ (Mg) into the medium before mitochondria is shown to the right of the trace in experiments with ADP and CsA (A, B). Arrows (A, B) show additions after mitochondria of 100 μM Ca^{2+} (Ca^{2+}), 500 μM ADP (ADP), 1 μM CsA (CsA), and 3 mM Tris-PO_4 (P_i). Numbers above traces indicate oxygen consumption rates (n atom O/min per mg of protein). Typical traces for three independent experiments are shown. Numbers on abscissa axis show experiments: free of additions (1) or with Ca^{2+} alone (2). In other cases (3–7), as in panels A and B, after Ca^{2+} addition, following reagents were added into the medium: 1 μM CsA (3), 500 μM ADP (4), ADP + CsA (5), 7.5 μM RR (6), and 3 mM Tris-PO_4 (7). Asterisks show significant differences from the control, experiments with Ca^{2+} alone ($P < .05$).

Table 4.14 Effects of Ca^{2+}, ADP, CsA, Mg^{2+}, RR, and P_i on oxygen consumption rates (n atom O/min per mg of protein) in succinate-energized rat liver mitochondria.

	Media							
	75 mM $TlNO_3$ and 250 mM sucrose				75 mM $TlNO_3$ and 125 mM KNO_3			
Additions	4_0	+ Ca^{2+}	$3U_{DNP}$	RCR_{DNP}	4_0	+ Ca^{2+}	$3U_{DNP}$	RCR_{DNP}
Free of Ca^{2+}	58 ± 2 (9)		197 ± 9 (9)*	3.42 ± 0.22 (9)	59 ± 2 (9)		122 ± 8 (9)*	2.08 * ± 0.12 (9)
Control	54 ± 3 (9)	59 ± 2 (9)	168 ± 11 (9)	3.39 ± 0.29 (9)	56 ± 2 (9)	47 ± 3 (9)	68 ± 13 (9)	1.19 ± 0.22 (9)
Mg^{2+}	54 ± 3 (3)	53 ± 4 (3)	166 ± 25 (3)*	3.09 ± 0.42 (3)	62 ± 2 (3)	57 ± 2 (3)	107 ± 7 (3)*	1.74 * ± 0.15 (3)
CsA	62 ± 9 (3)	50 ± 3 (3)	168 ± 9 (3)	2.81 ± 0.30 (3)	52 ± 4 (3)	46 ± 2 (3)	83 ± 11 (3)	1.58 * ± 0.15 (3)
ADP	63 ± 4 (3)	67 ± 2 (3)	173 ± 14 (3)	2.78 ± 0.29 (3)	58 ± 4 (3)	58 ± 4 (3)	109 ± 14 (3)*	1.74 * ± 0.23 (3)
ADP + Mg^{2+}	47 ± 4 (3)	47 ± 4 (3)*	183 ± 22 (3)	3.65 ± 0.29 (3)	52 ± 3 (3)	48 ± 5 (3)*	91 ± 16 (3)	1.72 * ± 0.24 (3)
ADP + CsA + Mg^{2+}	51 ± 3 (3)	41 ± 4 (3)*	176 ± 25 (3)*	3.49 ± 0.31 (3)	52 ± 2 (3)	43 ± 5 (3)*	93 ± 11 (3)*	2.16 * ± 0.21 (3)
P_i	65 ± 5 (3)	76 ± 3 (3)*	141 ± 10 [72 ± 3] (3)*	0.94 ± 0.02 (3)*	62 ± 1 (3)	59 ± 3 (3)*	22 ± 3 (3)*	0.36 ± 0.05 (3)*
RR	59 ± 3 (3)	59 ± 3 (3)*	201 ± 18 (3)*	3.42 ± 0.15 (3)	57 ± 1 (3)	69 ± 2 (3)*	133 ± 15 (3)*	2.33 * ± 0.19 (3)

Additions	Media							
	75 mM $TlNO_3$ and 125 mM $NaNO_3$				75 mM $TlNO_3$ and 125 mM NH_4NO_3			
	4_0	+ Ca^{2+}	$3U_{DNP}$	RCR_{DNP}	4_0	+ Ca^{2+}	$3U_{DNP}$	RCR_{DNP}
Free of Ca^{2+}	60 ± 2 (8)		98 ± 4 (8)*	1.63 * ± 0.04 (8)	59 ± 2 (8)		117 ± 7 (8)*	2.01 * ± 0.11 (8)
Control	54 ± 3 (9)	51 ± 2 (9)	69 ± 13 (9)	1.14 ± 0.21 (9)	58 ± 2 (8)	55 ± 2 (8)	78 ± 8 (8)	1.35 ± 0.15 (8)
Mg^{2+}	62 ± 1 (3)	46 ± 5 (3)	91 ± 3 (3)*	1.48 * ± 0.06 (3)	59 ± 4 (3)	51 ± 4 (3)	88 ± 4 (3)	1.42 ± 0.19 (3)
CsA	60 ± 2 (3)	46 ± 3 (3)	88 ± 11 (3)	1.46 * ± 0.05 (3)	60 ± 2 (3)	47 ± 4 (3)*	80 ± 5 (3)*	1.34 ± 0.05 (3)
ADP	63 ± 2 (3)	51 ± 3 (3)	101 ± 6 (3)*	1.59 * ± 0.07 (3)	61 ± 3 (3)	60 ± 2 (3)*	117 ± 9 (3)*	1.92 * ± 0.20 (3)
ADP + Mg^{2+}	60 ± 2 (3)	44 ± 4 (3)*	90 ± 6 (3)*	1.52 * ± 0.06 (3)	49 ± 2 (3)	46 ± 1 (3)*	95 ± 4 (3)*	1.84 ± 0.17 (3)
ADP + CsA + Mg^{2+}	53 ± 2 (3)	38 ± 3 (3)*	90 ± 5 (3)*	1.63 * ± 0.12 (3)	46 ± 1 (3)	41 ± 2 (3)*	83 ± 3 (3)	1.81 * ± 0.10 (3)
P_i	66 ± 4 (3)	71 ± 5 (3)	33 ± 2 (3)*	0.49 ± 0.03 (3)*	66 ± 4 (3)	76 ± 4 (3)	30 ± 7 (3)*	0.45 ± 0.09 (3)
RR	65 ± 3 (3)	65 ± 3 (3)	140 ± 13 (3)*	2.16 * ± 0.12 (3)	62 ± 1 (3)	62 ± 1 (3)	130 ± 9 (3)*	2.09 ± 0.13 (3)

Asterisk (*) indicate significant differences from the control, experiments free of $TlNO_3$ ($P < .05$).

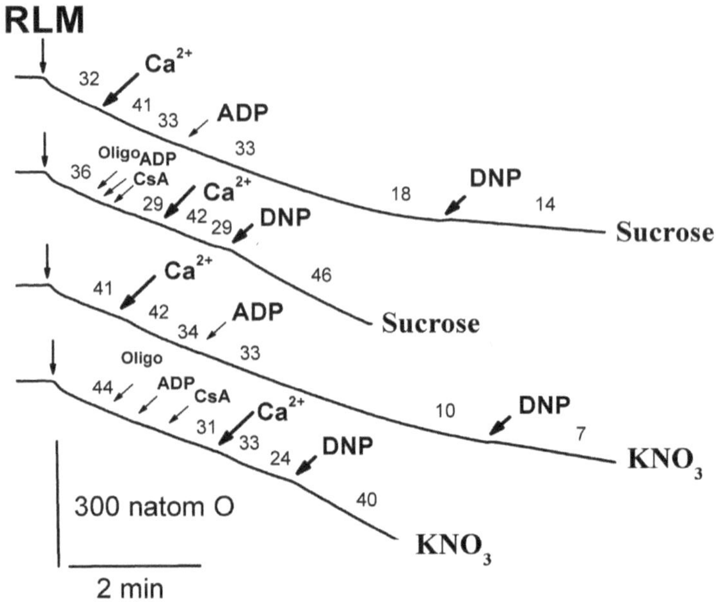

Figure 4.26 Oxygen consumption rates of rat liver mitochondria energized by glutamate and malate in the presence of ADP and CsA. Mitochondria (1.5 mg/mL) were added into 400 mOsm medium containing 75 mM $TlNO_3$, 5 mM glutamate, 5 mM malate, and 3 μg/mL oligomycin along with 250 mM sucrose (Sucrose) or 125 mM KNO_3 (KNO_3). Other additions and designations are as in Fig. 4.25. Typical traces for three independent experiments are shown.

2002, 2004b; Belyaeva and Korotkov, 2003; Dorta et al., 2003). However, this did not occur in the presence of MPTP inhibitors (ADP, CsA, Mg^{2+}), and the effect of these substances was enhanced when they were simultaneously present in the medium. Similarly, MPTP inhibitors (ADP, CsA, Mg^{2+}, Li^{+}) significantly reduced the swelling of calcium-loaded RLM energized with succinate (Fig. 4.21 and Tables 4.5, 4.6, and 4.11) or glutamate and malate (Figs. 4.22 and 4.23 and Tables 4.7 and 4.8) (Korotkov and Saris, 2011). And in this case, the simultaneous presence of ADP and CsA in the medium showed the maximal effect. In the media containing $TlNO_3$ used (Fig. 4.22 and Table 4.12), swelling of calcium-loaded RLM energized by glutamate and malate did more inhibit in experiments with CsA than with ADP. An opposite pattern observed in similar succinate experiments, and ADP showed more effectiveness than CsA (Fig. 4.21 and Table 4.11). This finding may indicate the involvement of calcium-binding sites near the ADP-dependent ANT domain in the opening of thallium-induced MPTP (Halestrap and Brenner, 2003). Increased swelling (Figs. 4.21 and 4.22 and Tables 4.11 and 4.12), decreased

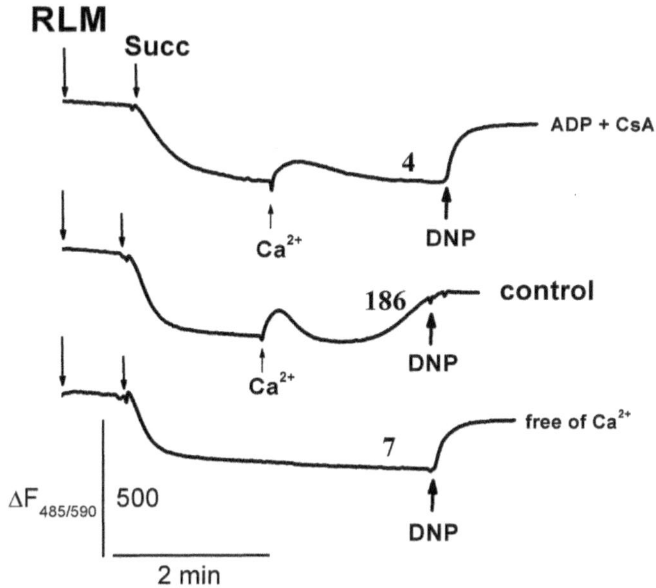

Figure 4.27 Changes of safranin O fluorescence (a.u.) in suspension of rat liver mitochondria. Mitochondria (0.5 mg/mL of protein) were added into 400 mOsm medium containing 20 mM $TlNO_3$, 110 mM sucrose, 125 mM KNO_3, 5 mM Tris-NO_3 (pH 7.3), 1 mM Tris-PO_4, 2 μM rotenone, 3 μM safranin O, and 1 μg/mL oligomycin. Arrows indicate additions of mitochondrial (RLM), 5 mM succinate (Succ), 75 μM Ca^{2+} (Ca^{2+}), and 30 μM DNP (DNP). The ordinate shows change in fluorescence ($\Delta F_{485/590}$), and the abscissa shows time (min). The control is experiments with Ca^{2+} alone. Numbers above the traces show the fluorescence rate change after Ca^{2+} injection (arb.units/min). Typical traces for three independent experiments are shown.

$3U_{DNP}$ state respiration (Figs. 4.25 and 4.26 and Table 4.14), and $\Delta\Psi_{mito}$ decline (Fig. 4.27 and Table 4.15) showed minimal magnitude in experiments with calcium-loaded rat liver mitochondria, injected in the nitrate media with $TlNO_3$ in the presence of MPTP inhibitors (Mg^{2+}, CsA, ADP) (Korotkov and Lapin, 2003; Korotkov and Saris, 2011; Korotkov et al., 2015b). The maximum decrease in the mitochondria swelling in experiments with the simultaneous presence of ADP and CsA (Figs. 4.21 and 4.22 and Tables 4.11 and 4.12) may be associated (Korotkov and Saris, 2011) with some cooperative interactions between various constituent parts of the pore with which these inhibitors interact, namely: CyP-D and the ANT calcium-binding sites (Halestrap and Davidson, 1990; Novgorodov et al., 1992; Halestrap and Brenner, 2003).

The maximal decrease of the succinate-energized RLM swelling in media with $TlNO_3$ (Fig. 4.21 and Table 4.11) (Korotkov and Saris, 2011) did observe in experiments with 2 mM ADP. This result can also explain by

Table 4.15 Effect of Ca^{2+}, ADP, and CsA on decrease of safranin O fluorescence rate (arb.units/min) in suspension of energized rat liver and rat heart mitochondria.

	5 mM glutamate + 5 mM malate		5 mM succinate + 2 μM rotenone	
	Sucrose	KNO_3	KNO_3	
Additions in the medium before mitochondria	**RLM**	**RLM**	**RLM**	**RHM**
Free of Ca^{2+}	13 ± 5 (3)*	5 ± 3 (3)	7 ± 2 (5)*	27 ± 7 (4)*
Control (after additions Ca^{2+})	165 ± 8 (3)	121 ± 11 (3)	186 ± 13 (5)	97 ± 9 (4)
CsA	8 ± 3 (3)*	2 ± 1 (3)		61 ± 5 (3)*
ADP	2 ± 1 (3)*	2 ± 1 (3)		82 ± 7 (3)
ADP + CsA	8 ± 4 (3)*	1 ± 1 (3)	4 ± 2 (5)*	65 ± 11 (3)*

Asterisk (*) indicate significant differences from the control, experiments free of $TlNO_3$ ($P < .05$).

the ability of ADP to reduce the affinity of Ca^{2+} to calcium-binding sites located near the ANT (Halestrap and Davidson, 1990; Halestrap et al., 1997; Halestrap and Brenner, 2003). The above data assume that Tl^{+} can stimulate the opening of CsA-inhibited and ADP-dependent MPTP in the inner membrane of calcium-loaded mitochondria. This phenomenon results in high-amplitude swelling of energized mitochondria in nitrate media, a decrease in their respiration in states 3 and $3U_{DNP}$, as well as a noticeable decrease in $\Delta\Psi_{mito}$.

In experiments with energized mitochondria, inorganic phosphate accelerated the matrix K^{+} entry accompanied by swelling of these organelles and stimulation of their state 4 respiration (Azzi and Azzone, 1966; Barrera and Gomez-Puyou, 1975; Jung and Brierley, 1981; Tedeschi, 1981; Toth et al., 1991). Stimulated by Cd^{2+} ions, state 4 respiration and swelling of mitochondria increased significantly in the presence of phosphate (Koike et al., 1991; Belyaeva et al., 2002, 2004b). The probability of the MPTP opening in a low-conductance state increased in the presence of inorganic phosphate (Ichas and Mazat, 1998; Garlid and Paucek, 2003). For this reason, ionic permeability of the inner membrane increased in experiments with Cd^{2+} (Belyaeva et al., 2002). In our experiments with calcium-loaded RLM, their swelling (Fig. 4.21 and Table 4.11) and decrease in state $3U_{DNP}$ respiration (Fig. 4.25 and Table 4.14) increased in the $TlNO_3$ media containing phosphate. The thallium-induced effects decreased in experiments with MPTP inhibitors (ADP, CsA, Mg^{2+}), but they enhanced in the presence of the phosphate, the pore modulator (Figs. 4.21–4.27 and Tables 4.11–4.15). Thus, these results are essential evidence of the MPTP pore induction by Tl^{+} ions in the inner mitochondrial membrane in experiments with calcium-loaded mitochondria.

4.5 The role of calcium sites and potassium channels of the inner membrane in the Tl^{+}-induced MPTP opening

4.5.1 The Tl^{+}-induced MPTP and calcium sites

According to recent data (Fan et al., 2018), the mitochondrial calcium uniporter consists of two closely interacting protein dimers that form a central pore diameter of about 4.5 Å. Another mechanism of Ca^{2+} transport into the matrix is the fast pathway (RaM), which is similar in many respects to the previous one, but its activation requires much lower Ca^{2+} concentrations. The activity of these channels sharply decreases in the presence of noncompetitive inhibitors of ruthenium red and Ru360 (Paillard et al., 2018), as well as competitive inhibitors of Y^{3+}, La^{3+}, Pr^{3+}, Ga^{3+}, and Ni^{2+} (Lehninger and Carafoli, 1971; Harris and Zaba, 1977; Bragadin and Viola, 1997). The primary way to remove Ca^{2+} from the matrix is the electrogenic Ca^{2+}/Na^{+} exchanger, which catalyzes exchanging the medium three Na^{+} ions for the matrix one Ca^{2+} ion. Another way of pumping Ca^{2+} out of mitochondria is the electrically neutral Ca^{2+}/H^{+} exchanger, which (according to the latest data) catalyzes (under physiological conditions) the exchange of the medium two protons for the Ca^{2+} matrix one ion (De Stefani et al., 2016). At the same time, with a significant calcium overload of mitochondria, the stoichiometry of this exchanger changes due to the MPTP opening in the inner membrane, which we will refer the next section.

Ca^{2+}, Sr^{2+}, Ba^{2+}, and Mn^{2+} ions are known to transport into the matrix via the mitochondrial calcium uniporter (Caplan and Carafoli, 1965; Carafoli, 1965; Lehninger and Carafoli, 1971; Vasington et al., 1972; Reed and Bygrave, 1974a; Hughes and Exton, 1983; Allshire et al., 1985; Lukács and Fonyó, 1985; Zoratti and Szabò, 1995; Bragadin and Viola, 1997; Dong et al., 2009; Gunter and Sheu, 2009; Gunter et al., 2010). The effect of these divalent cations and La^{3+} on the transport of Ca^{2+} into mitochondria depends on their crystal-chemical radius and hydration enthalpy (Reed and Bygrave, 1974a; Tedeschi, 1981; Dubyts'kyĭ and Vovkanych, 2000). Ruthenium red (RR) is this uniporter noncompetitive inhibitor (Vainio et al., 1970; Moore, 1971; Vasington et al., 1972; Reed and Bygrave, 1974b; Heaton and Nicholls, 1976; Saris and Järvisalo, 1977; Dong et al., 2009). Reacting with low-affinity calcium sites located on the cytoplasmic side of the IMM, Sr^{2+}, Ba^{2+}, Mn^{2+}, and Ni^{2+}, as well as Y^{3+}, La^{3+}, Nd^{3+}, Pr^{3+}, and Gd^{3+}, competitively inhibit Ca^{2+} transport into mitochondria (Lehninger and Carafoli, 1971; Moore, 1971; Reed and Bygrave, 1974a,b; Heaton and Nicholls, 1976; Harris and Zaba, 1977; Ligeti et al., 1981; Saris and Bernardi, 1983; Jung and Brierley, 1984; Allshire et al., 1985; Diwan, 1985; Konji et al., 1985; Lukács and Fonyó, 1985; Bragadin and Viola, 1997; Gunter and Sheu, 2009; Tsai et al., 2014). The highest inhibition of Ca^{2+} transport into mitochondria did find in experiments with La^{3+} and RR (Reed and Bygrave, 1974a,b; Dubyts'kyĭ and Vovkanych, 2000; Dong et al., 2009). Interaction of the above cations, as well as Ca^{2+}, Mg^{2+}, H^{+}, and Li^{+} with these sites, reduced the MPTP opening in the inner membrane (Bernardi et al., 1992, 1993; Zoratti and

Szabò, 1995; Ichas and Mazat, 1998; Shalbuyeva et al., 2006, 2007; Halestrap, 2009).

The interaction of Ca^{2+} with Ca^{2+} sites located on the matrix side stimulates the MPTP opening in the inner membrane (Bernardi et al., 1993; Zoratti and Szabò, 1995; Gunter and Sheu, 2009; Halestrap, 2009). Occupation of these sites by Sr^{2+}, Mn^{2+}, and Ba^{2+} ions inhibited the MPTP opening (Bernardi et al., 1992, 1993; Zoratti and Szabò, 1995; Ichas and Mazat, 1998; Halestrap, 2009). The research found that Sr^{2+} promoted ATP-dependent contraction of mitochondria preswollen in the presence of Ca^{2+} or phosphate (Caplan and Carafoli, 1965). Ni^{2+} inhibited calcium-induced swelling of RLM in potassium media (Ligeti et al., 1981). Previously, we found that Sr^{2+}, Mn^{2+}, and RR prevented the decrease in respiration of rat liver mitochondria in 3 or $3U_{DNP}$ states and inhibited the opening of MPTP in the inner membrane in experiments with Cd^{2+} (Skul'skiĭ et al., 1988; Korotkov and Skul'skiĭ, 1996; Belyaeva et al., 2001). Taking these facts into account, we investigated the possible involvement of the IMM external and internal calcium sites in the MPTP induction in experiments with calcium-loaded RLMs in media containing $TlNO_3$.

It can see in Fig. 4.28 that Ca^{2+}-induced swelling of energized rat liver mitochondria markedly reduced in the presence of competitive inhibitors of the mitochondrial Ca^{2+} uniporter (Y^{3+}, La^{3+}, Ni^{2+}) (Korotkov et al., 2014b). The noncompetitive inhibitor of this uniporter (ruthenium red) had a similar effect. Also, Sr^{2+} and Mn^{2+} ions relate to the uniporter competitive inhibitors. The fact is that they can compete with Ca^{2+} for binding with the matrix-faced specific calcium sites of the inner membrane and thus prevent Ca^{2+}-induced MPTP opening (Bernardi et al., 1992). At the same time, these inhibitors significantly prevented the decrease of both respiration in $3U_{DNP}$ state (Fig. 4.29) and $\Delta\Psi_{mito}$ (Fig. 4.30) in experiments with calcium-loaded RLM.

Manganese and nickel salts were found to have weakened cadmium-induced damage to various organs of rat cells and prevented potassium loss by these cells in experiments in vivo (Goering and Klaassen, 1985). It was previously found that inhibition of MPTP by Sr^{2+} and Mn^{2+} results in their interaction with the matrix-facing calcium sites (Drahota et al., 1969; Jurkowitz et al., 1983; Bernardi et al., 1992; Zoratti and Szabò, 1995; Korotkov and Skul'skiĭ, 1996; Belyaeva et al., 2001). Ions of bivalent metals (Sr^{2+}, Ni^{2+}, Ba^{2+}, Mn^{2+}) and RR delayed Ca^{2+} release from the mitochondrial matrix (Rigoni et al., 1980; Ligeti et al., 1981; Jurkowitz et al., 1983; Saris and Bernardi, 1983; Jung et al., 1984; Allshire et al., 1985; Konji

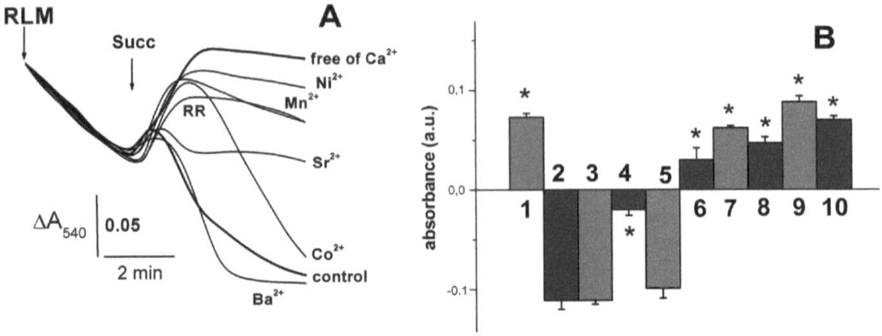

Figure 4.28 Change in absorbance (ΔA_{540}) of rat liver mitochondria suspension in the presence of ruthenium red and bivalent and trivalent metals. Mitochondria (1.5 mg/mL of protein) were injected into 400 mOsm medium containing 75 mM $TlNO_3$, 125 mM KNO_3, 5 mM Tris-NO_3, 2 μM rotenone, 50 μM Ca^{2+}, and 1 μg/mL oligomycin. Arrows indicate addition of mitochondria (RLM) and 5 mM succinate (Succ). To the right of traces (A), 500 μM bivalent metals (Sr^{2+}, Ba^{2+}, Mn^{2+}, Co^{2+}, Ni^{2+}) or 7.5 μM ruthenium red (RR) were added into the medium before mitochondria. Control (A) is experiments with Ca^{2+} alone. Typical traces are shown for three independent experiments. On the ordinate axis (B): the absorbance change (ΔA_{540}) 3 min after adding succinate into the medium, as in panel A. The numbers on the abscissa indicate experiments free of Ca^{2+} (1) and with 50 μM Ca^{2+} (2). In other cases (3–10) along with Ca^{2+}, 5 mM Mg^{2+} (3), 0.5 mM Sr^{2+} (4), 0.5 mM Ba^{2+} (5), 0.5 mM Mn^{2+} (6), 0.5 mM Ni^{2+} (7), 12.5 μM Y^{3+} (8), 12.5 μM La^{3+} (9), and 7.5 μM RR (10). Asterisks indicate significant differences from control (2), experiments with Ca Ca^{2+} alone ($P < .05$).

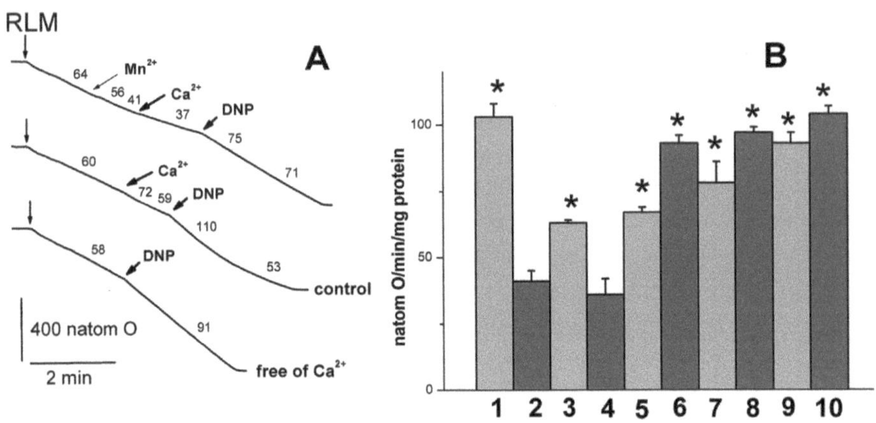

Figure 4.29 Effect of ruthenium red, bivalent and trivalent metal ions on oxygen consumption rates in succinate-energized rat liver mitochondria. The medium is similar to that shown in Fig. 4.28. *Arrows* indicate mitochondrial addition (RLM), 50 μM Ca^{2+} (Ca), 500 μM Mn^{2+} (Mn^{2+}) and 30 μM DNP (DNP). Oxygen consumption rates (n atom O/min per mg protein) are shown by numbers above traces (A), which represent data from three independent experiments. On the ordinate axis (B): oxygen consumption rates of mitochondria being in $3U_{DNP}$ state. Numerals on the abscissa indicate concentration of metals, which are correspondingly similar to those indicated in Fig. 4.28B. Asterisks show significant differences from control, experiments with Ca^{2+} alone ($P < .05$).

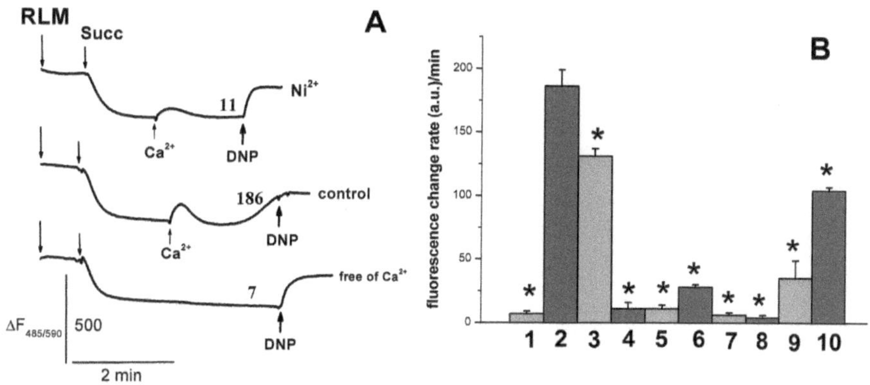

Figure 4.30 Intensity of safranin O fluorescence (a.u.) in rat liver mitochondria suspension in the presence of ruthenium red, ions of divalent and trivalent metals. Mitochondria (0.5 mg/mL of protein) were added into medium containing 20 mM $TlNO_3$, 110 mM sucrose, 125 mM KNO_3, 2 μM rotenone, 3 μM safranin O, and 1 mg/mL oligomycin. *Arrows* show addition of mitochondria (RLM), 5 mM succinate (Succ), 50 μM Ca^{2+} (Ca^{2+}), and 30 μM DNP (DNP). To the right of the traces (panel A) is shown the presence of 200 μM Ni^{2+} (Ni^{2+}) added into the medium before mitochondria. Numbers above traces (A) or along the ordinate axis (B) show the rate (arb.units/min) of the safranin O fluorescence change 1 minute before DNP injection. Typical traces are shown for three independent experiments. The data of experiments free of Ca^{2+} (1) or ones with Ca^{2+} alone (2) are shown by numerals on the abscissa axis, as in panel A. In other cases (3–10): 0.5 mM Ba^{2+} (3), 0.5 mM Mn^{2+} (4), 0.5 mM Ni^{2+} (5), 0.5 mM Co^{2+} (6), 12.5 μM Y^{3+} (7), 12.5 μM La^{3+} (8), 7.5 μM RR (9), or 250 μM EGTA (10) were correspondingly added into the medium before mitochondria and Ca^{2+}. Asterisks show significant differences from the control, experiments with Ca^{2+} alone ($P < .05$).

et al., 1985; Lukács and Fonyó, 1985; Saris and van den Bosch, 1988; Gunter and Sheu, 2009; Gunter et al., 2010).

The MPTP opening in the inner membrane, mitochondrial respiration decrease, and increase in the IMM ionic permeability in experiments with Ca^{2+}, Cd^{2+}, or Ca^{2+} with PAO were less pronounced in the presence of Sr^{2+} and Mn^{2+} (Drahota et al., 1969; Haworth and Hunter, 1979; Bernardi et al., 1992; Zoratti and Szabò, 1995; Skul'skiĭ et al., 1988; Korotkov and Skul'skiĭ, 1996; Ichas and Mazat, 1998; Belyaeva et al., 2001, 2002; Gunter and Sheu, 2009). At the same time, the effect of Mg^{2+} and Ba^{2+} was absent, or it was much weaker in similar experiments. Cd^{2+}-induced release of K^+ ions from isolated rat hepatocytes decreased in the presence of Mn^{2+}, Sr^{2+}, and Mg^{2+} (Stacey and Klaassen, 1981). The MPTP opening in the RLM inner membrane followed by Ca^{2+} release from the matrix was less pronounced in the presence of some divalent metals (Hughes and Exton,

1983). This effect increased in the series $Co^{2+} < Ni^{2+} < Mg^{2+}$, $Ba^{2+} < Mn^{2+}$, Sr^{2+} and it was associated with the interaction of these cations with matrix-facing specific calcium-binding sites (Caplan and Carafoli, 1965; Bernardi et al., 1992; Zoratti and Szabò, 1995; Ichas and Mazat, 1998; Halestrap, 2009). Sr^{2+} increased the contraction of mitochondria pre-swollen in a medium with Ca^{2+} and Pi, blocked tBHP-induced swelling of RLM, and prevented a calcium-induced decrease in $\Delta\Psi_{mito}$ (Caplan and Carafoli, 1965; Bernardi and Pietrobon, 1982; Kushnareva and Sokolove, 2000; García et al., 2009). Uncouplers-stimulated release of Ca^{2+} ions from the matrix decreased in the presence of Ni^{2+} and Ba^{2+} (Ligeti et al., 1981; Lukács and Fonyó, 1985). Sr^{2+} and Mn^{2+} ions prevented an increase in swelling (Fig. 4.28), a decrease in both $\Delta\Psi_{mito}$ (Fig. 4.30) and state $3U_{DNP}$ respiration (Fig. 4.29), which was observed in experiments with calcium-loaded RLM in the medium with $TlNO_3$ and KNO_3.

Inhibition of mitochondrial respiration by Cd^{2+} ions was found to have decreased in the presence of Mn^{2+}, Co^{2+}, Ni^{2+}, and RR (Hunter and Ford, 1955; Korotkov and Skul'skiĭ, 1996; Belyaeva et al., 2001). Inhibition by Ni^{2+} and Co^{2+} ions of an increase in swelling (Fig. 4.28A), a decrease in both $\Delta\Psi_{mito}$ (Fig. 4.30B) and respiration in $3U_{DNP}$ state (Fig. 4.29B) in experiments with calcium-loaded succinate-energized RLM indicates that these ions inhibit the transport of Ca^{2+} into the matrix of these organelles. On the other hand, effects of Cd^{2+}, Ca^{2+} with phosphate, or Ca^{2+} with PAO on mitochondria and the MPTP opening induction in the inner membrane by these agents were unchanged in experiments with Ba^{2+} (Caplan and Carafoli, 1965; Skul'skiĭ et al., 1988; Bernardi et al., 1992; Korotkov and Skul'skiĭ, 1996; Belyaeva et al., 2001). Ba^{2+} did not prevent the MPTP opening in a medium with $TlNO_3$ (Figs. 4.28–4.30) because of the increase in swelling of succinate-energized and calcium-loaded RLM, first, $\Delta\Psi_{mito}$ decline, second, and the decrease of state $3U_{DNP}$ respiration, third, remained the same in similar experiments with Ba^{2+} (Figs. 4.28–4.30). This result can explain by the weaker binding of Ba^{2+} to specific calcium sites on the matrix side (Bernardi et al., 1992).

Depolarization of calcium-loaded RLMs is a prerequisite for the MPTP opening in the inner membrane (Zoratti and Szabò, 1995; Halestrap, 2009). Tl^{+} showed a depolarizing effect in experiments with isolated mitochondria (Saris et al., 1981; Korotkov, 2009). The observed decrease in the found results (Figs. 4.28–4.30) due to the opening of thallium-induced MPTP in the inner membrane of calcium-loaded RLM in experiments with Sr^{2+} and

Mn^{2+} is consistent with similar results, where the presence of these metals in a medium reduced the likelihood of opening the mitochondrial pore induced by valinomycin. FCCP, CCP, PAO or Cd^{2+} (Drahota et al., 1969; Jurkowitz et al., 1983; Bernardi et al., 1992; Korotkov and Skul'skiĭ, 1996; Belyaeva et al., 2001). So, it suggests that the observed inhibition of thallium-induced MPTP in experiments with calcium-loaded RLMs may result from Sr^{2+} and Mn^{2+} competition with Ca^{2+} ions for calcium sites located on the matrix side. Ions of Y^{3+} and La^{3+} stimulated de-energized rat liver mitochondria swelling in calcium-free media containing $TlNO_3$ (Fig. 4.41). This swelling increased with increasing $TlNO_3$ concentration in the medium. The mitochondrial contraction did observe after succinate injection, which significantly decreased in increasing the concentration of $TlNO_3$ in experiments with La^{3+}. The energized mitochondria contraction in a medium with 75 mM $TlNO_3$ (Fig. 4.31B) rose noticeably in the presence of MPTP inhibitors (ADP, CsA, Mg^{2+}). This contraction in the experiments with ADP is the same as that found in experiments free of La^{3+}. Similar results did obtain in experiments with Y^{3+} (Korotkov et al., 2014b).

The binding of La^{3+} to the IMM external and internal calcium sites and this cation affinity to the calcium uniporter was much higher than that of

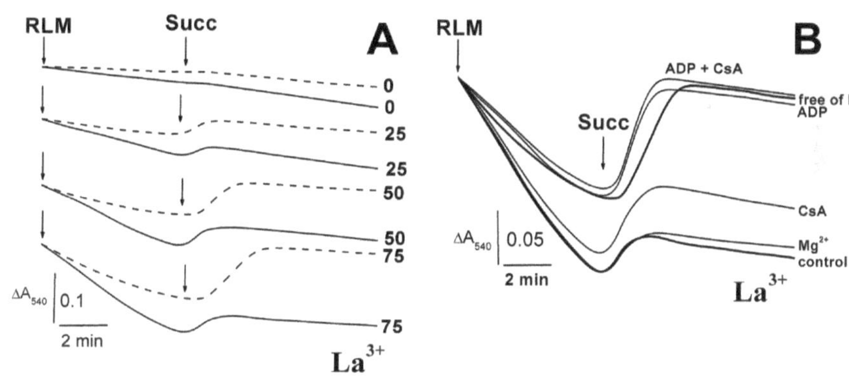

Figure 4.31 Effect of La^{3+} on swelling of rat liver mitochondria. Mitochondria (1.5 mg/mL of protein) were added into 400 mOsm medium containing 0–75 mM $TlNO_3$ (A) or 75 mM $TlNO_3$ (B), as well as 125 mM KNO_3, 5 mM Tris-NO_3 (pH 7.3), 2 μM rotenone, 50 μM La^{3+}, and 1 μg/mL oligomycin. Dotted traces show experiments free of La^{3+}. Numbers near traces show concentration of $TlNO_3$ (mM) in the medium. Arrows indicate addition of rat liver mitochondria (RLM) and 5 mM succinate (Succ). Panel B to the right of traces shows additions before mitochondria into the medium 500 μM ADP (ADP), 5 mM Mg^{2+} (Mg^{2+}), and 1 μM CsA (CsA). Control (B), experiments with La^{3+} alone. Typical traces are shown for three independent experiments.

Ca^{2+} itself (Reed and Bygrave, 1974a). Binding of La^{3+} to the IMM external and internal calcium sites and this cation affinity to the calcium uniporter was much higher than that of Ca^{2+} itself (Reed and Bygrave, 1974b; Saris and Järvisalo, 1977). At the same time, La^{3+} ions did not wholly displace RR from the IMM external sites having high affinity to Ca^{2+} (Reed and Bygrave, 1974b). Effects of Cd^{2+} and Pb^{2+} on rat liver mitochondria were manifested in an increase in swelling and state 4 respiration, in a decrease of respiration in 3 and $3U_{DNP}$ states, as well as in stimulation of Ca^{2+} uptake by energized organelles (Kapoor et al., 1985; Korotkov and Skul'skiĭ, 1996; Korotkov et al., 1998; Belyaeva et al., 2001). The presence of calcium uniporter inhibitors (La^{3+} and RR) significantly reduced these effects of Cd^{2+} and Pb^{2+}. An increase in swelling, a decrease both in mitochondrial respiration in $3U_{DNP}$ state and in $\Delta\Psi_{mito}$ were was much smaller in the presence of Y^{3+}, La^{3+}, Pr^{3+}, Nd^{3+} and Ni^{2+} in our experiments with calcium-loaded RHM (Shemarova et al., 2010, 2013; Korotkov et al., 2016b, 2019). Thus, we have shown a weakening in the effects of Tl^{+} ions on swelling, respiration, and $\Delta\Psi_{mito}$ in experiments with calcium-loaded RLM added into media with $TlNO_3$. An increase in mitochondrial swelling and a decrease in state $3U_{DNP}$ respiration and $\Delta\Psi_{mito}$ were much less pronounced in the presence of competitive inhibitors (Y^{3+}, La^{3+}, Ni^{2+}) and noncompetitive inhibitor RR (Figs. 4.28–4.30). It can be assumed that these results are associated with a decrease in Ca^{2+} uptake into mitochondria, which leads to low saturation of matrix-specific calcium sites and, as a result, reduces the thallium-induced MPTP opening in the inner membrane of calcium-loaded RLM.

La^{3+} increased swelling of nonenergized RLM in a medium with KCNS and stimulated respiration of these organelles in state 4_0 in a sucrose medium (Reed and Bygrave, 1974a). Thus, increased swelling of non-energized RLM did find in experiments with calcium-free medium containing $TlNO_3$, KNO_3, Y^{3+} or La^{3+} (Fig. 4.31A). This finding suggests that this Tl^{+} effect can be associated with Y^{3+} or La^{3+} induction of the IMM passive potassium permeability in adding the similar action of Tl^{+} itself (Figs. 4.11 and 4.12) (Brierley et al., 1970; Korotkov and Saris, 2011; Korotkov et al., 2014b). The interaction of La^{3+} and Gd^{3+} with both external and internal calcium sites of the IMM can induce the MPTP opening accompanied by swelling of mitochondria, a decrease of $\Delta\Psi_{mito}$, and release of cytochrome *c* into the intermembrane space (Reed and Bygrave, 1974a; Dong et al., 2009; Zhao et al., 2014). CsA did inhibit all these effects (Dong et al., 2009; Zhao et al., 2014). Differences in the impact of La^{3+} and Ca^{2+} are due to the higher affinity of La^{3+} to membrane sites responsible for mitochondrial swelling (Reed and Bygrave, 1974a).

We recently found that state 4_0 respiration of rat heart mitochondria and the IMM ion permeability increased in the presence of Ni^{2+} and La^{3+} (Shemarova et al., 2010, 2013). ADP inhibited La^{3+}-induced potassium permeability in CCP uncoupled bovine heart mitochondria (Jung et al., 1984). MPTP inhibitors (ADP, CsA, Mg^{2+}) in the medium with $TlNO_3$ and KNO_3 reduced markedly swelling of succinate-energized mitochondria in experiments with Ca^{2+} alone (Fig. 4.21) or Y^{3+} and La^{3+} free of Ca^{2+} (Fig. 4.31B). It can conclude that Y^{3+} and La^{3+} can induce the pore opening in the RLM inner membrane (Korotkov et al., 2014b). Similar to experiments with Cd^{2+}, the pore opening did not require the presence of Ca^{2+} in the medium (Korotkov and Skul'skiĭ, 1996; Korotkov et al., 1998; Belyaeva et al., 2001). Calcium ions did not additionally induce swelling of energized RLM in medium containing $TlNO_3$ and KNO_3 in the presence of Y^{3+} or La^{3+} (Korotkov et al., 2014b). This circumstance is probably because La^{3+} and Y^{3+} at low concentrations more strongly than Ca^{2+} bound to the IMM external calcium sites. That is why these trivalent cations reduced thereby effects of calcium on mitochondria (Lehninger and Carafoli, 1971; Reed and Bygrave, 1974a,b). If concentrations of La^{3+} and Y^{3+} were high (Fig. 4.31B), they could interact with the matrix side calcium sites and, like Ca^{2+} (Section 2.4.), induce the pore opening in the medium with $TlNO_3$.

Thus, presented results (Figs. 4.28–4.30) showed that inhibitors of the IMM external calcium sites (Li^+, Y^{3+}, La^{3+}, Co^{2+}, Ni^{2+}, RR) preventing Ca^{2+} entry into the matrix, as well as competing for matrix-facing calcium sites Sr^{2+} and Mn^{2+} which reduced the thallium-induced MPTP opening in the inner membrane of succinate-energized and calcium-loaded RLM (Korotkov and Saris, 2011; Korotkov et al., 2014b). Marked inhibition of effects in calcium-loaded mitochondria manifested in the medium with $TlNO_3$ and KNO_3, as a result. These effects were the swelling increase (Fig. 4.28), state $3U_{DNP}$ respiration decrease (Fig. 4.29), and the IMM potential decline (Fig. 4.30). Fig. 4.32 shows schematically interaction of calcium inhibitors under have studied with external and internal calcium sites of the inner mitochondrial membrane in the pore regulatory part containing ANT (Halestrap and Brenner, 2003), but free of PiC and ATP synthase structures (Halestrap, 2010; Halestrap and Richardson, 2015).

Designations: ANT: adenine nucleotide translocase, ETC: electron transport chain, MCU: mitochondrial calcium uniporter, MPU:

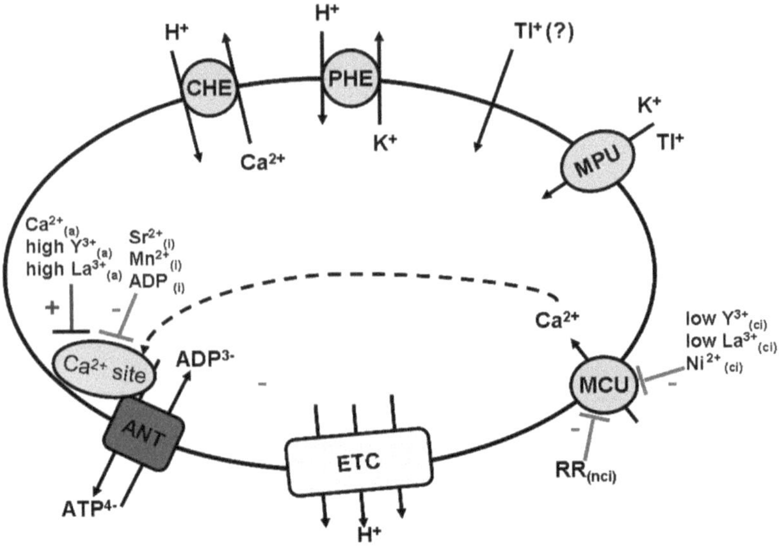

Figure 4.32 Inhibition of the thallium-induced MPTP pore in the inner membrane of rat liver mitochondria by calcium channel blockers.

mitochondrial potassium uniporter, PHE: mitochondrial K^+/H^+ antiporter, CHE: mitochondrial Ca^{2+}/H^+ antiporter, Ca^{2+}-site—calcium site activating MPTP opening, (a) MPTP inducer, (i) MPTP inhibitor, (ci) competitive inhibitor of mitochondrial calcium uniporter, (nci) noncompetitive inhibitor of mitochondrial calcium uniporter.

So, increased mitochondrial swelling and decreased both respiration and $\Delta\Psi_{mito}$, resulting in the Tl^+-induced pore opening in the inner membrane, were noticeably leveled in experiments with calcium uniporter inhibitors (RR, Y^{3+}, La^{3+}, Ni^{2+}), or with Sr^{2+} and Mn^{2+} competing with Ca^{2+} for specific binding sites on the matrix side (Korotkov et al., 2014b). Thus, our experiments revealed an essential role in the calcium sites' occupancy in this pore opening in the inner membrane.

4.5.2 The Tl^+-induced MPTP and potassium channels

One of the results of the mitochondrial calcium load is the opening in their inner membrane of calcium-sensitive potassium channels with high ($mitoBK_{Ca}$), medium (IK_{Ca}), and low (SK_{Ca}) conductance (Laskowski et al., 2016; Krabbendam et al., 2018). Low-molecular-weight protein and α-subunits mainly form these channel structures (Li and Gao, 2016). In addition, they contain β and γ auxiliary subunits that regulate their biophysical and pharmacological properties. MitoBK_{Ca} can modulate by NS1619 and inhibit paxillin (Pax), (+)-tubocurarine, charybdotoxin, and iberiotoxin. At the same time, the main flow of K^+ ions enters mitochondria through ATP-dependent potassium channels (mitoK_{ATP}). These channels consist of four Kir6.x subunits that form the pore channel and four additional proteins that bind to sulfonylurea (Garlid and Halestrap, 2012). The channels are inhibited by glibenclamide and 5-hydroxyodecanoate (5-HD) and modulated by diazoxide (Dia) and pinacidil (Pin).

Earlier studies showed that inhibition of mitoK_{ATP} or mitoBK_{Ca} contributed to calcium accumulation in the mitochondrial matrix (see diagram in Fig. 4.33; Ardehali and O'Rourke, 2005; Kang et al., 2007; Kupsch et al., 2007; Costa and Garlid, 2008). This result was an increase in ROS production by mitochondria and the MPTP opening in their inner membrane. On the other hand, the mitoK_{ATP} modulators (Diazoxide (Dia), pinacidil (Pin), nicorandil) prevented the inner membrane MPTP

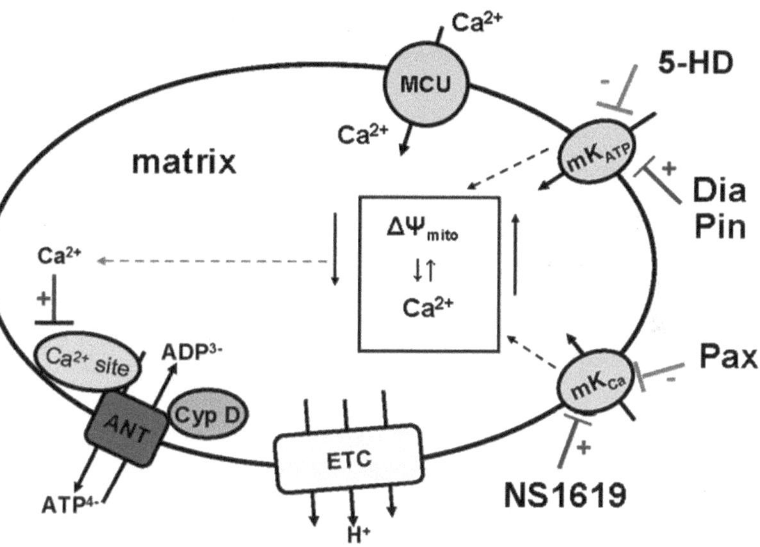

Figure 4.33 Influence of the activity of potassium channels on the inner membrane potential and voltage-dependent calcium uptake in mitochondria. Abbreviations: *5-HD*, 5-hydroxodecanoate; $\Delta\Psi_{mito}$, the inner membrane potential; *ANT*, adenine nucleotide translocase; *Ca^{2+}-site*, MPTP-activating calcium site; *Dia*, diazoxide; *ETC*, electron transport chain; *mK_{ATP}*, mitochondrial ATP-dependent channel; *mK_{Ca}*, mitochondrial Ca^{2+}-activated BK-type potassium channel; *MCU*, mitochondrial calcium uniporter; *Pax*, paxillin; *Pin*, pinacidyl. The pore regulatory part with the ANT is shown (Halestrap and Brenner, 2003).

opening due to the matrix Ca^{2+} uptake decrease (Korge et al., 2002; Costa et al., 2006a; Kupsch et al., 2007; Mio et al., 2014). On the opposite, potassium channels' opening protects mitochondrial functions from calcium overload (Holmuhamedov et al., 1999; Crestanello et al., 2000).

Experiments with isolated mitochondria have shown that transport of Tl^+ into the matrix of energized mitochondria occurs with the direct participation of $mitoK_{ATP}$ (Nikitina and Glazunov, 2003; Wojtovich et al., 2010; Queliconi et al., 2011) and $mitoBK_{Ca}$ (Wojtovich et al., 2013).

The swelling of calcium-loaded rat liver mitochondria in a medium with $TlNO_3$ and KNO_3 (Fig. 4.34) (Korotkov et al., 2015a) increased in the presence of 5-HD ($mitoK_{ATP}$ inhibitor) or paxillin ($mitoBK_{Ca}$ inhibitor). In experiments with diazoxide (Dia) and pinacidil (Pin), this effect decreased. Fig. 4.35 and Tables 4.16 and 4.17 show the effect of inhibitors (5-HD and paxillin) and modulators (Dia, Pin, and NS1619) of these potassium channels on respiration of succinate-energized rat liver and rat heart mitochondria (Korotkov et al., 2013, 2015a). Inhibitors enhanced calcium-induced decrease in state $3U_{DNP}$ respiration of mitochondria. On

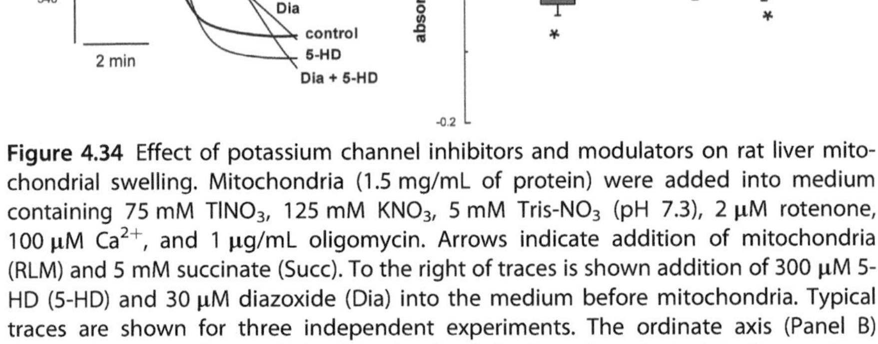

Figure 4.34 Effect of potassium channel inhibitors and modulators on rat liver mitochondrial swelling. Mitochondria (1.5 mg/mL of protein) were added into medium containing 75 mM $TlNO_3$, 125 mM KNO_3, 5 mM Tris-NO_3 (pH 7.3), 2 μM rotenone, 100 μM Ca^{2+}, and 1 μg/mL oligomycin. Arrows indicate addition of mitochondria (RLM) and 5 mM succinate (Succ). To the right of traces is shown addition of 300 μM 5-HD (5-HD) and 30 μM diazoxide (Dia) into the medium before mitochondria. Typical traces are shown for three independent experiments. The ordinate axis (Panel B) shows absorption change (ΔA_{540}) 3 min after injection of succinate into the medium, as in panel A. Numerals on the abscissa indicate the additions into the medium containing 100 μM Ca^{2+} before mitochondria: control (1), 300 μM 5-HD (2), 30 μM Dia (3), Dia + 5-HD (4), 100 μM Pin (5), Pin + 5-HD (6), 30 μM NS1619 (7), 10 μM Pax (8), and NS1619 + Pax (9). Asterisks show significant differences ($P < .05$) from control, experiments with Ca^{2+} alone.

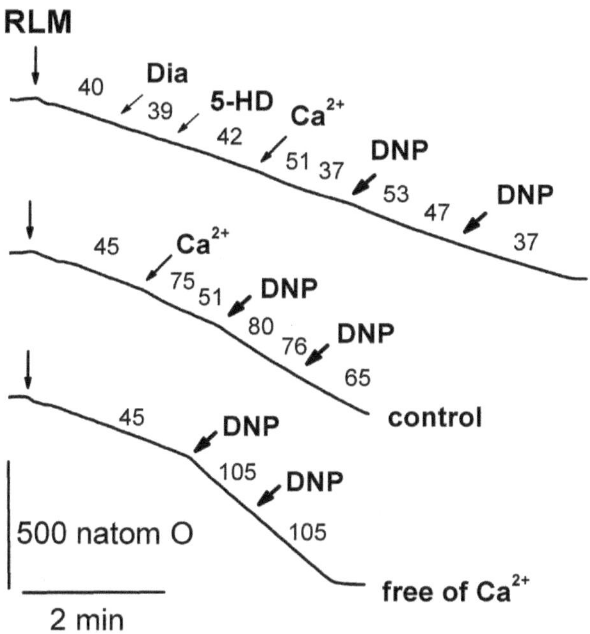

Figure 4.35 Effect of diazoxide, 5-hydroxydecanoate and Ca^{2+} on oxygen consumption rates in succinate-energized rat liver mitochondria. The medium is as in Fig. 4.34, but it was supplemented by 5 mM succinate before mitochondria. Numbers above traces show oxygen consumption rates (n atom O/min per mg protein). After mitochondrial injection, 30 μM Dia (Dia), 300 μM 5-HD (5-HD), 40 μM Ca^{2+} (Ca^{2+}), and 15 μM DNP (DNP) were added into the medium. Typical traces are shown for three independent experiments.

Table 4.16 Effect of inhibitors and modulators of potassium channels on oxygen consumption rates (n atom O/min per mg of protein) and RCR_{DNP} value in succinate-energized rat liver mitochondria.

Supplements after mitochondria	4_0	$3U_{DNP}$	RCR_{DNP}
Free of Ca^{2+}	46 ± 1 (3)	104 ± 4 (3)*	2.23 ± 0.11 (3)*
Ca^{2+}	49 ± 2 (3)	63 ± 5 (3)	1.27 ± 0.12 (3)
Ca^{2+} + 5-HD	53 ± 2 (3)	40 ± 3 (3)*	0.84 ± 0.05 (3)*
Ca^{2+} + Dia	44 ± 3 (3)	57 ± 4 (3)	1.75 ± 0.07 (3)*
Ca^{2+} + Dia + 5-HD	44 ± 2 (3)	67 ± 15 (3)	1.35 ± 0.15 (3)
Ca^{2+} + Pin	43 ± 1 (3)	69 ± 2 (3)	1.56 ± 0.19 (3)
Ca^{2+} + Pin + 5-HD	46 ± 2 (3)	66 ± 10 (3)	1.38 ± 0.08 (3)
Ca^{2+} + NS1619	42 ± 1 (3)	29 ± 6 (3)*	0.62 ± 0.08 (3)*
Ca^{2+} + Pax	46 ± 1 (3)	50 ± 6 (3)*	1.16 ± 0.13 (3)
Ca^{2+} + NS1619 + Pax	42 ± 2 (3)	30 ± 8 (3)*	0.59 ± 0.13 (3)
Dia	44 ± 2 (3)	75 ± 2 (3)#	1.75 ± 0.07 (3)#
Pin	46 ± 1 (8)	104 ± 4 (3)	2.31 ± 0.19 (3)
5-HD	51 ± 1 (7)#	110 ± 4 (3)	2.24 ± 0.17 (3)
NS1619	50 ± 1 (3)#	71 ± 4 (3)#	1.56 ± 0.19 (3)#

Asterisk (*) indicate significant differences from the control, experiments free of $TlNO_3$ ($P < .05$). The hash signs (#) significant differences between experiments in the medium with nitrates from those that were carried out in the medium with sucrose for the same concentration of $TlNO_3$ ($P < .05$).

Table 4.17 Effect of inhibitors and modulators of potassium channels on oxygen consumption rates (n atom O/min per mg of protein) and RCR_{DNP} value in succinate-energized rat heart mitochondria.

Supplements after mitochondria	4_0	$3U_{DNP}$	RCR_{DNP}
Free of Ca^{2+}	65 ± 2 (5)	101 ± 5 (5)*	1.55 ± 0.07 (5)*
Ca^{2+}	56 ± 4 (4)	72 ± 2 (4)	1.25 ± 0.04 (4)
Ca^{2+} + Dia	57 ± 5 (4)	63 ± 4 (4)*	1.32 ± 0.07 (4)
Ca^{2+} + Dia + 5-HD	54 ± 2 (4)	58 ± 4 (4)*	1.19 ± 0.09 (4)
Ca^{2+} + Pin	61 ± 5 (4)	72 ± 6 (4)	1.68 ± 0.18 (4)*
Ca^{2+} + Pin + 5-HD	58 ± 2 (4)	63 ± 5 (4)*	1.23 ± 0.07 (4)

Asterisk (*) indicate significant differences from the control, experiments free of $TlNO_3$ ($P < .05$).

the other hand, Dia and Pin did not significantly affect this mitochondrial respiration, regardless of calcium in the medium.

The medium and additions of Ca^{2+}, ADP and DNP are the same as in Fig. 4.35. Number of experiments performed is indicated in parentheses. Designations in the table: "4_0"—state 4 respiration before ADP addition into the medium, "$3U_{DNP}$"—state $3U_{DNP}$ respiration after 30 μM DNP injection into the medium, "RCR_{DNP}"—RCR_{DNP} is equal the ratio, state $3U_{DNP}$/state 4_0. Asterisks indicate significant differences from the control, experiments with Ca^{2+} alone ($P < .05$). The hash signs shows significant differences between calcium-free experiments with Dia, Pin, 5-HD, and NS-1619 from the experiments free of Ca^{2+} and additions ($P < .05$).

Mitochondria (1 mg/mL of protein) were added into medium containing 25 mM $TlNO_3$, 125 mM KNO_3, 100 mM sucrose, 5 mM Tris-NO_3 (pH 7.3), 5 mM succinate, 3 mM Tris-PO_4, and 3 mM $Mg(NO_3)_2$. After mitochondria, 75 μM Ca^{2+} and 10 μM DNP were sequentially added into the medium. Experimental conditions and additions are as in Fig. 4.35. The number of tests carried out is indicated in parentheses. Asterisks (*) indicate significant differences from the control, experiments with Ca^{2+} alone ($P < 0.05$).

Inhibitors of $mitoK_{ATP}$ (5-HD) and $mitoBK_{Ca}$ (Pax) decreased the matrix calcium retention and increased the calcium-induced decrease in $\Delta\Psi_{mito}$ in rat liver and rat heart mitochondria (Figs. 4.36 and 4.37). The former manifested itself in an increase of concentration of Ca^{2+} (the dye Fluo-5N fluorescence) in mitochondrial suspension due to increased Ca^{2+} release from the matrix (Fig. 4.36) (Korotkov et al., 2015a). Ca^{2+} release from mitochondria in experiments with paxillin was observed already after the first calcium pulse (Fig. 4.36). However, the Ca^{2+} release with 5-HD

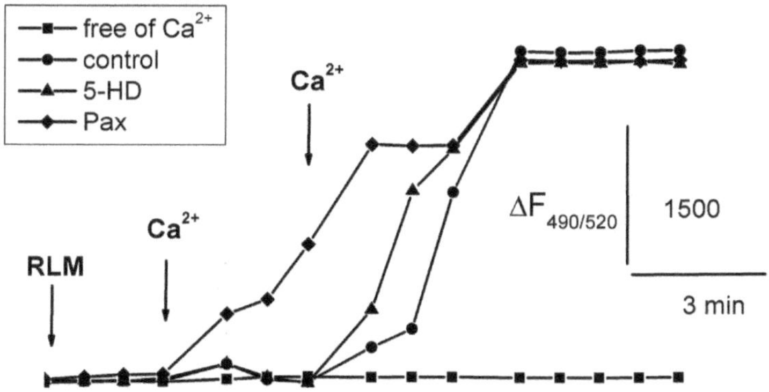

Figure 4.36 Fluorescence change of Fluo-5N ($\Delta F_{480/520}$) in a suspension of succinate-energized RLM in the presence of potassium channel inhibitors (Pax and 5-HD). Mitochondria (0.5 mg/mL of protein) were injected into 400 mOsm medium containing 20 mM $TlNO_3$, 110 mM sucrose, 125 mM KNO_3, 5 mM Tris-NO_3 (pH 7.3), 1 mM Tris-PO_4, 3 μM safranin O, 5 mM succinate, 1 μM Fluo-5N, 2 μM rotenone, and 1 mg/mL oligomycin. The arrows show the addition of mitochondria (RLM) and two additions of 2.5 μM Ca^{2+} (Ca^{2+}).

was after the second pulse but more intensively than in the control one (without additions).

As indicated above, the release of Ca^{2+} from the matrix, a decrease in $\Delta\Psi_{mito}$, and an increase in swelling of mitochondria accompany the MPTP opening in the inner membrane (Zoratti and Szabò, 1995; Ichas and Mazat, 1998; Halestrap, 2009). The complex of 30- and 70-kD succinate dehydrogenase parts with mABC1 protein (mitochondrial ATP-binding cassette protein-1) and proteins involved in the formation of MPTP (ANT, ATP synthase, PiC) synthesized earlier (Ardehali and O'Rourke, 2005). This complex catalyzed the transport of K^+ across lipid bilayers and exhibited characteristics similar to those of $mitoK_{ATP}$. As in the case of mitoKATP, this potassium transport was activated by diazoxide and inhibited by ATP, 5-HD, and glibenclamide (Ardehali and O'Rourke, 2005). $MitoK_{ATP}$ modulators (Dia, ATPenin A5, sodium trioxonitrate) or $mitoBK_{Ca}$ modulator (NS1619) inhibited the ATP-induced decrease of voltage-dependent Tl^+ transport into energized mitochondria (Wojtovich et al., 2010, 2013; Queliconi et al., 2011). However, $mitoK_{ATP}$ inhibitors (5-HD, glibenclamide, fluoxetine) or $mitoBK_{Ca}$ inhibitor (paxillin) eliminated the effect of these modulators in similar experiments.

Increased matrix calcium concentration, decreased oxidative phosphorylation efficiency, and released cytochrome *c* into the intermembrane

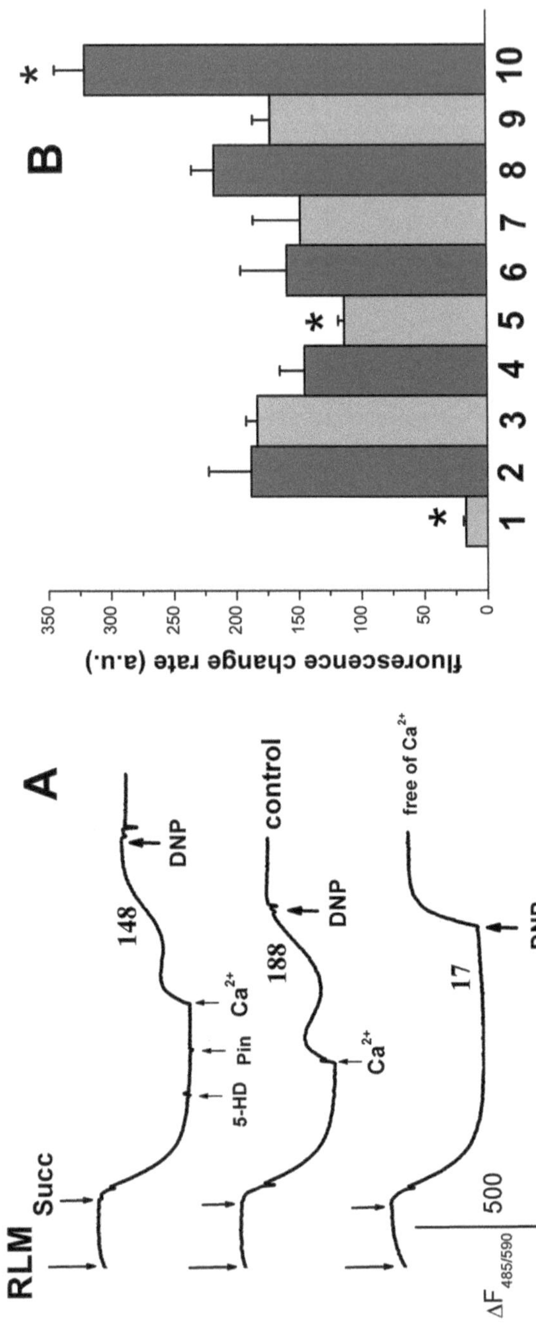
A
RLM
Succ
5-HD
Pin
Ca2+
148
DNP
control
188
17
free of Ca2+
ΔF485/590
500
2 min
B
fluorescence change rate (a.u.)
0
50
100
150
200
250
300
350
1
2
3
4
5
6
7
8
9
10
*

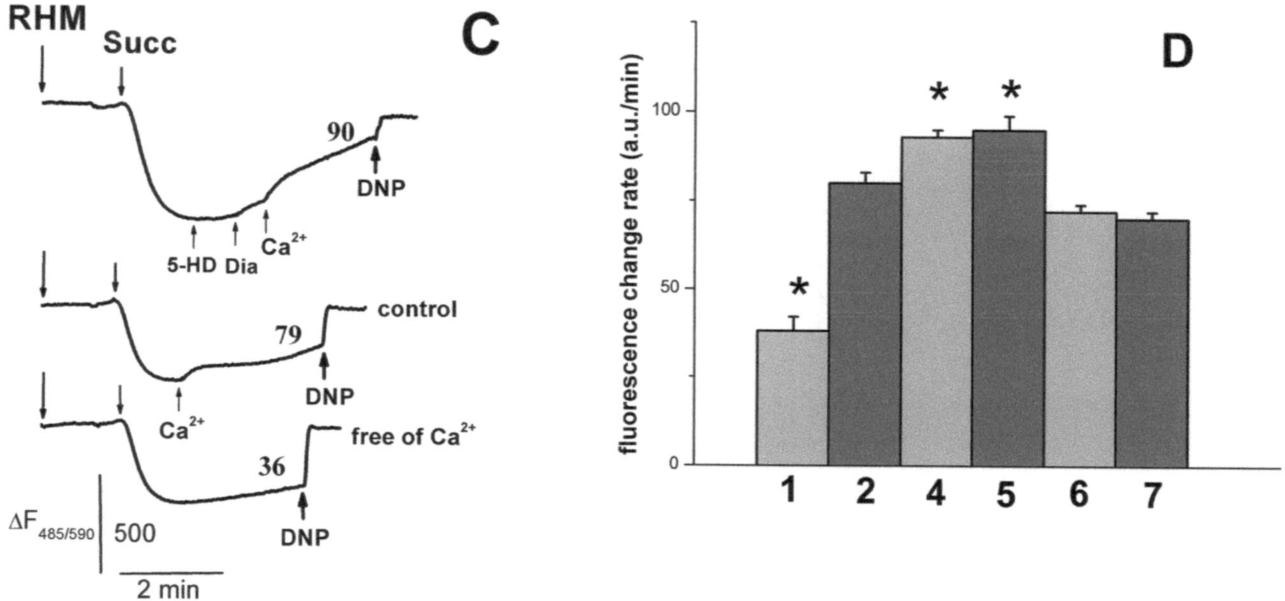

Figure 4.37 Change in safranin O fluorescence ($\Delta F_{485/590}$) in a suspension of succinate-energized rat liver (A, B) and rat heart (C, D) mitochondria in the presence of inhibitors and modulators of mitochondrial potassium channels. Mitochondria (0.5 mg/mL of protein) were injected into 400 mOsm medium containing 20 mM $TlNO_3$ and 110 mM sucrose (A, B), or 15 mM $TlNO_3$ and 120 mM sucrose (C, D) along with 125 mM KNO_3, 5 mM tris-NO_3 (pH 7.3), 1 mM tris-PO_4, 3 μM safranin O, 2 μM rotenone, and 1 mg/mL oligomycin. Arrows show additions rat liver mitochondria (RLM), rat heart mitochondria (RHM), 5 mM succinate (Succ), 35 μM Ca^{2+} (A) or 40 μM Ca^{2+} (C), 300 μM 5-HD (5-HD), 30 μM Dia (Dia), 100 μM Pin (Pin), and 40 μM DNP (DNP). Numbers above traces (A, C) indicate calcium-induced changes in safranin O fluorescence rate (arb.units/min). Typical traces are shown for three independent experiments. The ordinate axis (B, D) shows the fluorescence change rate (arb.units/min.) after adding Ca^{2+} into the medium. Numerals on the abscissa (B, D) mark experiments free of Ca^{2+} (1), or with Ca^{2+} injected into the medium after mitochondria (2). In other cases (3–10) before Ca^{2+} the next reagents were injected as in panels A and C: 300 μM 5-HD (3), 30 μM Dia (4), Dia + 5-HD (5), 100 μM Pin (6), Pin + 5-HD (7), 30 μM NS1619 (8), 10 μM Pax (9), and NS1619 + Pax (10). Asterisks show significant differences ($P < .05$) from the control, experiments with Ca^{2+} alone.

space have a place in control experiments with an isolated rat heart after ischemia/reperfusion (Wang et al., 2001; Pasdois et al., 2007). Diazoxide attenuated these effects, but they became like the control with the mito-K_{ATP} inhibitor (5-HD). MitoK_{ATP} modulators (Dia, Pin, nicorandil) reduced accumulation of Ca^{2+} in the matrix and inhibited in vitro MPTP opening in mitochondria of rat heart, rat brain, and rat liver, as well as N2A cells and myocytes of adult rats (Korge et al., 2002, 2011; Hausenloy et al., 2004; Rousou et al., 2004; Ardehali and O'Rourke, 2005; Costa et al., 2006a,b; Dahlem et al., 2006; Kupsch et al., 2007; Nuñez et al., 2011; Mio et al., 2014). MitoK_{ATP} inhibitors (5-HD, glibenclamide) leveled the effects above modulators, and mitochondria calcium overload became the same as in experiments without modulators. Diazoxide reduced FCCP-induced swelling in succinate-energized and calcium-loaded rat heart mitochondria (Costa et al., 2006b). Effect of diazoxide eliminated by 5-HD, and the mitochondria swelling became the same as in experiments with Ca^{2+} alone. Closure of BK-type calcium-activated potassium channels (mitoBK_{Ca}) promoted MPTP opening in rat brain mitochondria (Cheng et al., 2008).

Our data (Korotkov et al., 2013, 2015a) suggested that if mitoK_{ATP} is closed, the probability of thallium-induced MPTP opening in Ca^{2+} loaded mitochondria will increase compared to control experiments with Ca^{2+} alone. Indeed, the decrease in state $3U_{DNP}$ respiration and $\Delta\Psi_{mito}$ (Figs. 4.35 and 4.36 and Tables 4.17 and 4.18) were more robust in the presence of 5-HD in experiments with calcium-loaded mitochondria than control experiments with Ca^{2+} alone. At the same time, the mitochondria resistance to calcium overload decreased even more (Fig. 4.37), and these organelles' swelling became even more pronounced (Fig. 4.34). These mitochondrial parameters were the same in similar experiments with mitoK_{ATP} modulators (diazoxide and pinacidil) and 5-HD (Figs. 4.34—4.36 and Tables 4.16 and 4.17; Korotkov et al., 2015a).

The medium, additions and notation are identical to Fig. 4.40A. The number of independent experiments performed is indicated in parentheses. The table shows change in absorbance (Fig. 4.40A) 3 min after addition of 100 μM Ca^{2+} into the medium. Number of experiments is indicated in brackets. Asterisks show significant differences from control 2 (experiments with thiol reagents and Ca^{2+}) from the control, experiments with Ca^{2+} alone ($p < 0.05$). The hash signs mark significant differences between experiments with thiol reagents, Ca^{2+} and MPTP inhibitors (ADP, CsA, NEM) from the control 2 ($p < 0.05$). The opening of potassium channels (mitoK_{ATP} and mitoBK_{Ca}) accompanies the $\Delta\Psi_{mito}$ decrease (Fig. 4.33),

Table 4.18 Change in absorbance (ΔA_{540}) in suspension of succinate-energized rat liver mitochondria in the presence of Ca^{2+} and thiol reagents.

Additions before mitochondria	2 μM PAO	50 μM tBHP	50 μM diamide	5 μM DIDS
Control (Ca^{2+} alone)	−0.197 ± 0.006 (7)*	−0.203 ± 0.010 (9)*	−0.209 ± 0.011 (7)*	−0.288 ± 0.007 (3)*
Control 2 (Ca^{2+} + reagent)	−0.255 ± 0.006 (7)*	−0.229 ± 0.011 (8)*	−0.231 ± 0.007 (6)*	−0.314 ± 0.012 (3)*
ADP	−0.078 ± 0.007 (9)#	−0.178 ± 0.015 (5)#	−0.189 ± 0.022 (6)#	−0.132 ± 0.007 (3)#
CsA	−0.248 ± 0.011 (3)	−0.220 ± 0.034 (3)	−0.250 ± 0.003 (3)	−0.276 ± 0.018 (3)#
NEM	−0.157 ± 0.030 (5)#	−0.119 ± 0.027 (3)#	−0.099 ± 0.019 (4)#	−0.235 ± 0.039 (3)#
ADP + NEM	−0.028 ± 0.003 (5)#	−0.023 ± 0.006 (3)#	−0.015 ± 0.002 (3)#	−0.065 ± 0.028 (3)#
CsA + NEM	−0.070 ± 0.015 (4)#	−0.111 ± 0.016 (3)#	−0.064 ± 0.007 (3)#	−0.122 ± 0.008 (3)#
ADP + CsA	−0.032 ± 0.006 (3)#	−0.019 ± 0.005 (3)#	−0.028 ± 0.007 (4)#	−0.015 ± 0.003 (3)#
ADP + CsA + NEM	−0.013 ± 0.002 (3)#	−0.013 ± 0.001 (3)#	−0.019 ± 0.004 (3)#	−0.046 ± 0.011 (4)#

Asterisk (*) indicate significant differences from the control, experiments free of $TlNO_3$ ($P < .05$).
The hash signs (#) significant differences between experiments in the medium with nitrates from those that were carried out in the medium with sucrose for the same concentration of $TlNO_3$ ($P < .05$).

which reduces the driving force of Ca^{2+} transport into the matrix by the energy-dependent uniport system (Korge et al., 2002; Mio et al., 2014). In turn, this prevents mitochondrial calcium overload and reduces MPTP opening in cells and isolated mitochondria (Rousou et al., 2004; Ardehali and O'Rourke, 2005; Dahlem et al., 2006; Cheng et al., 2010; Nishida et al., 2010). This finding may explain some delay in swelling of Ca^{2+}-loaded mitochondria after the succinate addition into the medium containing Dia (Fig. 4.34A).

Another reason for the above phenomenon may be partial inhibition of succinate dehydrogenase by diazoxide (Schäfer et al., 1969; Portenhauser et al., 1971; Lim et al., 2002; Holmuhamedov et al., 2004). Studies have found that diazoxide, regardless of Ca^{2+} in the medium, reduced respiration of succinate-energized rat heart and rat liver mitochondria in states 3 and 3U (uncoupled state) (Lim et al., 2002; Korotkov et al., 2006; Riess et al., 2008). Therefore, the observed decrease in Ca^{2+}-induced swelling of succinate-energized mitochondria or the delay in the contraction of these organelles in a calcium-free medium in experiments with diazoxide (Fig. 4.34A) may associate with succinate dehydrogenase inhibition, which manifests itself as states 3 and $3U_{DNP}$ respiration decrease (Tables 4.16 and 4.17) (Korotkov et al., 2015a).

An uncoupled respiration decrease in heart mitochondria found in the presence of $mitoBK_{Ca}$ channel modulator NS1619 (Heinen et al., 2007a,b, 2008). We found a decrease in swelling (Fig. 4.34B) and state $3U_{DNP}$ respiration (Fig. 4.35 and Table 4.16) in experiments with calcium-loaded RLMs in the presence of NS1619. On the other hand, some activation of RLM respiration in states 4_0 and 4_1 (Fig. 4.35 and Table 4.16) together with a decrease in $\Delta\Psi_{mito}$ (Fig. 4.36A and B) found in experiments with NS1619 could have been resulted from an acceleration of Tl^+ transport through the IMM under participating $mitoBK_{Ca}$ (Tl^+ inflow into the matrix) and mitochondrial K^+/H^+ exchanger (Tl^+ outflow from the matrix) (Korotkov, 2009; Wojtovich et al., 2013).

The opening of MPTP in the inner membrane of rat brain mitochondria increased in the presence of $mitoBK_{Ca}$ inhibitor iberiotoxin since a decrease in $\Delta\Psi_{mito}$, caused by this pore opening in experiments with safranin O, occurred already after three 50 μM Ca^{2+} pulses. In contrast, the opening required four such pulses in the control experiment (free of iberiotoxin) (Cheng et al., 2008). In our experiments, the $mitoBK_{Ca}$ blocker paxillin increased swelling (Fig. 4.34B) and decreased calcium retention of the rat liver mitochondria (Fig. 4.37). This result suggests that

Pax increases the Tl^+-induced MPTP opening because most of mitoBK_{Ca} channels are in a closed state. The decrease in respiration of rat liver mitochondria in $3U_{DNP}$ state and RCR_{DNP} value (Fig. 4.35B and Table 4.16) was even more noticeable in experiments with Pax and NS1619. In this case, the decrease in $\Delta\Psi_{mito}$ (Fig. 4.36) and the increase in swelling (Fig. 4.44) became even stronger after adding Ca^{2+} into the medium compared with the control experiments with Ca^{2+} alone. We found that the closure of potassium channels (mitoK_{ATP} and mitoBK_{Ca}) in the presence of the corresponding inhibitors (5-HD and paxillin) increased the thallium-induced MPTP opening in the inner membrane of calcium-loaded mitochondria. Because the increased mitochondrial swelling, decreased state 3UDNP respiration and the inner membrane potential decline (Figs. 4.34–4.37 and Tables 4.17 and 4.18) became even more significant after injection of Ca^{2+} into the medium than similar experiments without these inhibitors.

Thus, our experiments found that inhibition of mitochondrial ATP-dependent or Ca-stimulated potassium channels contributed to a decrease in the matrix calcium retention and accelerated the Tl^+ -induced MPTP opening in the inner membrane of calcium-loaded rat liver mitochondria (Korotkov et al., 2015a). These data are in good agreement with the results of experiments with isolated mitochondria and cells, in which these inhibitors increased the probability of MPTP opening in the inner membrane (Korge et al., 2002; Costa et al., 2006a,b; Kupsch et al., 2007; Mio et al., 2014).

4.6 The dependence of the Tl^+-induced MPTP opening in the inner membrane of rat liver mitochondria on the adenine nucleotide translocase conformation and the phosphate symporter state

4.6.1 Tl^+-induced MPTP opening and the adenine nucleotide translocase

MPTP is known to be regulated by matrix adenine nucleotides (ADP, ATP), $\Delta\Psi_{mito}$, and ANT ligands (CAT, BKA) (Halestrap and Brenner, 2003; Halestrap, 2010). There is growing evidence that the binding of thiol reagents to SH groups localized near ANT is the main reason why these reagents promote the MPTP opening in the inner membrane (Halestrap et al., 1997, 2002; García et al., 2007). The MPTP opening was concluded to depend on the ANT conformation. The likelihood of this event increased in the presence of oxidative stress inducers (tBHP, diamide), thiol reagent (PAO), or CAT that stabilize this translocase in "c" conformation (Petronilli et al., 1994; Castilho et al., 1996; McStay et al., 2002; Halestrap and Brenner, 2003; Halestrap, 2010; Peng and Jou, 2010). At the same time, an increase in affinity for Ca^{2+} of specific calcium sites located on the matrix side was observed to be due to a decrease in ADP-binding to the IMM regions (located near the translocase) that reduce this affinity. Experiments with MSL and DIDS being monofunctional and bifunctional thiol reagents showed that the ability of the ANT thiol groups to react with these two reagents depends on the ANT conformation (Bernardes et al., 1994, 1997; Lapidus and Sokolove, 1994; Kowaltowski et al., 1997). The effect of these reagents on mitochondria depended on the concentration of extramitochondrial Ca^{2+} (Bernardes et al., 1994; Castilho et al., 1996; Nantes et al., 2011). In contrast, the MPTP opening in the inner membrane was inhibited by ADP, NEM, eosin maleimide, and thiol dimaleimides, which stabilize the ANT "m" conformation (Majima et al., 1993; Petronilli et al., 1994; Kowaltowski et al., 1997; Hashimoto et al., 1999; McStay et al., 2002; Halestrap and Brenner, 2003; Nantes et al., 2011). So, the swelling increase and $\Delta\Psi_{mito}$ decline induced by thiol reagents (tBHP, diamide, arsenite, menadione) in calcium-loaded mitochondria were attenuated by the above substances reducing the affinity of the inner membrane specific calcium sites to Ca^{2+}.

The MPTP opening was shown to accompany high-amplitude swelling after interaction of thiol reagent (Ag^{+}, p-chloromercuriphenyl sulfonate, MSL, PAO) with free thiol groups of the inner membrane proteins (about 11 nmol per mg of protein) (Riley and Lehninger, 1964; Brierley et al., 1968; García et al., 2007). Thallium toxic effect on living organisms was associated with forming various ligand bonds between Tl^{+} and thiol groups of cellular proteins (Mulkey and Oehme, 1993). To open the thallium-induced MPTP was needed to clarify the ANT conformation role. For this purpose, the effects of thiol reagents (PAO, DIDS, tBHP, diamide) and CAT fixing the ANT "c" conformation studied on respiration, swelling, and $\Delta\Psi$mito of calcium-loaded mitochondria injected into medium containing 75 mM $TlNO_3$ and 125 mM KNO_3. One can see in Fig. 4.38A and B that PAO and tBHP free of Ca^{2+} stimulated markedly swelling of succinate energized RLM. If $TlNO_3$ did replace in this medium with an equimolar amount of KNO_3, these effects noticeably decreased, especially

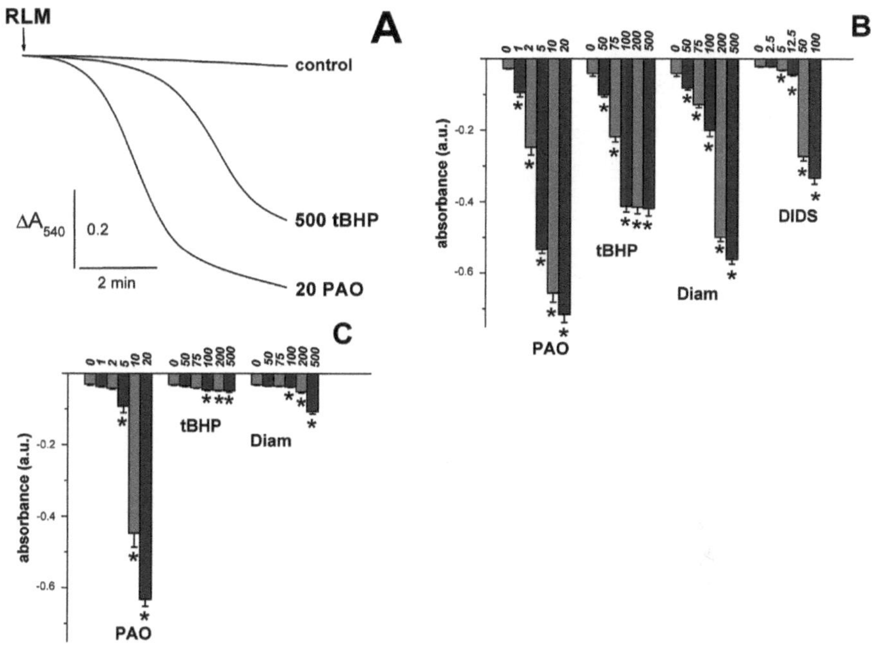

Figure 4.38 The effects of thiol reagents on the absorbance change (ΔA_{540}) in suspension of succinate-energized rat liver mitochondria. Mitochondria (1.5 mg/mL of protein) were added into medium containing 75 mM $TlNO_3$ and 125 mM KNO_3 (A, B) or 200 mM KNO_3 (C) along with 5 mM succinate, 2 μM rotenone, and 1 μg/mL oligomycin. Arrows indicate (A) addition of mitochondria into the medium (RLM). Numbers to the right of the traces indicate concentrations (μM) of reagents (PAO, tBHP) injected into the medium before mitochondria. Typical traces are shown for three independent experiments. The ordinate is the absorbance change 7 min after mitochondria addition into the medium as in panel A. Numbers on the abscissa (B, C) show the concentration of thiol reagents (μM). Asterisks indicate significant differences from the control ("0" on the abscissa) in experiments free of thiol reagents ($P < .05$).

in experiments with tBHP and diamide (Fig. 4.38C) (Korotkov et al., 2015c, 2016a,b).

The Ca^{2+}-induced swelling of energized RLM markedly reduced in experiments with PAO, tBHP, diamide, and DIDS in the presence of ADP and N-ethylmaleimide (Figs. 4.39A, C, and 4.40 and Table 4.18; Korotkov et al., 2015c; 2016a,b), which stabilize the ANT "m" conformation (McStay et al., 2002; Halestrap and Brenner, 2003; Nantes et al., 2011). However, the inhibitory effect of ADP decreased with increasing concentrations of these reagents (Fig. 4.39D). The swelling of succinate-

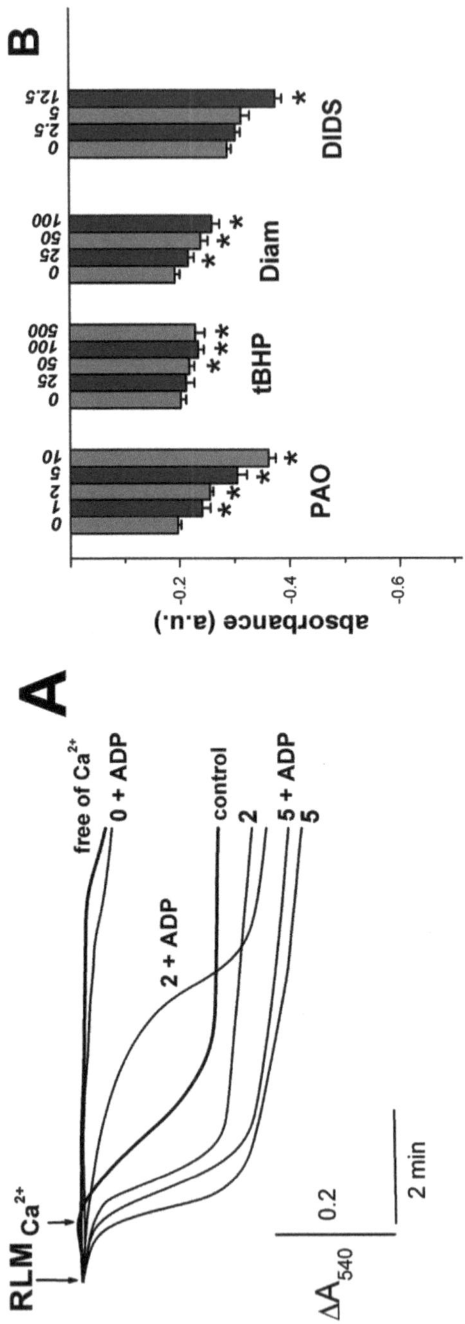
A
RLM
Ca^{2+}
free of Ca^{2+}
0 + ADP
2 + ADP
control
2
5 + ADP
5
ΔA_{540}
0.2
2 min
B
absorbance (a.u.)
-0.2
-0.4
-0.6
PAO
tBHP
Diam
DIDS
0
1
2
5
10
0
25
50
100
500
0
25
50
100
0
2.5
5
12.5

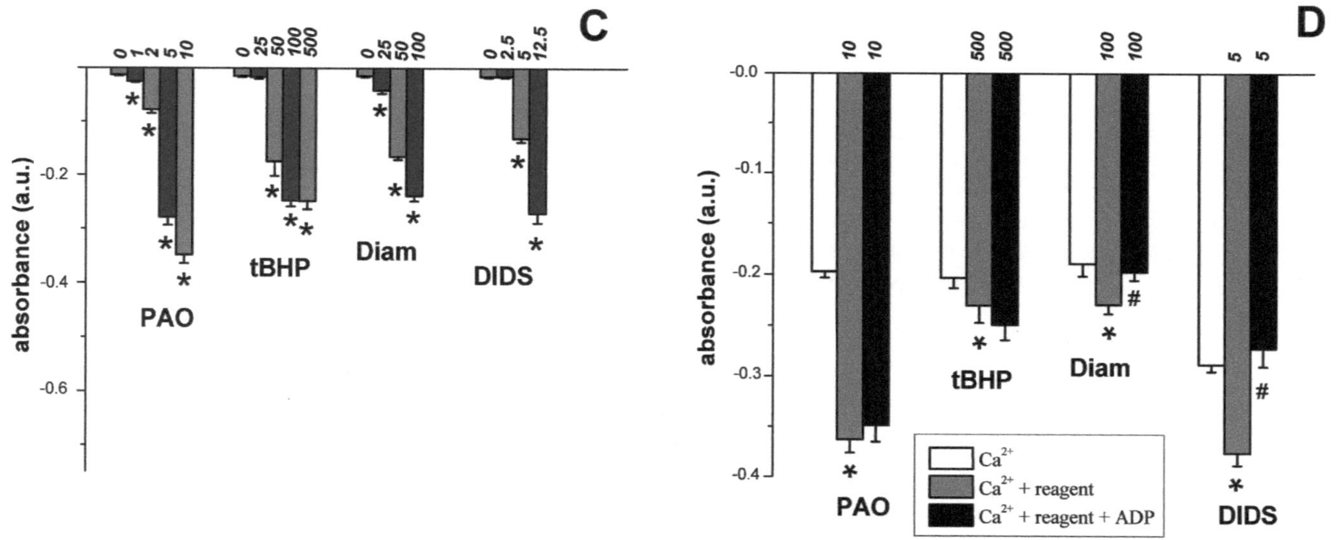

Figure 4.39 Absorbance change (ΔA_{540}) in suspension of succinate-energized rat liver mitochondria in the presence of thiol reagents, Ca^{2+} and ADP. Medium and designations as in Fig. 4.38. Additionally (A, C, D), 500 μM ADP was added into the medium before mitochondria. Arrows (A) show additions of mitochondria (RLM) and 100 μM Ca^{2+} (Ca^{2+}) into the medium. Numbers to the right of traces (A) show the injection of PAO (μM) and 500 μM ADP in the medium before mitochondria. Typical traces are shown for three independent experiments. Control (A–D)—experiments with Ca^{2+} alone. The ordinate axis (B–D) shows the absorbance change 3 min after the addition of Ca^{2+} into the medium as in the panel A. Asterisks (B–D) show significant differences from control, experiments with Ca^{2+} alone ($P < .05$). The number sign (D) marks significant differences between the experiments with Ca^{2+}, ADP, and the reagents from the control.

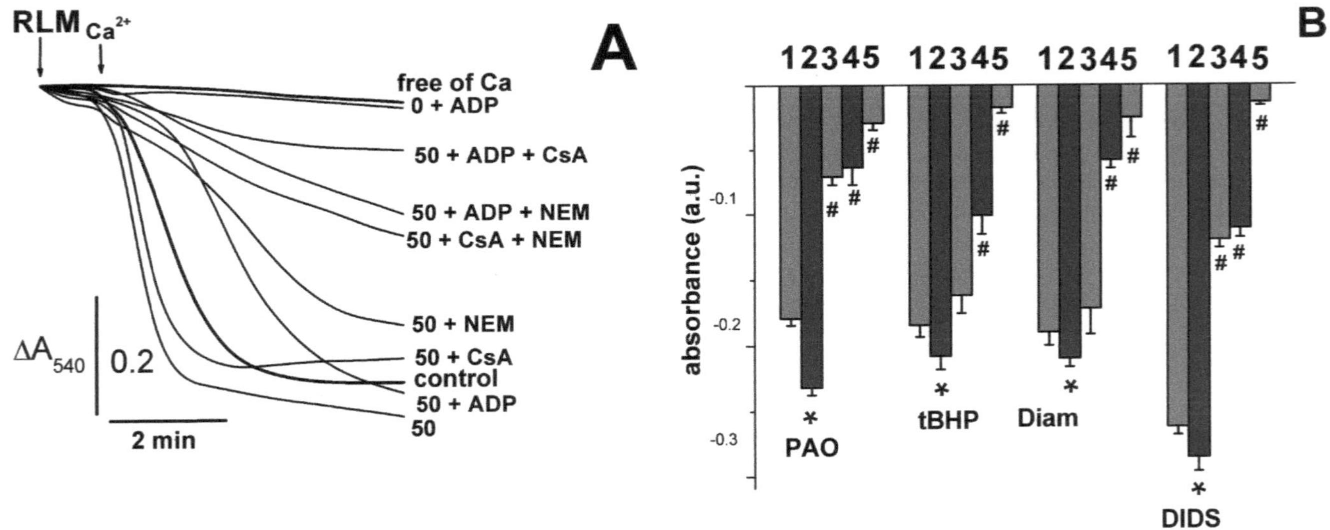

Figure 4.40 Effect of thiol reagents on absorption changes (ΔA_{540}) in a suspension of calcium-loaded and succinate-energized rat liver mitochondria.The medium is as in Fig. 4.49. Arrows (A) show mitochondria injection (RLM) and 100 μM Ca^{2+} (Ca^{2+}) into the medium. Numbers to the right of traces (A) indicate concentration (μM) of tBHP added into the medium before mitochondria. To the right of traces (A), additions before mitochondria are marked: 500 μM ADP (ADP), 1 μM CsA (CsA), and 50 μM N-ethylmaleimide (NEM). Control (A, B) are experiments with Ca^{2+} alone. Before mitochondria (columns 2–5, panel B), the next reagents were added into the medium before mitochondria: 2 μM PAO (PAO), 50 μM tBHP (tBHP), 50 μM diamide (diamide), and 5 μM DIDS. On the ordinate axis: change in absorbance as in panel A 3 min after adding Ca^{2+} to the medium. Numerals on the abscissa axis (B) are Ca^{2+} alone (1), thiol reagent + Ca^{2+} free of inhibitors (2). Columns 3–5 are the same as 2, but 500 μM ADP (3), 50 μM NEM + 1 μM CsA (4), and ADP + CsA (5) were added into the medium before mitochondria. Asterisks show significant differences between experiments with Ca^{2+} along with thiol reagents from the control research ($P < .05$). Typical traces are shown for three independent experiments. Number signs mark significant differences between experiments with Ca^{2+}, thiol reagent and inhibitors from experiments with Ca^{2+} and thiol reagent ($P < .05$).

energized RLM caused by thiol reagents and Ca^{2+} markedly reduced in experiments with NEM and ADP or NEM and CsA (Fig. 4.40 and Table 4.18). This swelling further increased in experiments with high concentrations of NEM (250−500 μM) and CAT (Fig. 4.41 and Table 4.19A), but BKA noticeably inhibited this swelling increase (Table 4.19B). Similarly, ADP, BKA, and NEM inhibited calcium-induced swelling of rat heart mitochondria in a medium containing $TlNO_3$ and NH_4NO_3 (Fig. 4.42).

The medium and procedure for carrying out experiments free of Ca^{2+} shown as in Fig. 4.41, and experiments with 100 μM Ca^{2+} did as in Fig. 4.40. The following reagents were added before mitochondria: 500 μM ADP (ADP), 1 μM CsA (CsA), 50 μM NEM (NEM), 10 μM CAT (CAT) and 10 μM BKA (BKA). For experiments free of Ca^{2+} (A), the change in absorbance (ΔA_{540}) was shown 6 minutes after additions of mitochondria into the medium (Fig. 4.41), and for experiments with Ca^{2+} (A, B), they did 3 minutes after adding Ca^{2+} into the medium (Fig. 4.40). Number of experiments is indicated in brackets. Asterisks (A) indicate significant differences between the results of experiments with 250−500 μM NEM from control, experiments free of NEM and Ca^{2+} ($P < .05$). The hash signs (A) mark significant differences from control 2, experiments with Ca^{2+} alone ($P < .05$). The gralfa signs (@) (B) indicated

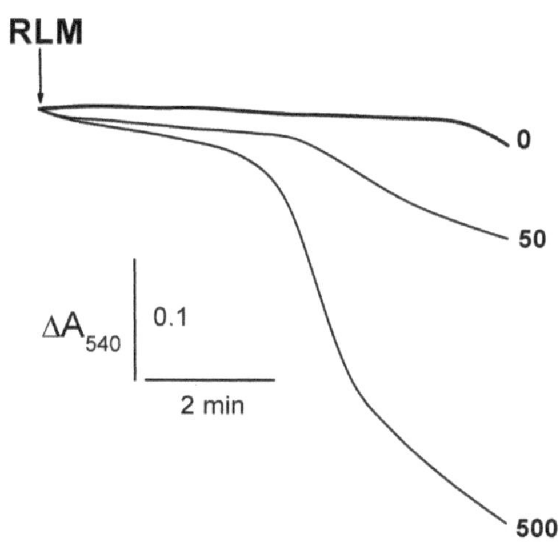

Figure 4.41 Effect of N-ethylmaleimide on swelling of succinate-energized rat liver mitochondria. The medium and designations as in Fig. 4.40. Numbers to the right of traces indicate addition of N-ethylmaleimide (μM) into the medium before mitochondria. Typical traces are shown for three independent experiments.

Table 4.19 Change in absorbance (ΔA_{540}) in suspension of calcium-loaded and succinate-energized rat liver mitochondria in the presence of N-ethylmaleimide, ADP, CsA, BKA, and carboxyatractyloside.

A

Before mitochondria NEM (μM)	Free of Ca^{2+}	+100 μM Ca^{2+}	
	Free of ADP	Free of ADP	+500 μM ADP
0	−0.009 ± 0.001 (3) control	−0.192 ± 0.008 (4) control 2	−0.014 ± 0.002 (4)#
50	−0.093 ± 0.007 (3)*	−0.095 ± 0.009 (3)#	−0.021 ± 0.004 (3)#
250		−0.206 ± 0.006 (3)	−0.220 ± 0.008 (3)
500	−0.298 ± 0.009 (3)*	−0.288 ± 0.007 (4)#	−0.265 ± 0.023 (4)

B

Before mitochondria	+100 μM Ca^{2+}	Before mitochondria	+100 μM Ca^{2+}
Ca^{2+}	−0.196 ± 0.008 (5) control 3	Ca^{2+} + CsA	−0.215 ± 0.015 (4)
Ca^{2+} + ADP	−0.025 ± 0.003 (5)@	Ca^{2+} + ADP + CAT	−0.136 ± 0.026 (4)@
Ca^{2+} + CAT	−0.204 ± 0.005 (4)	Ca^{2+} + BKA	−0.034 ± 0.006 (3)@

Asterisk (*) indicate significant differences from the control, experiments free of $TlNO_3$ ($P < .05$).
The hash signs (#) significant differences between experiments in the medium with nitrates from those that were carried out in the medium with sucrose for the same concentration of $TlNO_3$ ($P < .05$).
The gralfa signs (@) (B) indicated significant differences between experiments with Ca^{2+} and reagents (ADP, CsA, CAT, BKA) from control 3, experiments with Ca^{2+} alone ($P < .05$).

significant differences between experiments with Ca^{2+} and reagents (ADP, CsA, CAT, BKA) from control 3, experiments with Ca^{2+} alone ($P < .05$).

States 3 and $3U_{DNP}$ respiration and respiratory control values (RCR and RCR_{DNP}) became even smaller in the presence of thiol reagents (PAO, tBHP, diamide, DIDS) fixing ANT in "c" conformation as compared to the control experiments in experiments with Ca^{2+} alone (Fig. 4.43 and Table 4.20) (Korotkov et al., 2015c; 2016a,b). In contrast, decrease in $3U_{DNP}$ state respiration of rat liver (Fig. 4.44 and Table 4.21) and rat heart (Fig. 4.45) mitochondria in the experiments with Ca^{2+} was less noticeable in the presence of MPTP inhibitors (ADP and NEM), stabilizing the ANT "m" conformation (Korotkov et al., 2013).

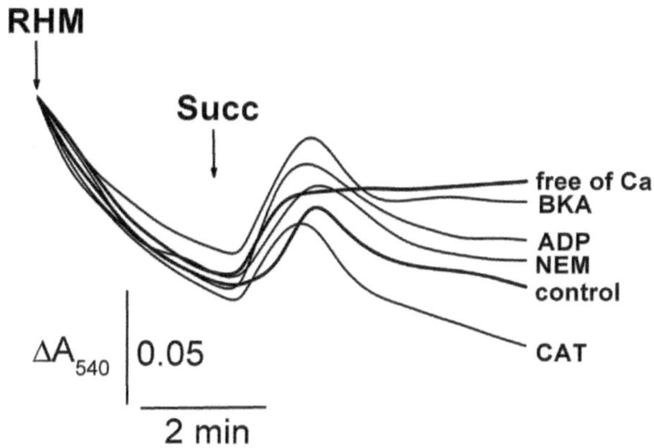

Figure 4.42 Effect of ADP, N-ethylmaleimide, bongkrekic acid and carboxyatractyloside on the swelling of rat heart mitochondria. Rat heart mitochondria (1 mg/mL of protein) were added to a medium containing 25 mM $TlNO_3$, 100 mM sucrose, 3 mM Tris-PO_4, 300 μM Ca^{2+}, 2 μM rotenone, and 1 μg/mL oligomycin. Arrows indicate addition of mitochondria (RHM) and 5 mM succinate (Succ). To the right of the traces are marked additions into the medium before mitochondria: 500 μM ADP (ADP), 10 μM CAT (CAT), 50 μM NEM (NEM), and 10 μM BKA (BKA). Control is experiments with one Ca^{2+} alone. Typical traces are shown for three independent experiments.

Media, additions, and the order of experiments are as in Fig. 4.43. The number of experiments is indicated in brackets. Asterisks show significant differences from control, experiments free of thiol reagents ($P < .05$). The hash signs marks indicate experiments with glutamate and malate energized mitochondria in the presence of DIDS.

Media, additions and the order of experiments are as in Fig. 4.44. The number of experiments is indicated in brackets. The number of experiments is indicated in brackets. Designations in the table: "+ Ca^{2+}"" - state 4_0 respiration after adding of 100 μM Ca^{2+} into the medium, "state $3U_{DNP}$ + Ca^{2+}" - mitochondrial respiration after consistent injection of 100 μM Ca^{2+} and 30 μM DNP into the medium after mitochondria. Asterisks (*) show significant differences from the control, experiments with Ca^{2+} alone and free of thiol reagents ($P < .05$).

Media, additions and the order of experiments are as in Fig. 4.45. After mitochondria and before Ca^{2+} (where indicated), reagents were sequentially added into the medium: 500 μM ADP (ADP), 1 μM CsA (CsA), 50 μM NEM (NEM), 10 μM BKA (BKA), 10 μM CAT (CAT), 75 μM Ca^{2+} (Ca^{2+}) and 15 μM DNP (DNP). Other designations are as in Table 4.21 except for "+ Ca^{2+}"—state 4_0 respiration after adding of 75 μM

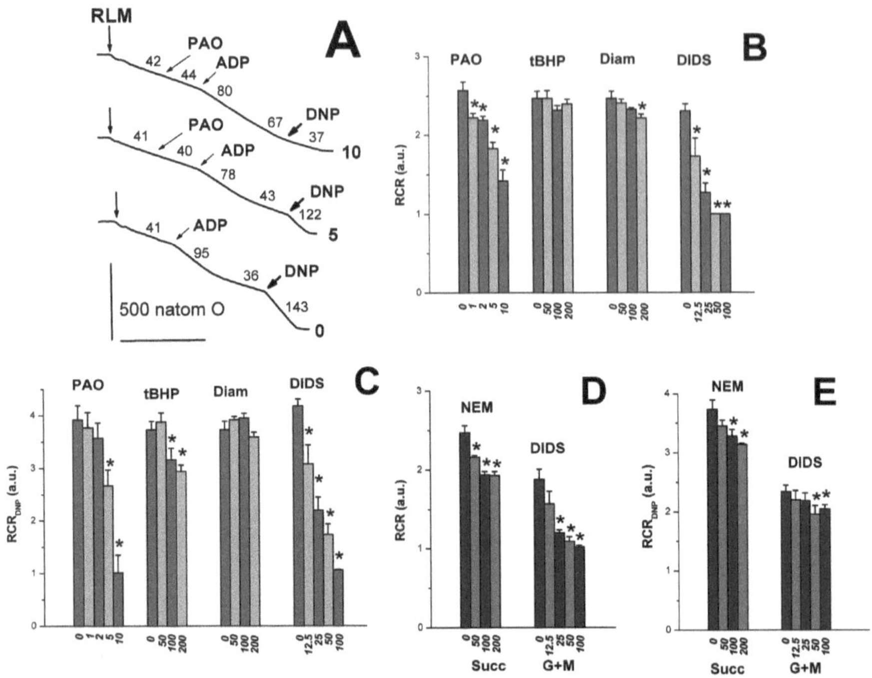

Figure 4.43 Effect of thiol reagents on oxygen consumption rates and RCR and RCR_{DNP} values in experiments with energized rat liver mitochondria. Mitochondria (1.5 mg/mL of protein) were added into 400 mOsm medium containing 25 mM $TlNO_3$, 100 mM sucrose, 125 mM KNO_3, 5 mM Tris-NO_3 (pH 7.3), 3 mM Tris-PO_4, and 3 mM $Mg(NO_3)_2$ along with 5 mM succinate (Succ) and 2 μM rotenone (panels AE) or 5 mM glutamate and 5 mM malate (G + M) (panels DE). *Arrows* indicate mitochondrial addition of mitochondria (RLM), 130 μM ADP (ADP) and 30 μM DNP (DNP). Numbers to the right of traces (A) indicate the PAO addition (μM). Numbers above the traces show oxygen consumption rates (n atom O/min per mg protein). Typical traces for three independent experiments are shown. The ordinates (B, D) and (C, E) show RCR and RCR_{DNP} values. The abscissa axis (B, C) shows concentration of thiol reagents (μM) in the medium. Asterisks show significant differences from control ("0" under the abscissa), experiments free of thiol reagents ($P < .05$).

Ca^{2+} into the medium. Asterisks show significant differences from the control, experiments with Ca^{2+} alone ($P < .05$).

The addition of Ca^{2+} into the medium containing 20 mM $TlNO_3$ reduced $\Delta\Psi_{mito}$ of the RLM (Fig. 4.46A and C, control) (Korotkov et al., 2013, 2016a). This Ca^{2+} effect enhanced significantly in thiol reagents (PAO, tBHP, diamide, DIDS). Still, it decreased substantially in similar experiments with the same reagents in the presence of ADP and CsA (Fig. 4.46A and B). Similar inhibition of the calcium-induced decrease in

Table 4.20 Effect of thiol reagents on oxygen consumption rates (n atom O/min per mg of protein) in energized rat liver mitochondria.

Additions after mitochondria	State 3	State 4	State $3U_{DNP}$
Control	92 ± 2 (3)	37 ± 4 (3)	145 ± 2 (3)
1 μM PAO	84 ± 5 (3)	38 ± 3 (3)	142 ± 12 (3)
2 μM PAO	88 ± 6 (3)	40 ± 3 (3)	143 ± 13 (3)
5 μM PAO	82 ± 6 (3)*	45 ± 1 (3)*	120 ± 17 (3)*
10 μM PAO	82 ± 6 (3)*	59 ± 4 (3)*	58 ± 18 (3)*
Control	97 ± 1 (3)	39 ± 2 (3)	146 ± 3 (3)
50 μM tBHP	89 ± 3 (3)*	36 ± 2 (3)	140 ± 3 (3)
100 μM tBHP	89 ± 2 (3)*	38 ± 2 (3)	122 ± 12 (3)*
200 μM tBHP	87 ± 4 (3)*	37 ± 2 (3)	108 ± 11 (3)*
Control	97 ± 1 (3)	39 ± 2 (3)	146 ± 3 (3)
50 μM diamide	89 ± 2 (3)*	37 ± 2 (3)	145 ± 9 (3)
100 μM diamide	82 ± 2 (3)*	35 ± 1 (3)	138 ± 6 (3)
200 μM diamide	86 ± 2 (3)*	39 ± 1 (3)	140 ± 4 (3)
Control	47 ± 3 (3)	25 ± 2 (3)	58 ± 4 (3)
12.5 μM DIDS#	31 ± 2 (3)*	20 ± 2 (3)	44 ± 4 (3)*
25 μM DIDS#	28 ± 1 (3)*	24 ± 1 (3)	52 ± 6 (3)
50 μM DIDS#	24 ± 1 (3)*	22 ± 2 (3)	43 ± 3 (3)*
100 μM DIDS#	21 ± 1 (3)*	21 ± 2 (3)	43 ± 2 (3)*
Control	69 ± 4 (3)	33 ± 4 (3)	137 ± 10 (3)
12.5 μM DIDS	55 ± 7 (3)*	32 ± 3 (3)	96 ± 3 (3)*
25 μM DIDS	39 ± 4 (3)*	30 ± 1 (3)	66 ± 6 (3)*
50 μM DIDS	30 ± 3 (3)*	30 ± 3 (3)	52 ± 3 (3)*
100 μM DIDS	26 ± 2 (3)*	26 ± 2 (3)	27 ± 2 (3)*
Control	97 ± 1 (3)	39 ± 2 (3)	146 ± 3 (3)
50 μM NEM	88 ± 2 (3)*	41 ± 1 (3)	140 ± 2 (3)
100 μM NEM	79 ± 1 (3)*	41 ± 1 (3)	133 ± 2 (3)*
200 μM NEM	65 ± 2 (3)*	34 ± 1 (3)*	106 ± 2 (3)*

Asterisk (*) indicate significant differences from the control, experiments free of $TlNO_3$ ($P < .05$).

$\Delta\Psi_{mito}$ found in experiments with succinate-energized RHM in the presence of ADP and bongkrekic acid (Fig. 4.46C and D) stabilizing the ANT "m" conformation.

The tBHP in calcium-free media was found previously to have induced membrane permeability due to oxidation or crosslink of adjacent thiol groups, and also increased mitochondrial swelling, decreased both $\Delta\Psi_{mito}$ and oxidation of pyridine nucleotides, glutathione, and NADH (Bindoli et al., 1997; Salvi et al., 2007; Singh et al., 2011). Interaction of DIDS with membrane proteins increased the IMM ion permeability (Bernardes et al., 1994; Nantes et al., 2011). It was previously assumed that DIDS stimulation of the ionic permeability of the inner membrane could occur due to

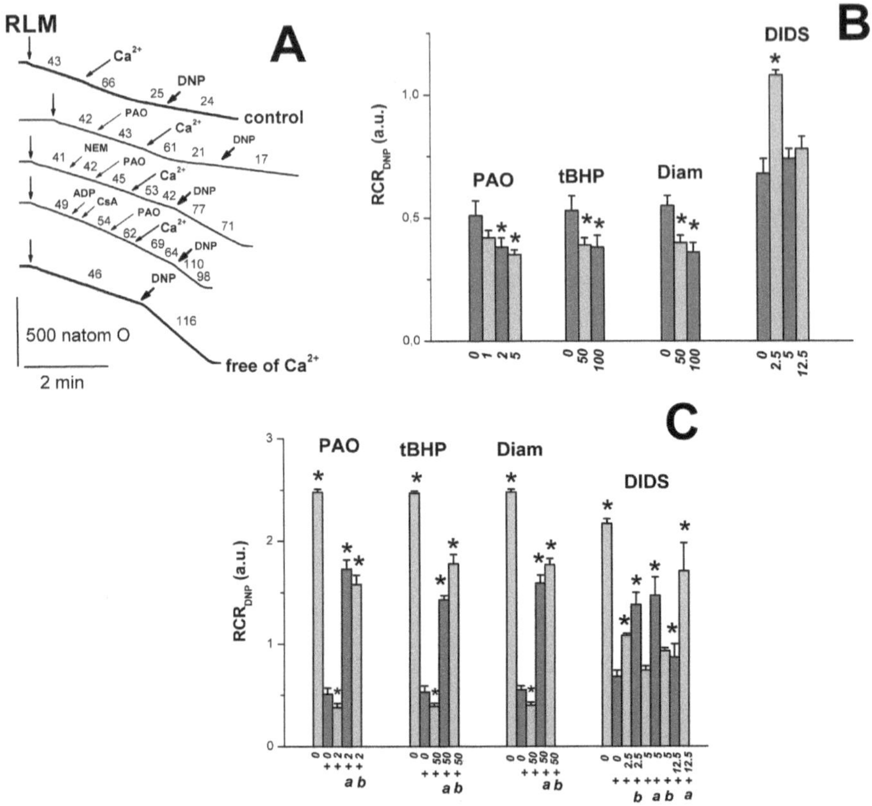

Figure 4.44 Effect of thiol reagents on oxygen consumption rates and RCR_{DNP} value in experiments with succinate-energized rat liver mitochondria. Mitochondria (1.5 mg/mL of protein) were added into medium containing 75 mM $TlNO_3$, 125 mM KNO_3, 5 mM succinate, 5 mM Tris-NO_3 (pH 7.3), 2 μM rotenone, and 1 mg/mL oligomycin. Arrows indicate (A) into the medium introduction of mitochondria (RLM), 100 μM Ca^{2+} (Ca^{2+}), 2 μM PAO (PAO), 50 μM NEM (NEM), 500 μM ADP (ADP), and 1 μM CsA (CsA). Numbers above traces show oxygen consumption rates (n atom O/min per mg protein). Typical traces are shown for three independent experiments. The ordinates (B and C) show RCR_{DNP} values. In panels B and C, Ca^{2+} addition into the medium was everywhere 100 μM. Numbers below the abscissa indicate concentration (μM) of thiol reagents. The plus sign (C) indicates experiments with Ca^{2+}, and latin letters (a,b) indicate experiments with 500 μM ADP and 1 μM CsA (a) or 50 μM NEM (b) added after mitochondria. Asterisks show significant differences from the control, experiments ("0" under the abscissa without the plus sign), experiments with Ca^{2+} free of thiol reagents ($P < .05$).

Table 4.21 Effect of thiol reagents and Ca^{2+} on oxygen consumption rates (n atom O/min per mg of protein) in energized rat liver mitochondria.

Additions after mitochondria	State 4_0	$+ Ca^{2+}$	State $3U_{DNP} + Ca^{2+}$
Free of Ca^{2+}	46 ± 1 (3)		112 ± 3 (3)*
Control	48 ± 1 (3)	40 ± 2 (3)	24 ± 3 (3)
1 μM PAO	47 ± 2 (3)	42 ± 3 (3)	20 ± 2 (3)
2 μM PAO	47 ± 2 (3)	20 ± 1 (3)*	18 ± 2 (3)
5 μM PAO	44 ± 2 (3)	19 ± 2 (3)*	16 ± 1 (3)*
2 μM PAO + ADP + CsA	57 ± 3 (3)*	63 ± 1 (3)*	98 ± 7 (3)*
2 μM PAO + NEM	46 ± 1 (3)	19 ± 2 (3)*	73 ± 4 (3)*
Free of Ca^{2+}	45 ± 1 (3)		112 ± 2 (3)*
Control	47 ± 1 (4)	32 ± 4 (4)	24 ± 2 (4)
50 μM tBHP	43 ± 2 (3)	23 ± 3 (3)*	17 ± 2 (3)*
100 μM tBHP	45 ± 1 (3)	18 ± 2 (3)*	17 ± 2 (3)*
50 μM tBHP + ADP + CsA	43 ± 1 (3)	44 ± 1 (3)*	62 ± 3 (3)*
50 μM tBHP + NEM	44 ± 1 (3)	44 ± 4 (3)*	78 ± 4 (3)*
Free of Ca^{2+}	45 ± 1 (3)		112 ± 3 (3)*
Control	47 ± 2 (3)	32 ± 4 (3)	26 ± 2 (3)
50 μM diamide	39 ± 2 (3)*	21 ± 5 (3)*	16 ± 2 (3)*
100 μM diamide	42 ± 2 (3)	17 ± 1 (3)*	15 ± 2 (3)*
50 μM diamide + ADP + CsA	44 ± 1 (3)	42 ± 2 (3)*	70 ± 4 (3)*
50 μM diamide + NEM	45 ± 1 (3)	49 ± 2 (3)*	79 ± 4 (3)*
Free of Ca^{2+}	44 ± 4 (3)		96 ± 10 (3)*
Control	45 ± 4 (3)	33 ± 6 (3)	30 ± 3 (3)
2.5 μM DIDS	43 ± 2 (3)	43 ± 3 (3)	47 ± 2 (3)*
5 μM DIDS	43 ± 3 (3)	41 ± 2 (3)	32 ± 3 (3)
12.5 μM DIDS	42 ± 3 (3)	38 ± 4 (3)	33 ± 3 (3)
5 μM DIDS + ADP + CsA	44 ± 3 (3)	36 ± 2 (3)	65 ± 11 (3)*
12.5 μM DIDS + ADP + CsA	43 ± 5 (3)	34 ± 2 (3)	72 ± 9 (3)*
5 μM DIDS + NEM	44 ± 2 (3)	47 ± 3 (3)*	61 ± 6 (3)*
12.5 μM DIDS + NEM	45 ± 1 (3)	49 ± 2 (3)*	41 ± 1 (3)*

Asterisk (*) indicate significant differences from the control, experiments free of $TlNO_3$ ($P < .05$).

modification of the membrane individual proteins due to formation of disulfide bonds resulting from oxidation of thiols, but it is not associated with the thiol crosslinking (Bernardes et al., 1994; Castilho et al., 1996). Respiratory chain dysfunction is known to promote MPTP opening (Chávez et al., 1997). An even more significant decrease in state $3U_{DNP}$ respiration (Fig. 4.43 and Table 4.20) in experiments with thiol reagents (PAO, tBHP, diamide, and DIDS) or Cd^{2+} (Korotkov et al., 1998) indicates inhibition of the respiratory chain.

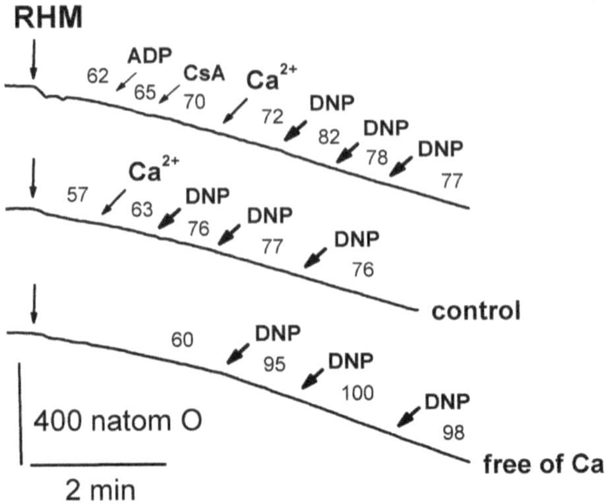

Figure 4.45 Effect of ADP and CsA on oxygen consumption rates in succinate-energized rat heart mitochondria. Mitochondria (1 mg/mL of protein) were injected into medium containing 25 mM $TlNO_3$, 125 mM KNO_3, 100 mM sucrose, 5 mM succinate, 5 mM Tris-NO_3 (pH 7.3), 3 mM Tris-PO_4, 3 mM $Mg(NO_3)_2$, 2 μM rotenone, and 1 μg/mL oligomycin. Reagents: 75 μM Ca^{2+} (Ca^{2+}), 500 μM ADP (ADP), 1 μM CsA (CsA), and three consecutive additions of 5 μM DNP (DNP) were injected after mitochondria.Typical traces are shown for three independent experiments.

Thus, it can be assumed that more noticeable swelling of succinate-energized RLM in medium containing $TlNO_3$ and KNO_3 in experiments with PAO, tBHP, diamide, and DIDS (Fig. 4.48) most likely results in the respiratory chain inhibition by these reagents (Fig. 4.43 and Table 4.20) rather than increasing the passive ionic permeability of the inner membrane. The inhibition of mitochondrial swelling in a medium containing ADP and NH_4NO_3 suggested the ANT involvement transporting H^+ and K^+ ions into the matrix (Klingenberg, 2008). It was shown previously that the ANT is in "c" conformation in the absence of ADP in a medium, and in this case, it can function as a nonselective channel that passes K^+ and H^+ ions to enter into the matrix (Panov et al., 1980; Halestrap and Davidson, 1990; Li et al., 2004; Panov, 2015). In this case, the addition of ADP into a medium transforms the translocase structure into "m" conformation, in which the IMM conductivity for H^+ and K^+ ions is very low (Panov, 2015). Swelling of nonenergized RLM in calcium-free media with $TlNO_3$ decreased in the presence of ADP in our experiments with Y^{3+} and La^{3+} (Fig. 4.31B), or only in experiments with Tl^+

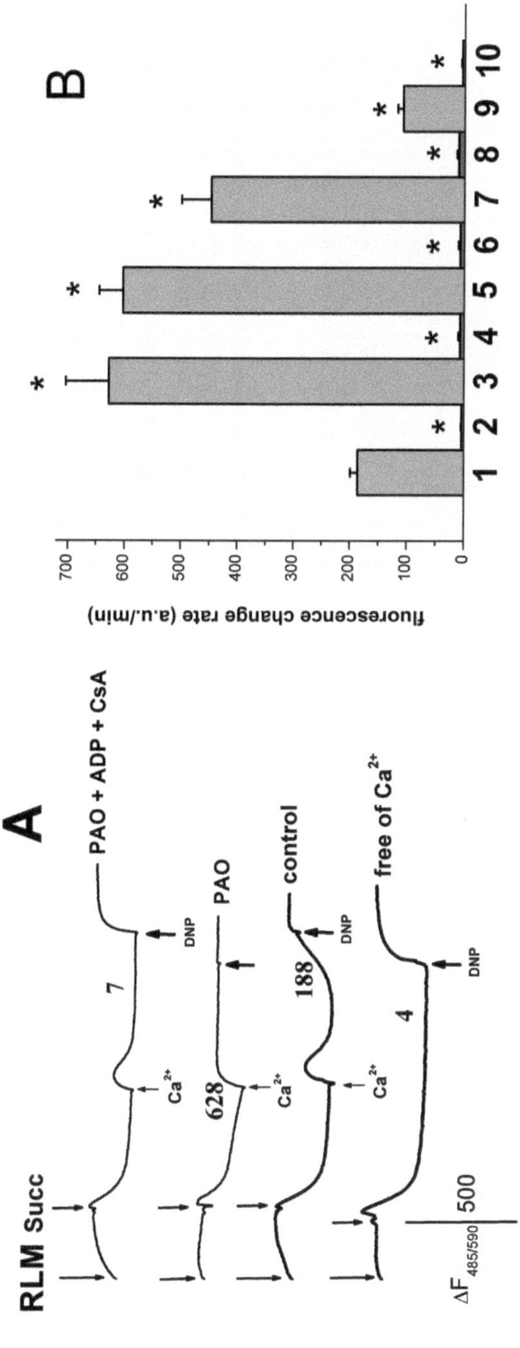
A
RLM Succ
PAO + ADP + CsA
PAO
control
free of Ca2+
Ca2+
DNP
7
628
188
4
ΔF485/590
500
2 min
B
fluorescence change rate (a.u./min)
0
100
200
300
400
500
600
700
1
2
3
4
5
6
7
8
9
10

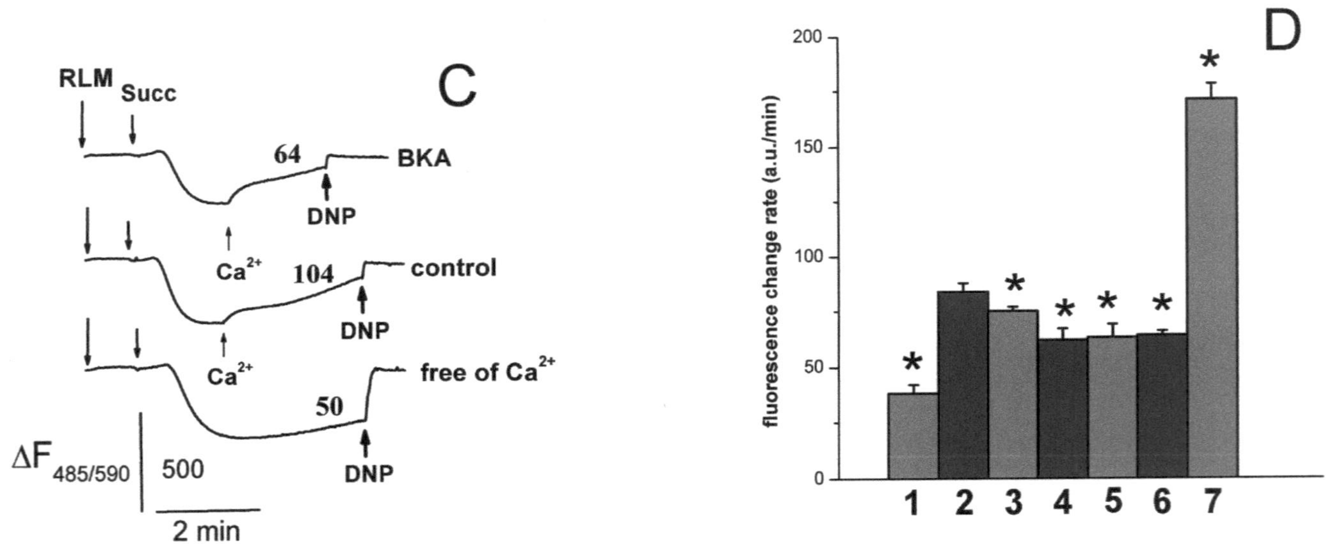

Figure 4.46 The fluorescence change of safranin O ($\Delta F_{485/590}$) in suspension of succinate-energized rat liver mitochondria in the presence of thiol reagents. Mitochondria (0.5 mg/mL of protein) of rat liver (A, B) or rat heart (C, D) were added into 400 mOsm medium containing 20 mM $TlNO_3$ and 110 mM sucrose (A, B) or 15 mM $TlNO_3$ and 120 mM sucrose (C, D) along with 125 mM KNO_3, 1 mM Tris-PO_4, 3 μM safranin O, 2 μM rotenone, and 1 mg/mL oligomycin. The numbers above the traces (A, C) or on the ordinate (B, D) show the calcium-induced rate of the fluorescence change (arb.units/min). Before mitochondria (where indicated), the medium was supplemented with 500 μM ADP (ADP), 1 μM CsA (CsA), 1 μM PAO (A), and 10 μM BKA (BKA). Typical traces for three independent experiments are shown. In panels A and C, arrows show the addition into the medium: rat liver mitochondria (RLM), rat heart mitochondria (RHM), 5 mM succinate (Succ), 40 μM DNP (DNP), as well as 35 μM Ca^{2+} (A) or 40 μM Ca^{2+} (C). Numerals alone panel B under the abscissa mark control experiments with Ca^{2+} alone (1). In other cases (2–10), before adding Ca^{2+} and mitochondria (as in panel A), the following additions were injected into the medium: ADP + CsA (2), 1 μM PAO (3), PAO + ADP + CsA (4), 50 μM tBHP (5), tBHP + ADP + CsA (6), 50 μM diamide (7), diamide + ADP + CsA (8), 2.5 μM DIDS (9), and DIDS + ADP + CsA (10). On panel D, numerals under the abscissa indicate experiments: free of Ca^{2+} (1), control experiments with Ca^{2+} alone and free of additions (2). In other cases (3–7), before adding Ca^{2+} and mitochondria (as in panel C), the following additions were injected into the medium: ADP (3), CsA (4), ADP + CsA (5), BKA (6), and 10 μM CAT (7). Asterisks indicate significant differences from control, experiments with Ca^{2+} alone ($P < .05$).

alone (Fig. 4.4). These data prove the ANT regulates the inner membrane permeability to Tl^{+} because the translocase exists in "m" conformation in the presence of ADP and oligomycin.

The second reason for this mitochondrial swelling may be the reaction of thiol reagents with SH groups of ANT (Petronilli et al., 2009; Nantes et al., 2011; Ricchelli et al., 2011). The interaction of thiol reagent PAO and oxidants (tBHP, diamide) was previously proven to induce intense swelling of mitochondria due to crosslinking of the ANT cysteine groups (Cys159 and Cys256). As a result, the binding of CyP-D to the IMM increased and decreased to ADP (Kowaltowski et al., 1997; Halestrap and Brenner, 2003). Indeed, swelling of succinate-energized RLM in a medium with KNO_3 alone significantly decreased in experiments with tBHP and diamide diamide. It was less pronounced in the presence of PAO (Fig. 4.40C) compared to similar experiments in the medium with $TlNO_3$ and KNO_3 (Fig. 4.40B) (Korotkov et al., 2015c, 2016a,b). This result is consistent with higher passive permeability of the IMM to Tl^{+} than to K^{+} and indicates an essential role of the state of the ANT thiol groups in regulating the inner membrane ionic permeability of energized mitochondria (Mitchell and Moyle, 1969; Brierley et al., 1970, 1977). It was believed that the interaction of DIDS with thiol groups of succinate dehydrogenase or H^{+}-ATP synthase was correspondingly one of the reasons for decreasing FCCP-uncoupled and state 3 respiration RLM energized by succinate or α-ketoglutarate (Bernardes et al., 1994, 1997). DIDS completely inhibited state 3 respiration of RLM in our experiments (Fig. 4.43 and Table 4.20) regardless of the substrate used. At the same time, as in experiments carried out by Bernardes et al. (1994, 1997), state $3U_{DNP}$ respiration was completely inhibited by 100 μM DIDS only in the case of succinate-energized mitochondria, but not in the presence of substrates of the I-th complex (glutamate with malate or α-ketoglutarate alone).

Earlier studies suggested that oxidation and crosslinking of thiol groups of the inner membrane proteins and their interaction with ROS generated by the respiratory chain contributes to the MPTP opening in the inner membrane (Bernardes et al., 1994; Petronilli et al., 1994; Kowaltowski et al., 1997). Effects of oxidative stress inducers (diamide, tBHP) and thiol reagent (PAO) on mitochondria was noted to lead to crosslinking the translocase Cys_{159} with Cys_{256} from the matrix side (Petronilli et al., 1994; McStay et al., 2002; Halestrap and Brenner, 2003; García et al., 2007; Nantes et al., 2011). Our experiments with RLM injected into the medium containing $TlNO_3$ with KNO_3 revealed that free thiols' content for the

IMM fraction decreased in series Diam < NEM, MSL < PAO (Korotkov, 2021). However, tBHP and DIDS did not affect the thiol content. This interaction promotes the MPTP opening due to a decrease in ADP-binding to the IMM specific sites and an increase in the CyP-D binding to this membrane (Petronilli et al., 1994; McStay et al., 2002; Halestrap and Brenner, 2003; García et al., 2007; Nantes et al., 2011). Interaction of mitochondria with CAT, tBHP, PAO, or diamide led to stabilization of the "c" conformation translocase, which significantly increased sensitivity of specific calcium sites to Ca^{2+} and increased the MPTP opening in the inner membrane (Castilho et al., 1996; McStay et al., 2002; Halestrap and Brenner, 2003; Klingenberg, 2008; Halestrap, 2010). Both a decrease of FCCP-uncoupled respiration or $\Delta\Psi_{mito}$ and an increase in mitochondrial swelling due to interaction with tBHP or PAO were markedly inhibited by EGTA in the presence of ADP, DTT, or ruthenium red in experiments with rat liver mitochondria, energized with succinate and loaded with calcium (Riley and Pfeiffer, 1985; Castilho et al., 1996; Kowaltowski and Castilho, 1997). ADP fixing the ANT "m" conformation decreased the sensitivity of the inner membrane calcium sites to Ca^{2+} (Halestrap and Brenner, 2003; Klingenberg, 2008; Halestrap, 2010). Induction of MPTP in experiments with Ca^{2+}, FCCP, and thiol reagents resulted in a decrease in the IMM specific sites' affinity to ADP and an increase in mitochondrial swelling in series tBHP or diamide < DIDS < PAO (Halestrap et al., 1997). The ability of ADP to inhibit the thallium-induced MPTP opening in the presence of these reagents decreased in a similar series tBHP or diamide < DIDS < PAO. Like series can see when comparing the corresponding concentrations of the reagents, there was a comparable swelling of mitochondria (Fig. 4.39) and a decrease in their respiration (Fig. 4.44 and Table 4.21) and $\Delta\Psi_{mito}$ (Fig. 4.46). Considering these results and data of Halestrap et al. (1997), we can be concluded that the opening of the Tl^{+}-induced MPTP in the inner membrane depends on the ANT conformational state.

Diamide compared to tBHP was shown to have caused the mixed disulfides' formation with protein-protein and protein-glutathione bonds. This formation did accompany by a transition of glutathione to its disulfide form in the following experiments: HEK and HeLa cells, cultured aortic endothelial cells, red blood cells, and cultures of neonatal rat cardiomyocytes and human platelets (Collison et al., 1986; Di Simplicio et al., 1996; Giustarini et al., 2000; Hansen et al., 2009; Lock et al., 2012). MPTP opening, the IMM structure disruption and $\Delta\Psi_{mito}$ decrease in experiments with Ca^{2+} and diamide or with Ca^{2+} and tBHP were the result of protein polymerization due to the formation of disulfide bonds between various proteins and appearance of high-molecular protein aggregates (Fagian et al., 1990; Castilho et al., 1996; Nantes et al., 2011).

Zn^{2+}-induced mitochondrial swelling increased even more in the presence of diamide, tBHP, and monobrombimane (Wudarczyk et al., 1999). diamide more strongly than tBHP affected calcium retention of rat liver mitoplasts (Fagian et al., 1990). Tert-butyl hydroperoxide induced RLM swelling only with exogenous Ca^{2+}, while diamide was effective even free of Ca^{2+} (Kushnareva and Sokolove, 2000). Thus, one can assume that more significant swelling of energized RLM in a medium with 75 mM $TlNO_3$ and 125 mM KNO_3 or 200 mM KNO_3 alone (Fig. 4.38B and C) in experiments with diamide in comparison with tBHP (Korotkov et al., 2015c) may result from more active oxidation of glutathione by diamide. The oxidation did find in experiments with isolated mitochondria and cell cultures (Petronilli et al., 1994; Halestrap and Brenner, 2003; Halestrap, 2010). Consequently, the formation of disulfide bridges between adjacent thiol groups of mitochondrial proteins, including ANT in reaction with diamide (Costantini et al., 2000; Nantes et al., 2011), can increase the IMM permeability to Tl^{+} ions (Korotkov et al., 2015b).

BKA and CAT were noted previously to fix ANT in "m" and "c" conformation, respectively (Halestrap and Brenner, 2003; Klingenberg, 2008; Halestrap, 2010). The opening of MPTP induced by Ca^{2+} and P_i was stimulated by CAT and inhibited by Mg^{2+}, ADP, and BKA (Zoccarato et al., 1981). The swelling of calcium-loaded rat liver and heart mitochondria in media containing $TlNO_3$ and ADP significantly increased in the presence of CAT (Fig. 4.42 and Table 4.19B), and it was comparable to the control experiments with Ca^{2+} alone. This result is in good agreement with previously obtained data that the opening of MPTP in calcium-loaded mitochondria was more active when ANT was in "c" conformation (Klingenberg, 2008; Halestrap, 2010). At the same time, a decrease in swelling of Ca^{2+}-loaded mitochondria in similar experiments with ADP, BKA, or NEM, as compared to the control (Figs. 4.41 and 4.42 and Table 4.19B), maybe the result in stabilizing the ANT "m" conformation (Halestrap and Brenner, 2003; Halestrap, 2010).

The inner membrane MPTP opening resulted in the mitochondrial swelling increase and $\Delta\Psi_{mito}$ decline induced by thiol reagents (tBHP, diamide, arsenite, menadione, DIDS, and PAO) fix the ANT "c" conformation. Opposite, NEM reacts with Cys^{56}, stabilizes the ANT "m" conformation, and decreases the affinity of IMM calcium-binding sites to Ca^{2+}, it all inhibits the opening, as a result (Petronilli et al., 1994; Kowaltowski et al., 1997; McStay et al., 2002; Halestrap and Brenner, 2003; García et al., 2007; Nantes et al., 2011). So, it can be assumed that the decrease in calcium-induced swelling of RLM in the presence of NEM in experiments with PAO, tBHP, and diamide (Figs. 4.40 and 4.41 and Tables 4.18 and 4.19) may be due to the ANT "m" conformation fixation.

Thiol reagents (PAO, tBHP, diamide, DIDS) additively reduced the concentration of free thiols in mitochondrial proteins (Kowaltowski et al., 1997). This finding suggested that these reagents can react with free thiol groups of the IMM proteins. Perhaps, for this reason, there was no decrease in calcium-induced mitochondrial swelling in experiments with NEM and DIDS (Table 4.18). The number of thiols in membrane proteins available for NEM increased markedly in the presence of ADP (Kowaltowski et al., 1997). We found that ADP and NEM injected into the medium additively inhibited thallium-induced MPTP opening in the inner membrane. ADP with NEM more potently prevented the increased swelling (Fig. 4.40 and Table 4.18) and the decreased state 3UDNP respiration (Fig. 4.44 and Table 4.21) in experiments with calcium-loaded mitochondria and thiol reagents (PAO, tBHP, diamide, DIDS) compared to similar experiments where ADP and NEM used separately. A similar result can be associated with both stabilizations of the ANT "m" conformation and the effect of NEM reducing the ability of these reagents to interact with free thiol groups on the outside IMM.

As is known, CsA forming complex with CyP-D inhibits binding of the latter to specific sites located on the IMM matrix side, and thus it inhibits the MPTP opening in the inner membrane (Halestrap and Brenner, 2003; Halestrap, 2010). CsA inhibited the mitochondrial swelling increase, Ca^{2+} exit from the matrix, CyP-D binding with the IMM, and $\Delta\Psi_{mito}$ decline, induced by thiol reagents (PAO, tBHP, diamide, DIDS, Cd^{2+}) (Broekemeier et al., 1989; Bernardes et al., 1994; Broekemeier and Pfeiffer, 1995; Kowaltowski and Castilho, 1997; Zazueta et al., 2000; McStay et al., 2002; Halestrap and Brenner, 2003; Belyaeva et al., 2004b; García et al., 2007). We showed early that the opening of thallium-induced MPTP in the inner membrane resulted in the swelling increase, state $3U_{DNP}$ respiration decrease, and $\Delta\Psi_{mito}$ decline in calcium-loaded RLM. In the presence of MPTP inhibitors (ADP, CsA), these effects were decreased in the series CsA < ADP << ADP + CsA. A similar pattern we observed in experiments with PAO, tBHP, diamide, and DIDS. The markedly greater mitochondrial swelling (Fig. 4.40) and the decrease in both $3U_{DNP}$ respiration (Fig. 4.44 and Table 4.21) and $\Delta\Psi_{mito}$ (Fig. 4.46), induced by these thiol reagents, were inhibited in experiments with calcium-loaded RLM with various combinations of ADP and CsA. Replacing ADP by NEM in similar experiments with CsA revealed a similar pattern (Figs. 4.40, 4.44–4.46 and Tables 4.18, 4.19, 4.21, 4.22). The results indicate that the additive effect in experiments with CsA is possible if the ANT "m" conformation did stabilize by either ADP or NEM (Halestrap and Brenner, 2003; Halestrap, 2010; Nantes et al., 2011). However, inhibition of Ca^{2+}-induced swelling in experiments with CsA and NEM was not as significant in the presence of DIDS (Fig. 4.40B and Table 4.18). It was comparable to the effect of ADP alone but not to the effect of ADP with CsA pair. This difference may be because PAO and tBHP, on the one hand, and DIDS, on the other hand, can react with different thiol groups of the IMM proteins (Kowaltowski et al., 1997) and induce various sulfhydryl-disulfide transitions, followed by polymerization of the membrane protein in the presence of Ca^{2+} (Bernardes et al., 1994).

Interacting with vicinal SH groups of the matrix-faced proteins, thiol reagents (PAO, tBHP, diamide, DIDS, 500 μM NEM) stabilize the MPTP structure in the open state. This interacting has resulted in an increase in the IMM ionic permeability, mitochondrial swelling, a reduction in the flow of electrons along the respiratory chain, a decrease in the content of reduced thiols in the matrix, a decline in $\Delta\Psi_{mito}$, oxidation of pyridine nucleotides, an increase in ROS production, formation of mitochondrial protein aggregates, and release of Ca^{2+}, Mg^{2+}, and K^+ from the matrix (Beatrice

Table 4.22 Effect of ADP, CsA, N-ethylmaleimide, bongkrekic acid, and carboxyatractyloside on oxygen consumption rates (n atom O/min per mg of protein) in succinate-energized rat heart mitochondria.

Additions after mitochondria	State 4_0	$+ Ca^{2+}$	State $3U_{DNP} + Ca^{2+}$	RCR_{DNP}
Free of Ca^{2+}	59 ± 2 (4)		89 ± 7 (4)*	1.51 ± 0.06 (4)*
Control	56 ± 2 (4)	58 ± 2 (4)	67 ± 3 (4)	1.20 ± 0.04 (4)
ADP	57 ± 2 (4)	57 ± 4 (4)	67 ± 3 (4)	1.16 ± 0.02 (4)
CsA	57 ± 1 (4)	57 ± 2 (4)	70 ± 2 (4)	1.22 ± 0.04 (4)
ADP + CsA	55 ± 1 (4)	58 ± 1 (4)	71 ± 1 (4)	1.31 ± 0.01 (4)*
BKA	55 ± 1 (4)	58 ± 2 (4)	75 ± 1 (4)*	1.38 ± 0.02 (4)*
NEM	54 ± 1 (4)	60 ± 2 (4)	73 ± 2 (4)	1.35 ± 0.04 (4)*
CAT	55 ± 2 (4)	55 ± 2 (4)	59 ± 1 (4)*	1.07 ± 0.03 (4)*

Asterisk (*) indicate significant differences from the control, experiments free of $TlNO_3$ ($P < .05$).

et al., 1980; Riley and Pfeiffer, 1985; Broekemeier et al., 1989; Bernardes et al., 1994; Broekemeier and Pfeiffer, 1995; Rigobello et al., 1995; Castilho et al., 1996; Bindoli et al., 1997; Kowaltowski et al., 1997; Nantes et al., 2011; Singh et al., 2011). EGTA inhibited mitochondrial swelling, a decline in $\Delta\Psi_{mito}$, and a decrease in formatting protein aggregates in experiments with DIDS and Ca^{2+} (Bernardes et al., 1994). As discussed above, the toxic effect of thallium on living organisms may be associated with the formation of Tl^{+} ligand bonds with the thiol groups of protein molecules. Therefore, the Tl^{+} effects increase on RLM in experiments with thiol reagents (PAO, tBHP, diamide, DIDS, NEM) regardless of the presence of Ca^{2+} in the medium (Figs. 4.38, 4.39, 4.41, 4.43, and 4.44, and Tables 4.18, 4.20, and 4.21) can be associated with the formation of Tl^{+} complexes with the IMM proteins modified by the reagents (Korotkov, 2021). The increased swelling of energized RLM did find in experiments with these thiol reagents, Ca^{2+}, and $TlNO_3$, compared with the control ones free these reagents (Fig. 4.39 and Table 4.18). Besides, the reduced mitochondrial volume retention is in calcium-free medium experiments (Fig. 4.38). So, these findings can account for the IMM permeability increase to Tl^{+} due to the membrane concentration reduction of free SH groups (Korotkov, 2021), which can be the result of the interaction of membrane proteins with these thiol reagents (Kowaltowski et al., 1997; McStay et al., 2002).

The ability of ADP to inhibit the MPTP opening in the inner membrane is not so noticeable in experiments with mitochondria pretreated with diamide and thiol reagents (PAO, 80 μM eosin maleimide, 500 μM NEM) due to the reaction of these reagents with the ANT Cys_{159} (McStay et al., 2002). The cadmium-induced MPTP opening increased in the presence of NEM, since Cd^{2+} forms a strong complex with SH groups of the translocase vicinal cysteines (Cys_{56} and Cys_{159}), in contrast to the thiol compounds we used, which react with Cys_{159} or Cys_{256} (Zazueta et al., 2000; Halestrap and Brenner, 2003; Belyaeva et al., 2004). Dithiothreitol reduced vicinal thiol groups previously oxidized with Cd^{2+}, crosslinked by tBHP or diamide, or conjugated with arsenite (Petronilli et al., 1994; Zazueta et al., 2000). Perhaps this is why the noticeable increase of state 4 respiration in experiments with cadmium-inhibited RLM occurs (Belyaeva and Korotkov, 2003; Korotkov et al., 2008b). Conformational changes in ANT can affect the reactivity of its thiol groups (Kowaltowski et al., 1997; McStay et al., 2002). MPTP opening induced by reagents (PAO, tBHP, diamide, menadione) was inhibited in the presence of low (5—20 μM)

NEM concentrations, while high NEM concentrations (500−1000 μM) themselves induced the pore opening in the IMM (Petronilli et al., 1994; Costantini et al., 1998). The ability of ADP to inhibit the MPTP opening markedly reduced in the presence of 500 μM NEM (McStay et al., 2002). So, high NEM did induce high-amplitude swelling succinate-energized RLM (Fig. 4.41). On the one hand, high NEM prevented ADP's ability to inhibit calcium-induced swelling (Table 4.19). These findings may be due to the translocase transition to "c" conformation from NEM reaction with the ANT thiol groups in experiments with succinate-energized RLM.

The fact that the conformation of the adenine nucleotide translocase plays an essential role in regulating the thallium-induced mitochondrial permeability transition pore in rat liver mitochondria's inner membrane showed in Fig. 4.47.

Summing up, this chapter's results should emphasize that the opening of thallium-induced MPTP has resulted in experiments with thiol reagents (PAO, tBHP, diamide, DIDS) and CAT, which fixed the ANT "c" conformation. However, stabilizing the ANT "m" conformation inhibited this pore opening in experiments with MPTP inhibitors (ADP, NEM, BKA) (Korotkov et al., 2016a,b). Therewith, ADP and CsA inhibited MPTP opening induced by (Cd^{2+}, Hg^{2+}, Pb^{2+}, Zn^{2+}, Cu^{2+}) (Miyahara and Utsumi, 1975; Chávez and Holguín, 1988; Zoratti and Szabó, 1995; Wudarczyk et al., 1999; Belyaeva et al., 2002; Dorta et al., 2003; Lee et al., 2005). More intensive the pore opening in the joint presence of Cd^{2+} and Ca^{2+} also inhibited by MPTP inhibitors (ADP, CsA, and Mg^{2+}) (Belyaeva and Korotkov, 2003; Belyaeva et al., 2004b). The observed dependence of the appearance of thallium-induced MPTP in the inner membrane of calcium-loaded rat liver mitochondria on the ANT conformation is consistent with similar results of earlier experiments with isolated mitochondria (Kowaltowski et al., 1997; Halestrap and Brenner, 2003).

4.6.2 Tl^+-induced MPTP opening and the phosphate symporter

The mitochondrial phosphate symporter (PiC), CyP-D, and some parts of ATP synthase were thought to be a major component of MPTP (Leung et al., 2008; Baines, 2009; Halestrap, 2010; Bernardi, 2013; Elrod and Molkentin, 2013). Hydrophilic monofunctional thiol reagent MSL prevented MPTP opening in the inner membrane of calcium-loaded mitochondria. As a result, there was a decrease in calcium-induced release of adenine nucleotides and ions (Ca^{2+}, Mg^{2+}, $H_2PO_4^-$) from the matrix of

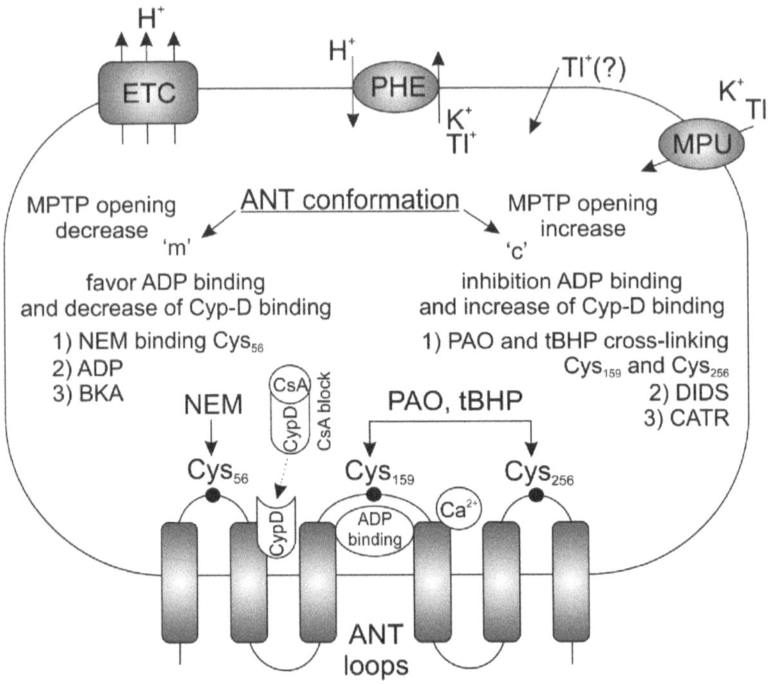

Figure 4.47 The role of some cysteines and the adenine nucleotide translocase conformation, as well as the matrix-facing specific sites binding to CyP-D, ADP and Ca^{2+} in the induction of thallium-induced MPTP in the inner membrane of rat liver mitochondria. Abbreviations: *ADP-binding*, high-affine ADP-binding site; *ANT*, adenine nucleotide translocase; *BKA*, bonkgrekic acid; *Ca^{2+}*, calcium site to induce the MPTP opening; *CATR*, carboxyatractyloside; *CsA*, MPTP blocker to bind to cyclophilin; D*CyP-D*, cyclophilin D; *Cys*, cysteine residues of the ANT structure; *DIDS*, 4,4′-diisothiocyanostilbene-2,2′-disulfonate; *ETC*, electron transport chain; *MPTP*, mitochondrial permeability transition pore; *MPU*, mitochondrial potassium uniporter; *NEM*, N-ethylmaleimide; *PAO*, phenylarsine oxide; *PHE*, mitochondrial K^+/H^+ antiporter; *tBHP*, tert-butyl hydroperoxide; *Tl^+(?)*, nonspecific transport pathway of Tl^+ across the inner membrane. Note: This CsA blocker is shown in the diagram as "CsA block."

these mitochondria (Rugolo et al., 1981; Rigoni et al., 1984; Wilson and Asimakis, 1987; Kowaltowski et al., 1997). By reacting with external thiol groups of phosphate symporter, MSL blocked the transport of P_i into mitochondria (Wehrle and Pedersen, 1981; Ligeti and Fonyo, 1984). However, according to other data, even 200 μM MSL did not affect the ANT activity in isolated RHM (Wilson and Asimakis, 1987). The above facts were the reason for our interest in studying the involvement of the phosphate symporter in the appearance of thallium-induced MPTP in the inner mitochondrial membrane.

In our experiments, MSL stimulated swelling of succinate-energized RLM in a medium containing $TlNO_3$ and KNO_3 (Fig. 4.48A and B) (Korotkov et al., 2018). This effect of MSL was more potent in experiments with RLM energized by glutamate and malate (Fig. 4.48B). Compared with the control (experiments free of MSL), succinate-induced contraction of RLM, pre-swollen in a medium with $TlNO_3$ and KNO_3, was increased in the presence of 5 μM MSL. However, there was contraction but no swelling of mitochondria when the concentration of MSL in the medium increased to 25–35 μM (Fig. 4.48C and D). A similar MSL-induced contraction did observe previously in experiments with energized rat liver and rat heart mitochondria, preswollen in media with KH_2PO_4 and KNO_3 (Azzone et al., 1976a; Brierley et al., 1977).

The MSL-induced swelling of energized mitochondria decreased significantly in the medium with 200 mM KNO_3 free of $TlNO_3$ (Fig. 4.48A). But in this case, the swelling was more significant in experiments with glutamate and malate than with succinate (Fig. 4.48B). This decrease in respiration of energized mitochondria suggests that MSL versus the thiol reagents our studied (PAO, DIDS) does not react most likely with thiol groups of the IMM affected its ion permeability. The latter does not contradict the data that the reaction of MSL with some inner membrane proteins did not increase its ion permeability (García et al., 2007). MSL of 2–20 μM (Fig. 4.49A) significantly increased the calcium-induced swelling in succinate-energized RHM, which ADP just entirely inhibited in the presence of 10 μM MSL (Fig. 4.49B) (Korotkov et al., 2018).

Fig. 49C shows that the RLM swelling in the medium with Ca^{2+} alone (control) accelerated slightly in experiments with 5 μM MSL (Fig. 4.49C). At the same time, it decreased with CsA and NEM and inhibited completely in experiments with ADP (Fig. 4.49C and D), regardless of the medium CsA and NEM presence (Table 4.23).

The changes in ΔA_{540} (see Fig. 4.49C) were detected within four minute interval after administration of 5 mM succinate to mitochondria and presented as Means ± SEM. The number of experiments showed in parentheses. P-values were accordingly calculated to experiments with 100 μM Ca^{2+} (a dash in the P value columns). Asterisks indicate that statistical difference between appropriate ΔA_{540} values is not statistically significant.

In experiments free of Ca $^{2+}$, respiration in 3 and $3U_{DNP}$ states of succinate-energized RLM, as well as values of RCR and RCR_{DNP}, were decreased significantly in the presence of 10–20 μM MSL (Fig. 4.50A and Table 4.24). In similar experiments with glutamate and malate, the

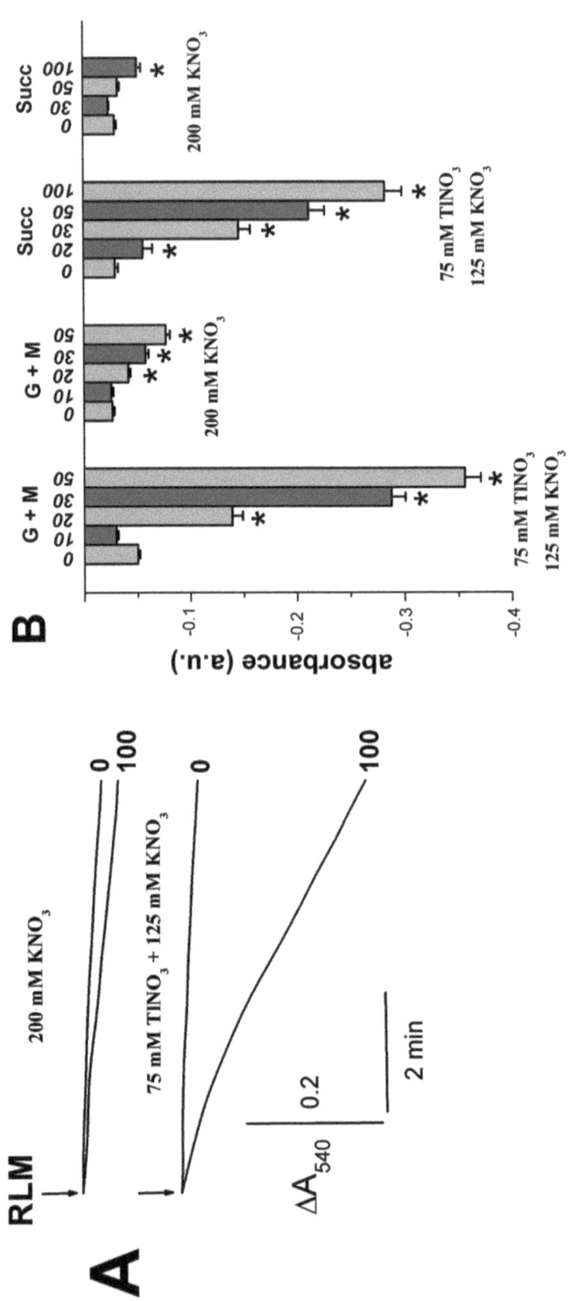
A
RLM
200 mM KNO$_3$
0
100
75 mM TlNO$_3$ + 125 mM KNO$_3$
0
100
ΔA_{540}
0.2
2 min
B
absorbance (a.u.)
-0.1
-0.2
-0.3
-0.4
G + M
0 10 20 30 50
75 mM TlNO$_3$
125 mM KNO$_3$
G + M
0 10 20 30 50
200 mM KNO$_3$
Succ
0 20 30 50 100
75 mM TlNO$_3$
125 mM KNO$_3$
Succ
0 30 50 100
200 mM KNO$_3$

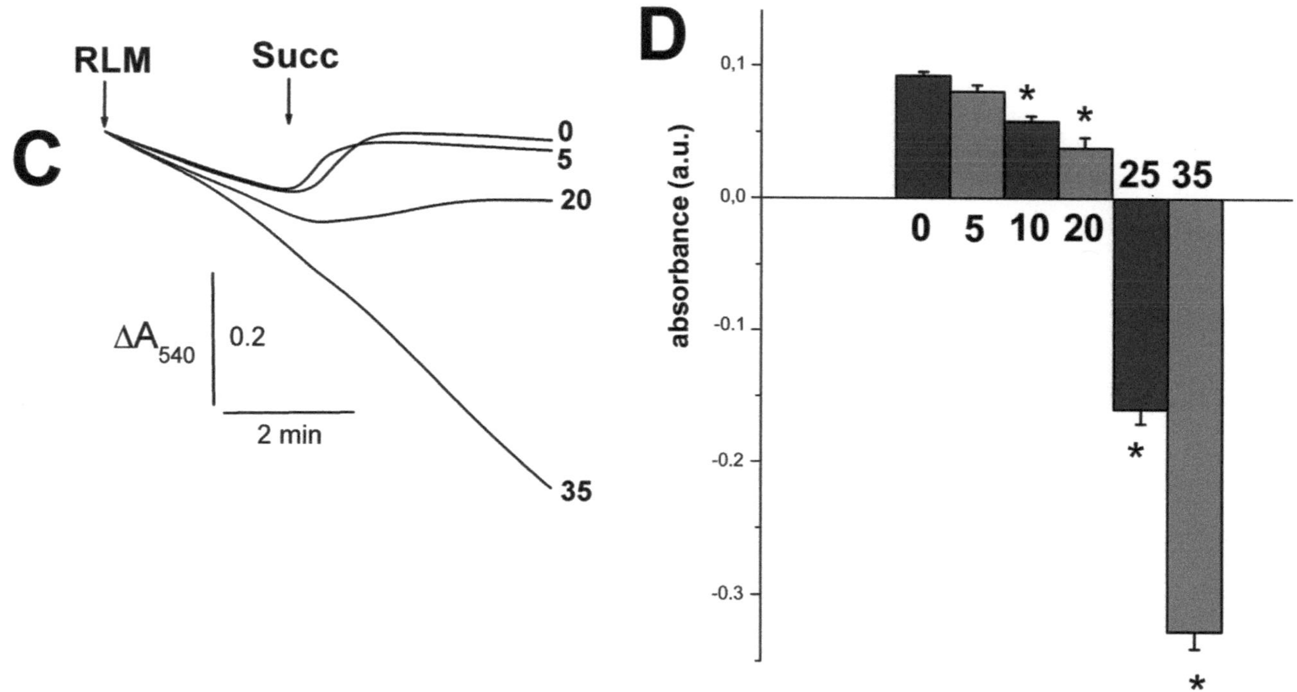

Figure 4.48 Effects of mersalyl (MSL) on swelling of rat liver mitochondria. Mitochondria (1.5 mg/mL of protein) were added into a medium containing 75 mM $TlNO_3$ and 125 mM KNO_3 (A, D) or 200 mM KNO_3 (A, B) along with 5 mM Tris-NO_3 (pH 7.3) and 1 μg/mL oligomycin. This medium also contained: 5 mM glutamate with 5 mM malate (G + M, panel B), or 5 mM succinate (Succ) with 2 μM rotenone (panels A and B). *Arrows* (A, C) show injection of mitochondria (RLM) and 5 mM succinate (Succ) into the medium. Numbers to the right of traces (A, C) indicate concentration of MSL (μM) added into the medium before mitochondria. Typical traces are shown for three independent experiments. The ordinate axis shows change in absorbance (ΔA_{540}) 7 min after mitochondria additions into the medium (B), as in panel A, or 4 min after succinate injection (D), as in panel C. Numbers on the abscissa (B, D) indicate concentration (μM) of MSL in the medium. Asterisks indicate significant differences from control ("0" near the abscissa axis), experiments free of MSL ($P < .05$).

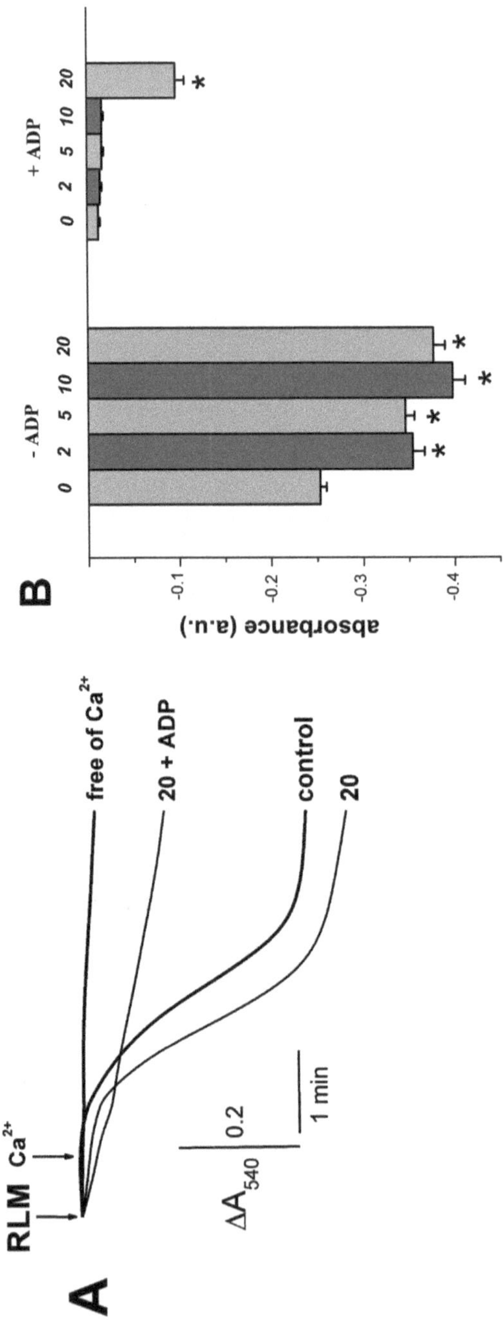
A
RLM
Ca^{2+}
free of Ca^{2+}
20 + ADP
control
20
ΔA_{540}
0.2
1 min
B
absorbance (a.u.)
-0.1
-0.2
-0.3
-0.4
- ADP
+ ADP
0 2 5 10 20
0 2 5 10 20
*

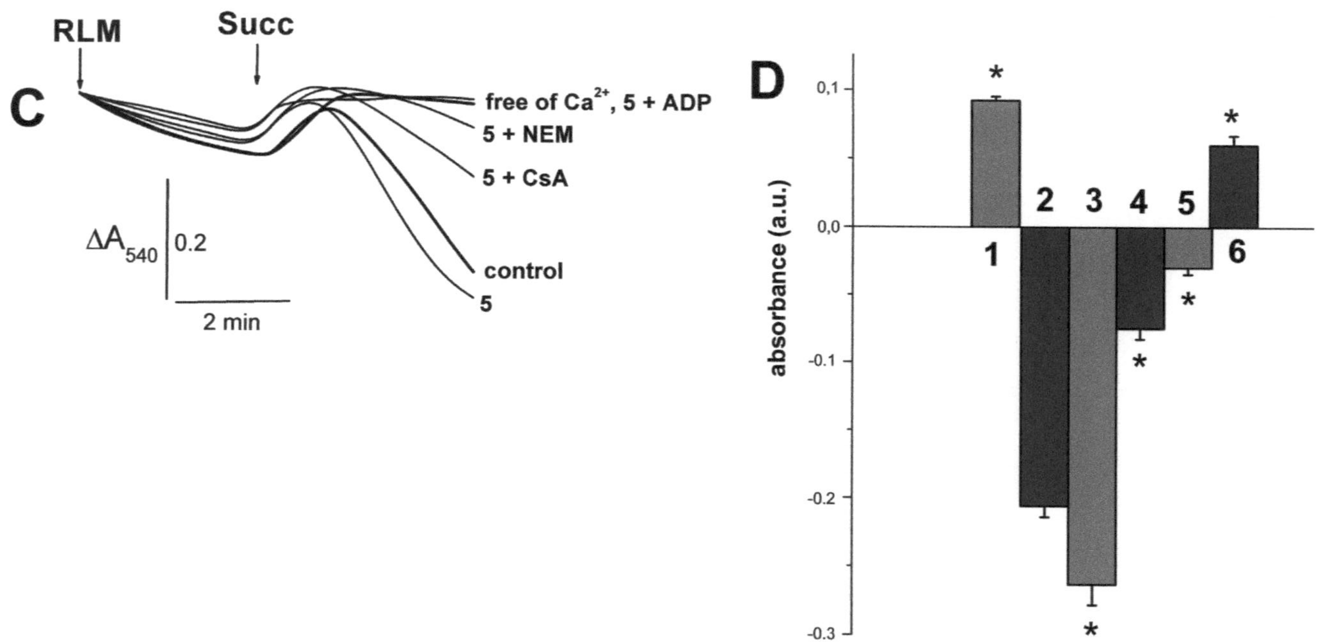

Figure 4.49 Influence of mersalyl (MSL), ADP, CSA and NEM on swelling of rat liver mitochondria. Mitochondria (1.5 mg/mL of protein) was added into medium containing 75 mM $TlNO_3$, 125 mM KNO_3, 5 mM Tris-NO_3 (pH 7.3), 5 mM succinate (A, B), 2 μM rothenon, and 1 μg/mL oligomycin. Up to mitochondria (where indicated), 0–35 μM MSL, 500 μM ADP (ADP), 1 μM CSA (CSA), and 50 μM NEM (NEM) were injected into the medium. *Arrows* (A, C) show addition into the medium of mitochondria (RLM), 100 μM Ca^{2+} (Ca^{2+}), and 5 mM succinate (Succ). Concentration (μM) of MSL introduced into the medium before mitochondria is shown to the right of traces (A, C). Typical traces are shown for three independent experiments. Control (A–D), experiments with Ca^{2+} alone. The ordinate axis (B, D) shows absorbance change (ΔA_{540}) 2 min after Ca^{2+} addition (B) or 4 min after succinate injection of (D) into the medium as in panels A and C, respectively. Numbers at the abscissa axis (B) show concentration of MSL in the medium. Numerals (D) indicate experiments free of MSL and Ca^{2+} (1), the control, with Ca^{2+} alone (2). In other cases (3–6), next chemicals before mitochondria were added into the medium along with 100 μM Ca^{2+}: 5 μM MSL (3), MSL and 1 μM CSA (4), MSL and 50 μM NEM (5), and MSL and 0.5 mM ADP (6). Asterisks (B, D) show significant differences from the control ($P < .05$).

Table 4.23 Change in absorbance (ΔA_{540}) in suspension of succinate-energized rat liver mitochondria in the presence of 5 μM mersalyl.

Supplements in the medium before mitochondria	ΔA_{540}
Ca^{2+}	−0.206 ± 0.008 (3)
5 мкМ MSL + Ca^{2+}	−0.263 ± 0.015 (3)*
5 мкМ MSL + Ca^{2+} + ADP	0.060 ± 0.007 (3) *
5 мкМ MSL + Ca^{2+} + ADP + CsA	−0.075 ± 0.008 (3)*
5 мкМ MSL + Ca^{2+} + NEM	−0.030 ± 0.005 (3)*
5 мкМ MSL + Ca^{2+} + ADP + NEM	0.070 ± 0.006 (3)*
5 мкМ MSL + Ca^{2+} + CsA + NEM	0.061 ± 0.009 (3)*

Asterisk (*) indicate significant differences from the control, experiments free of $TlNO_3$ ($P < .05$).

respiration of mitochondria in these states changed slightly in the presence of 5−20 μM MSL in the medium (Table 4.24) (Korotkov et al., 2018). State $3U_{DNP}$ respiration decrease found in experiments with calcium-loaded succinate-energized mitochondria (Fig. 4.50C and D and Table 4.25). However, this decrease was noticeable in experiments with 2.5−5 μM MSL, and it was absent if MPTP inhibitors (NEM and CsA), especially ADP, were added into the medium before mitochondria (Fig. 4.50C and D and Table 4.25).

Media, additions, and procedure in conducting experiments are as in Fig. 4.50A. Number of experiments is indicated in brackets. Asterisks show significant differences from control, experiments free of MSL ($P < .05$). Number signs indicate experiments with RLM energized by glutamate and malate.

A short-term decrease in $\Delta\Psi_{mito}$ did observe due to the energy-dependent uptake of Ca^{2+} in the control experiments (Figs. 4.20 and 4.41A). Noticeable safranin O fluorescence increase followed due to this dye release from mitochondria into the medium. In this case, the release of safranin O is due to a partial decrease in the IMM potential and the MPTP opening (Fig. 4.20) accompanied by intense mitochondrial swelling (Fig. 4.14 and Table 4.5) and a decline in their respiration rates in 4, 3, or 3UDNP states (Fig. 4.18 and Table 4.9; Korotkov and Saris, 2011). Similar results indicate a calcium-induced decrease in $\Delta\Psi_{mito}$ since the energized mitochondria contraction depending on the inner membrane potential did replace by an intense swelling of these organelles after their calcium load (Fig. 4.14). MSL significantly increased the Ca^{2+}-induced release of safranin O from mitochondria (Fig. 4.51A). However, the MSL effect did inhibit entirely in the presence of ADP with CsA. In this case, we can talk about

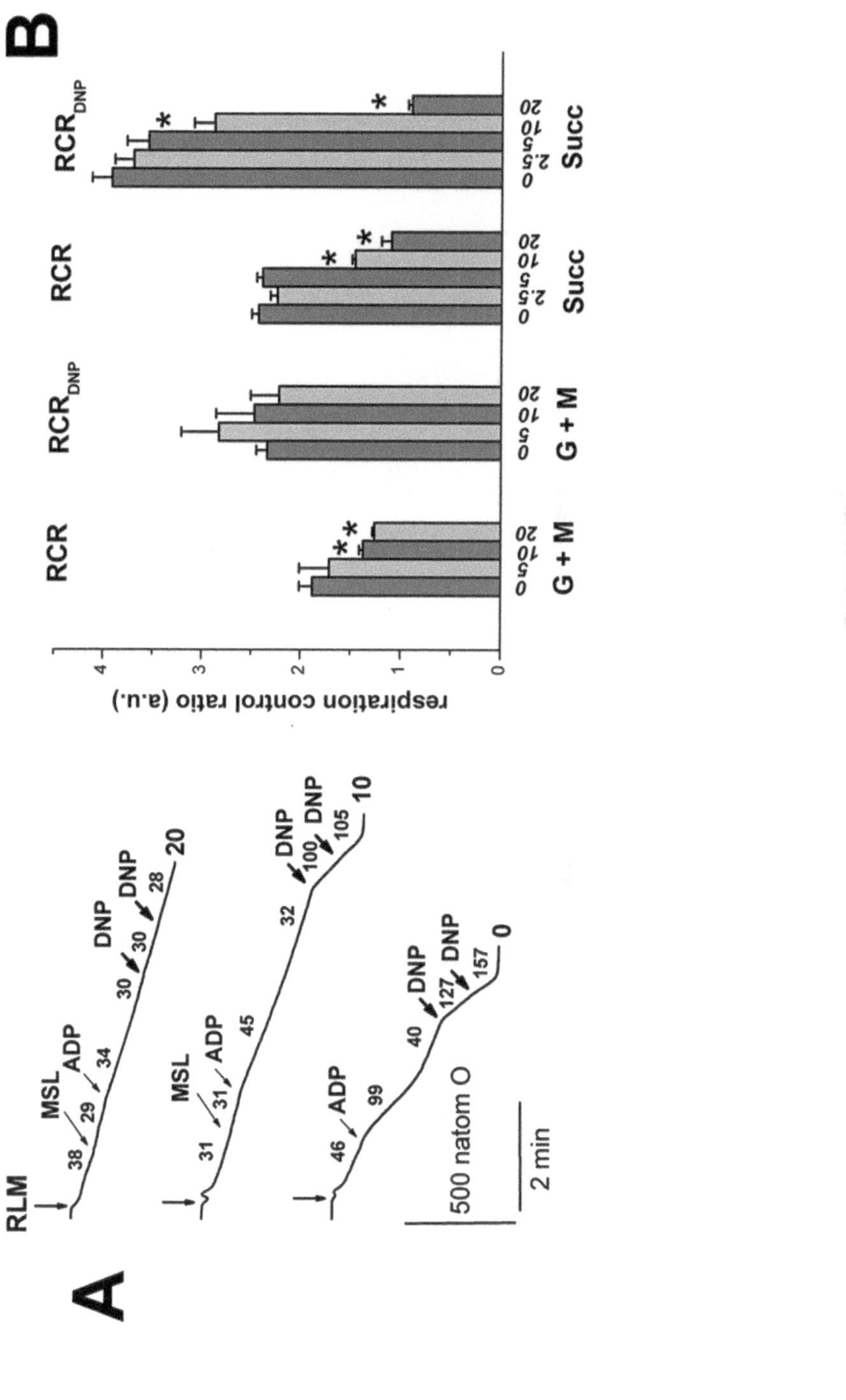
A
RLM
MSL
ADP
DNP
500 natom O
2 min
B
RCR
RCR$_{DNP}$
respiration control ratio (a.u.)
G + M
Succ

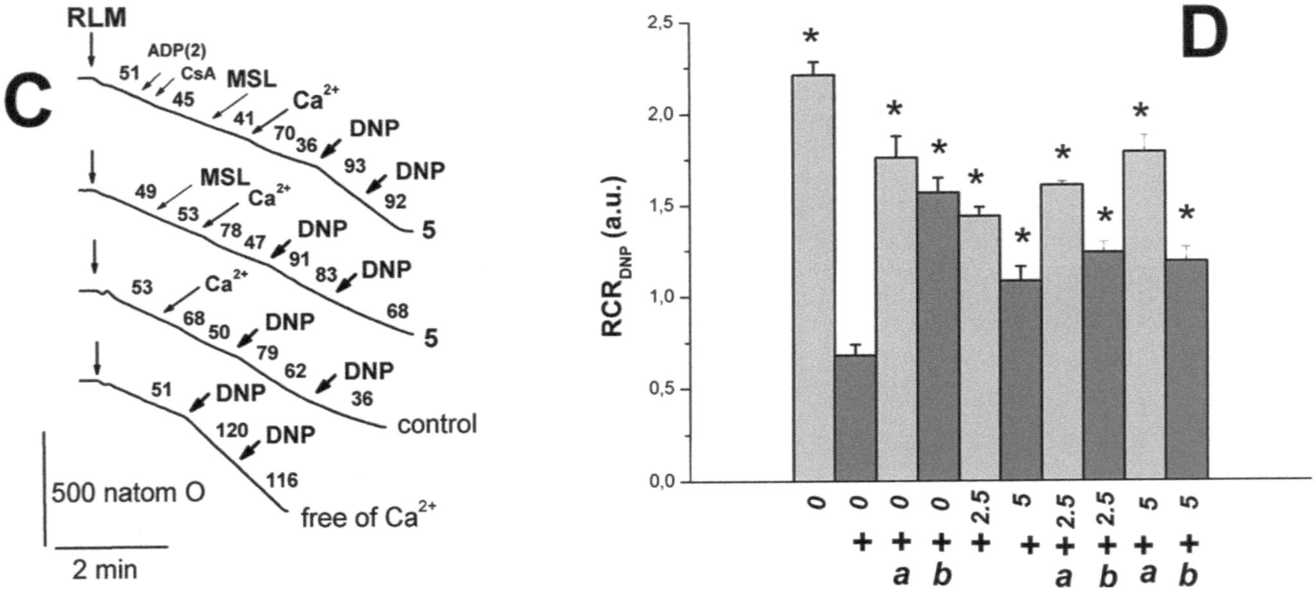

Figure 4.50 Effects of Ca^{2+} and mersalyl (MSL) on oxygen consumption rates of rat liver mitochondria. Mitochondria (1.5 mg/mL of protein) were added into a medium containing 125 mM KNO_3 and 5 mM Tris-NO_3 (pH 7.3), as well as 25 mM $TlNO_3$, 100 mM sucrose, 3 mM $Mg(NO_3)_2$, and 3 mM Tris-PO_4 (A, B) or 75 mM $TlNO_3$ and 1 mg/mL oligomycin (C, D). Before mitochondria, 5 mM glutamate and 5 mM malate (B) or 5 mM succinate and 2 μM rotenone (A–D) were added into the medium. Arrows (A, C) show additions after mitochondria of 100 μM Ca^{2+} (Ca^{2+}), 130 μM ADP (ADP), 500 μM ADP (ADP (2)), 15 μM DNP (DNP), 1 μM CSA (CSA), and 5–20 μM MSL. Digits to the right of traces (A, C) indicate MSL concentration (μM) in the medium. Numbers above traces show oxygen consumption rates (n atom O/min per mg of protein). Typical traces are shown for three independent experiments. Values of RCR and RCR_{DNP} are given along the ordinate axis (B, D). Addition of Ca^{2+} into the medium was 100 μM (C, D). Digits at the abscissa axis (B, D) indicate MSL concentration (μM). On panel D, the plus sign "+" shows experiments with Ca^{2+}, and latin letters note experiments with 500 μM ADP and 1 μM CsA ("**a**") or 50 μM NEM ("**b**"). Asterisks show significant differences from the control ("0" at the abscissa axis), experiments with Ca^{2+} and free of MSL ($P < .05$).

Table 4.24 Effects of mersalyl on oxygen consumption rates (n atom O/min per mg of protein) by energized rat liver mitochondria.

Additions after mitochondria	State 3	State 4	State $3U_{DNP}$
Control	47 ± 3 (3)*	25 ± 2 (3)	58 ± 4 (3)
5 μM MSL#	35 ± 2 (3)*	21 ± 3 (3)	59 ± 4 (3)
10 μM MSL#	30 ± 2 (3)*	22 ± 1 (3)	54 ± 7 (3)
20 μM MSL#	29 ± 1 (3)*	23 ± 1 (3)	50 ± 6 (3)
Control	80 ± 9 (3)	33 ± 4 (3)	130 ± 16 (3)
2.5 μM MSL	73 ± 2 (3)	33 ± 1 (3)	121 ± 7 (3)
5 μM MSL	76 ± 5 (3)	32 ± 1 (3)	113 ± 10 (3)
10 μM MSL	43 ± 1 (3)*	29 ± 1 (3)	85 ± 10 (3)*
20 μM MSL	31 ± 6 (3)*	28 ± 4 (3)	25 ± 3 (3)*

Asterisk (*) indicate significant differences from the control, experiments free of $TlNO_3$ ($P < .05$).

Table 4.25 Effects of MSL and Ca^{2+} on oxygen consumption rates (n atom O/min per mg of protein) in succinate-energized rat liver mitochondria.

Additions after mitochondria	State 4_0	+ Ca^{2+}	State $3U_{DNP}$ + Ca^{2+}
Free of Ca^{2+}	42 ± 2 (3)		94 ± 8 (3)*
Control	43 ± 3 (3)	44 ± 3 (3)	29 ± 2 (3)
ADP + CsA	44 ± 2 (3)*	43 ± 3 (3)*	78 ± 6 (6)*
NEM	46 ± 1 (3)*	50 ± 4 (3)*	73 ± 4 (6)*
2.5 μM MSL	38 ± 3 (3)	37 ± 3 (3)	55 ± 6 (3)*
5 μM MSL	41 ± 3 (3)	38 ± 6 (3)*	44 ± 4 (3)*
2.5 μM MSL + NEM	40 ± 2 (3)*	41 ± 4 (3)*	49 ± 4 (3)*
5 μM MSL + NEM	39 ± 3 (3)*	38 ± 3 (3)*	47 ± 5 (3)*
2.5 μM MSL + ADP + CsA	38 ± 3 (3)*	37 ± 5 (3)*	60 ± 3 (3)*
5 μM MSL + ADP + CsA	37 ± 4 (3)*	33 ± 2 (3)*	66 ± 6 (3)*

Asterisk (*) indicate significant differences from the control, experiments free of $TlNO_3$ ($P < .05$).

the restoration of $\Delta\Psi_{mito}$, since under these conditions, the swelling of calcium-loaded RLM was replaced by their contraction (Fig. 4.49C), as in experiments free of calcium and MSL (Fig. 4.21B). Herewith, calcium-induced decrease in $3U_{DNP}$ state respiration of these mitochondria markedly weakened in the presence of ADP and CsA (Fig. 4.50A), as in similar experiments free of MSL (Fig. 4.25B and Table 4.14).

The medium, additions, and experimental procedure are similar to those shown in 50C. The number of tests carried out is indicated in parentheses. Designations in the table: State 4_0—state 4 respiration before Ca^{2+} injection into the medium, + Ca^{2+}—respiration after 75 μM Ca^{2+} adding into the medium, State $3U_{DNP}$ + Ca^{2+}—state $3U_{DNP}$ uncoupled respiration

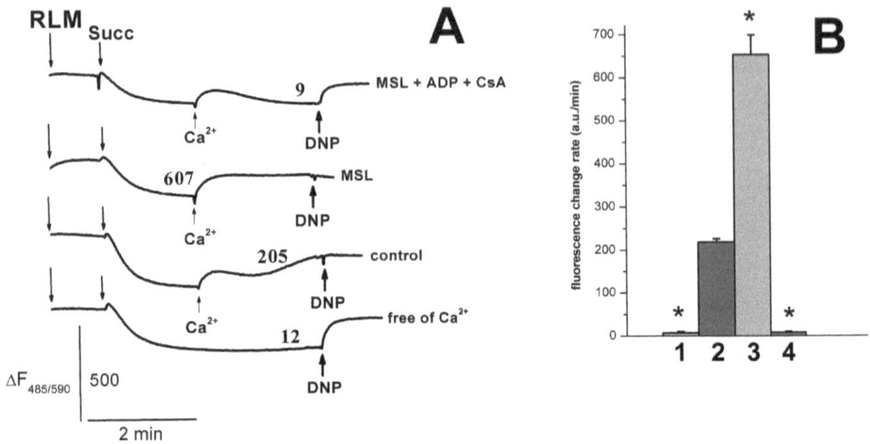

Figure 4.51 Effects of MSL on the safranin O fluorescence change ($\Delta F_{485/590}$) in suspension of succinate-energized rat liver mitochondria. The medium is as in Fig. 4.46A. The next reagents were added into the medium before mitochondria (where indicated): 2.5 μM MSL (MSL), 500 μM ADP (ADP), and 1 μM CsA (CsA). Numerals above the traces (A) or along the ordinate axis (B) show the fluorescence change rate after addition of Ca^{2+} (arb.units/min). Arrows (A) show the additions of RLM (RLM), 5 mM succinate (Succ), 35 μM Ca^{2+} (Ca^{2+}) and 40 μM DNP (DNP). Typical traces are shown for three independent experiments. The ordinate (B) shows the safranin fluorescence change rate (arb.units/min). Numerals alone the abscissa show the experiments: free of additions (1), the control, 75 μM Ca^{2+} addition after mitochondria (2), Ca^{2+} addition into the medium containing 2.5 μM MSL after mitochondria (3), Ca^{2+} addition into the medium containing 2.5 μM MSL, 500 μM ADP, and 1 μM CsA (4) after mitochondria. Asterisks indicate significant differences from the control, experiments with Ca^{2+} alone ($P < .05$).

after injection of 30 μM DNP followed the Ca^{2+} addition, as a Fig. 4.50C. Asterisks show significant differences ($P < .05$) from the control, experiments with Ca^{2+} alone.

Early studies found that increased influx of K^+ ions into the matrix led to intense swelling of succinate-energized RLM in the presence of MSL (Chávez, 1973; Diwan, 1985). Thus, the swelling increase in succinate-energized RLM in the medium containing MSL (Fig. 4.48) may be due to the induction by MSL of the IMM permeability to both K^+ and Tl^+. The modification of the inner membrane proteins caused by the interaction of thiol reagents (PAO, diamide) with the corresponding cysteines increased the MPTP opening (McStay et al., 2002; Halestrap and Brenner, 2003), but the ability of MSL to interact with the IMM proteins did not affect to this pore opening in this membrane (García et al., 2007). Perhaps, for this reason, the swelling of RLM in calcium-free media did not increase so significantly in the presence of 20 μM MSL (Fig. 4.48C) compared to noticeably more significant swelling in experiments with 20 μM FAO (Fig. 4.38). At the same time, the above effects of PAO became more potent in the presence of MSL, which manifested itself in leveling the inhibitory effects of CsA. García et al. (2007) suggested that protein dimerization caused by MSL did not affect probably proteins of the respiratory chain since MSL did not affect the ability of mitochondria to create a transmembrane electric potential on the inner membrane.

MSL inhibited respiration of succinate-energized mitochondria in 4, 3, and 3U states, but it did not affect ANT activity up to 200 μM (Wilsona and Asimakis, 1987; Le Quoc and Le Quoc, 1979; Samartsev et al., 2003). The contraction of succinate-energized RLM was more visible in the presence of 5 μM MSL (Fig. 4.48C). However, this MSL concentration did not affect these mitochondria respiration in 3 and $3U_{DNP}$ states (Fig. 4.50 and Table 4.24). At higher concentrations of MSL (20−35 μM), there was swelling but not contraction (Fig. 48C), which may be because MSL inhibited succinate transport into mitochondria (Le Quoc and Le Quoc, 1979). This assumption is in good agreement with the inhibition of respiration of succinate-energized RLM in 3 and $3U_{DNP}$ states in experiments with 20−35 μM MSL (Fig. 4.50 and Table 4.24) and DIDS (Table 4.20), which, like MSL, inhibits succinate transport into the matrix (Bernandes et al., 1997). MSL did not inhibit $3U_{DNP}$ state respiration (Table 4.24) in similar experiments with RLM energized by glutamate and malate (substrates of complex I), possibly because this complex is not inhibited by MSL (Wehrle and Pedersen, 1981). On the other hand, Pi transport was known to have inhibited by 5 μM MSL, which at this concentration blocked PiC by reacting with its external thiol groups (Wehrle and Pedersen, 1981; Ligeti and Fonyo, 1984). At this concentration of MSL, the contraction of energized and calcium-loaded mitochondria did accelerate compared to the calcium-free experiments (Fig. 4.48C). These data suggest that the phosphate symporter, but not the respiratory complex I, can be the main target in the observed effects of MSL upon the thallium-induced MPTP opening in the inner membrane of calcium-loaded RLM. High MSL concentrations (20−60 μM) were shown previously to have induced MPTP opening in experiments with rat heart mitochondria energized by glutamate with malate (Harris et al., 1979).

Ca^{2+} release from succinate-energized rat kidney mitochondria was inhibited by 5 μM Hg^{2+}, but not by 20 μM MSL. This circumstance was associated with greater availability of the IMM hydrophobic domains for the reaction with Hg^{2+} compared to MSL (Chávez and Holguín, 1988; Chávez et al., 1989). MSL inhibited Ca^{2+} uptake in the energized mitochondria from rat liver and rat heart (Harris and Zaba, 1977). Our results (Figs. 4.49–4.51 and Tables 4.23–4.25) did not confirm previously obtained data (Rugolo et al., 1981; Rigoni et al., 1984) that MSL can inhibit the opening of MPTP in the inner membrane calcium-loaded RLM. In this case, it seems that we should talk about a slight stimulation by MSL the Tl^{+}-induced MPTP in the inner membrane of calcium-loaded mitochondria. Since swelling of these mitochondria slightly increased in the presence of 20 μM MSL that does not penetrate the matrix (Fig. 4.49A) compared to the similar action of hydrophobic thiol reagent PAO, which already at 2 μM concentration significantly increased the mitochondria swelling (Fig. 4.39A). As shown above, the MPTP opening in the inner membrane was accompanied by **ΔΨ**mito decline, state 3 and 3UDNP respiration decrease with a simultaneous swelling increase of calcium-loaded RLM in media with $TlNO_3$. MSL more potently (Table 4.24) than DIDS (Table 4.21) prevented the decrease in state 3UDNP respiration of RLM loaded by Ca^{2+}. At the same time, respiration has just reduced more strongly in like experiments with other thiol reagents (PAO, tBHP, diamide) compared with experiments with calcium alone (Table 4.21).

Summing up the results of this chapter, it is necessary to emphasize the following. MSL, which does not penetrate mitochondria and blocks the cytoplasmic-facing thiol groups of the phosphate symporter, had a negligible effect on the MPTP opening in the inner membrane of calcium-loaded RLM in a medium with $TlNO_3$ and KNO_3 (Korotkov et al., 2018). Considering our data, we can assume that the binding of MSL to outer thiol groups of PiC did not have a noticeable effect on the Tl^{+}-induced MPTP opening in the inner membrane of calcium-loaded RLM. Therefore, it is necessary to investigate further the role of the matrix-facing PiC functional groups during this MPTP opening in the inner membrane.

Fig. 4.52 shows the generalized results of our studies of the mechanisms of toxic action of thallium on mitochondria.

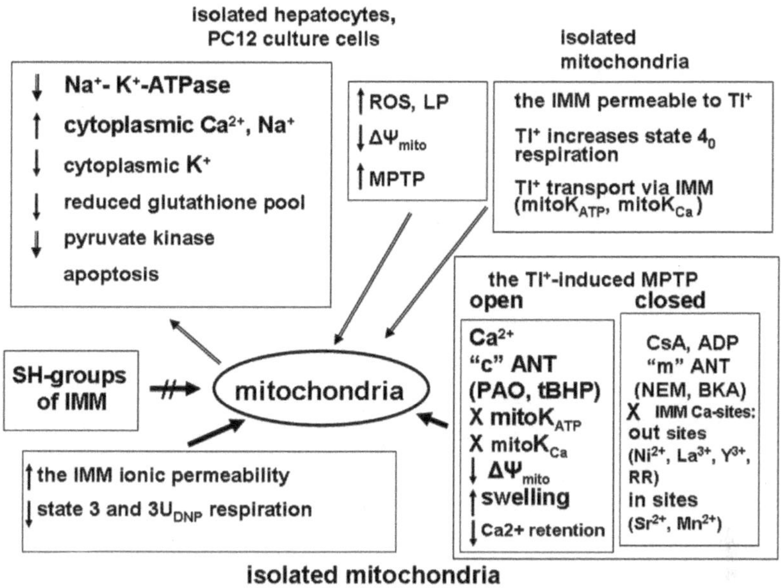

Figure 4.52 Biochemical mechanisms of thallium toxic effect on cells and mitochondria found in experiments *in vitro*. Legend: double vertical arrows indicate the Tl^+ -induced decrease in enzyme activity (Na^+-K^+-ATPase, pyruvate kinase) in experiments with cells in vitro. At the figure top, vertical arrows show changes in cytoplasmic concentrations of ions (Ca^{2+}, Na^+, K^+), induction of reactive oxygen species (ROS), and lipid peroxidation (LP) as well as the MPTP opening which were found previously after the action of Tl^+ on cells (isolated hepatocytes and PC12 cells) and mitochondria in experiments in vitro. In addition, the previously found ones are noted here: high passive permeability of the IMM to Tl^+, Tl^+-induced increase in state 4_0 respiration, and transport routes of Tl^+ into the matrix through $mitoK_{ADP}$ and $mitoBK_{Ca}$ channels. Downward sloping double arrows indicate previously obtained results on the effect of Tl^+ on mitochondria, and indirect effect on intracellular parameters, in experiments in vitro with cells and mitochondria. The figure lower part shows the results of our experiments with isolated rat liver and rat heart mitochondria in media with $TlNO_3$ and nitrates. The arrows show the Tl^+ -induced increase in the IMM passive ionic permeability, a decrease in mitochondrial respiration in states 3 and $3U_{DNP}$, a decrease in $\Delta\Psi_{mito}$, an increase in swelling, and a decrease in the mitochondria Ca^{2+} retention. Changes in mitochondrial parameters are shown depending on whether the IMM pore is open or closed. The pore is open: if the ANT is in the "c" conformation (in the presence of PAO, tBHP), if $mitoK_{ADP}$ and $mitoBK_{Ca}$ channels are closed (marked with "X"), intense mitochondrial swelling is present, $\Delta\Psi_{mito}$ decreases, and Ca^{2+} retention decreases. The pore is closed: if ANT is in the "m" conformation (in the presence of ADP, CsA, low NEM, BKA), both external (Ni^{2+}, La^{3+}, Y^{3+}, RR) and internal (Sr^{2+}, Mn^{2+}) of the IMM calcium sites are blocked. Bold oblique arrows schematically show the results of our experiments in studying the effect of Tl^+ on isolated mitochondria of rat liver and rat heart. The crossed-out bold arrow shows the absence of Tl^+ interaction with thiol groups of the inner mitochondrial membrane.

4.7 Conclusing remarks

This chapter shows the reasons and pathways of thallium penetration at different levels of biological organization (the whole organism, the organism systems and tissues, cells, intracellular structures, and in particular, mitochondria). Since the Tl^+ and K^+ effective radii are close in magnitude, this allows Tl^+ ions to use numerous potassium-dependent transport mechanisms initially intended to regulate K^+ transport of this essential biological element. As a rule, Tl^+ affinity for K^+-transport systems turns out to be an order of magnitude greater than that of K^+. Tl^+, in this case, is a kind of "Trojan horse" for cells, which, having entered the cell, distorts potassium-dependent processes in the cytoplasm and mitochondria and induces ROS production increase and the induction of the MPTP opening. At the same time, the easy polarizability of Tl^+ due to the peculiarities of its internal atomic structure (kainosymmetry, lanthanide contraction) enables Tl^+ to form numerous coordination bonds with various molecules that make up biological organisms. This Tl^+ feature, on the one hand, underlies this metal compounds' toxicity, and on the other hand, allows you to treat thallium poisoning, as in the case of Prussian blue, which forms strong complexes with Tl^+.

The strong interaction of heavy metals (Ag^+, Cd^{2+}, Hg^{2+}, Pb^{2+}, As^{3+}) with free thiol groups of mitochondria results in the respiratory enzymes' inhibition and the inner membrane ionic permeability increase. The interaction results in high-amplitude mitochondrial swelling, the inhibition of mitochondrial respiration in states 4, 3, and $3U_{DNP}$, $\Delta\Psi_{mito}$ decline, and Ca^{2+} matrix retention decrease in experiments with cells and mitochondria. Tl^+, like heavy metals, increases the inner membrane permeability of rat liver and heart mitochondria for K^+, H^+, and Na^+ ions. On the other hand, Tl^+ had little effect on mitochondrial respiratory enzymes' activity and $\Delta\Psi_{mito}$. Indeed, the reaction of mitochondrial respiration to ADP and DNP introduced into the medium with high Tl^+ concentrations and sucrose was persisted. One of the reasons for such a difference in the action of Tl^+ and these heavy metals may be the low interaction of Tl^+ with free thiol groups of the inner mitochondrial membrane. However, the decrease in mitochondrial respiration in states 3 and $3U_{DNP}$ in nitrate media with $TlNO_3$ may be associated with the mitochondrial volume increase caused by the IMM structure change. The decrease is probably not related to inhibiting the mitochondrial respiratory complexes in the heavy metal exposure case.

Lower concentrations of heavy metals (Cd^{2+}, Pb^{2+}, Hg^{2+}, Ag^{+}) were known to induce the MPTP opening in the inner membrane in media free of Ca^{2+}. In contrast, 10 times higher Ca^{2+} concentrations did require the MPTP opening. Such pore opening did accompany by high-amplitude swelling of mitochondria and $\Delta\Psi_{mito}$ decline. At the same time, Tl^{+} by itself did not induce the MPTP opening in the inner membrane in media with $TlNO_3$ and nitrates. The difference from other heavy metals was that the MPTP opening became possible only in calcium-loaded mitochondria. In this case, there was high-amplitude mitochondrial swelling, $\Delta\Psi_{mito}$ decline, and a more potent decrease of state 3, 4, $3U_{DNP}$ respiration compared to experiments in the calcium-free medium with $TlNO_3$. These effects became even more significant with an increase in the medium concentration of $TlNO_3$, which gave us to speculate the Tl^{+}-induced MPTP opening in the calcium-loaded mitochondria inner membrane. Thus, the MPTP opening became possible by acting on mitochondria either Tl^{+} in the presence of Ca^{2+} or some heavy metals alone (Cd^{2+}, Pb^{2+}, Hg^{2+}, Ag^{+}).

The MPTP inhibitors (ADP, CsA, BKA) partially inhibited this pore induced by heavy metals. However, these MPTP inhibitors abolished the above effects associated with the Tl^{+}-induced MPTP opening. Such action difference of inhibitors is possible due to the strong binding of heavy metals to the inner membrane thiol groups. The noticeable weakening of these metals' effects did associate with such pore opening, which did observe with the simultaneous presence of MPTP inhibitors and dithiothreitol, powerful SH groups' reductant. On the other hand, the common point was that ruthenium red (the calcium uniporter inhibitor), as well as Sr^{2+} and Mn^{2+} (competitive inhibitors of matrix-facing specific calcium sites), prevented the opening of MPTP induced by Cd^{2+} or Ca^{2+} or Tl^{+} (our experiments).

Thallium compared to bivalent heavy metals (Cd, Hg, Pb) was more toxic in experiments in vivo. However, the poisonous thallium effect on mitochondria was noticeably slower than the similar effect of these metals in experiments in vitro (our and literature data). Heavy metals' body penetrating induces synthesis of metallothionein proteins which bind with Cd^{2+}, Hg^{2+}, Pb^{2+}, and other heavy metal ions. This process underlies the detoxification processes for these heavy metals' but not thallium poisoning. Such a contradiction in the toxic impact of thallium and these metals can explain that metallothioneins containing active thiol groups, formed when heavy metals enter the body appear, do not bind strongly enough to Tl^{+} ions penetrated in the body.

In addition, the study showed that the adenine nucleotide translocase, phosphate symporter, potassium channels (mitoK_{ATP}, mitoBK_{Ca}) affect the Tl^+-induced MPTP opening in the inner membrane of calcium-loaded mitochondria. The pore closure occurred in experiments with MPTP inhibitors (ADP, BKA, CsA, Mg^{2+}, N-ethylmaleimide). Opposite, the pore opening was more intensive in the presence of MPTP modulators (PAO, CAT, tBHP, diamide).

Another reason for the thallium compounds' toxicity for living organisms' cells is well-known of crystal-chemical radii proximity of K^+ and Tl^+ ions. Despite this, the affinity of Tl^+ for most potassium cell transporters turned out to be an order of magnitude higher than that of K^+. This distinction in K^+ and Tl^+ ions' behavior is due to the electron structure difference between K and Tl atoms. Because of this, Tl^+ but not K^+ can quickly polarize, and Tl^+ but not K^+, for this reason, can form stronger chemical bonds with reactive electron-donor groups of biological molecules ($R{-}CO_2^-$, $R{-}OPO_3^{2-}$, $R{-}SO_3^-$, $R{-}SH$) with coordination numbers of 6–12. Tl^+ ion can replace K^+ in the active centers of potassium-dependent enzymes and change their activity in different directions. Tl^+ ions are known to easily replace K^+ in Prussian blue soluble forms, making it possible to use this substance to fight against human poisoning with thallium salts. At the same time, the use of various chelators and thiol compounds used effectively to reduce the heavy metal poisoning turned out to be entirely useless for toxicity reducing of thallium which, according to our data, revealed a low interaction with the IMM thiol groups.

Given that the radii Na^+, K^+, Rb^+, Cs^+, and Tl^+ are 0.98, 0.133, 0.149, 0.165, and 0.147 nm, respectively. Besides, effective Tl^+ radius (nm) increases with rising the coordination number and is equal to 0.164 (6), 0.173 (8), and 0.184 (12) (the number indicated in parentheses). The difference between K^+ and Tl^+ manifests itself even for their aqueous solutions. Most of the salts of potassium, rubidium, and cesium are known to be readily soluble in water. However, potassium, rubidium, and especially cesium perchlorates ($KClO_4$, $RbClO_4$, $CsClO_4$) are low soluble, in contrast to $TlClO_4$, which is moderately soluble in water. The metals' chlorides showed inverse solubility dependence. The hydration enthalpy for Tl^+ was even closer to Na^+ than for K^+.

Thus, this study revealed several Tl^+ toxic effects in experiments with mitochondria in vitro. These Tl^+ damage effects on mitochondria were the next: the increased inner membrane ion permeability, high-amplitude swelling, the decrease in mitochondrial respiration in states 4, 3, and $3U_{DNP}$, the inner membrane potential decline, and the Tl^+-induced MPTP opening, which got more visible in transiting of the adenine nucleotide translocase to "c" conformation. Thallium was found to deplete the pool of reduced glutathione and increase the production of ROS in cells, reduce the body's antioxidants defenses, and inhibit enzymes containing active thiol groups. The ROS production increase can probably relate to the reaction of Tl^+ with various cell structures and, in particular, with mitochondrial structures. This study allowed us to hypothesize that oxidizing thiol groups' Tl^{3+} ions can form in interacting Tl^+ with ROS produced by the mitochondrial respiratory chain complexes. This future knowledge could improve people's treatment of thallium compounds in the complex use of antioxidants, hemoperfusion, and hemodialysis.

The study of Tl^+ toxic effects on the mitochondria energetics and structure, along with the already discovered biochemical mechanisms, can explain the harmful thallium effects on living organisms and contribute to the search for new ways to prevent thallium poisoning. On the other hand, the used model in studying calcium-loaded mitochondria in the medium with $TlNO_3$ and KNO_3 can search for new MPTP modulators and inhibitors among various chemical and natural compounds. The evidence we obtained and the already discovered Tl^+ toxic mechanisms allow us to speculate that mitochondria are the critical intracellular target of thallium toxicity.

Acknowledgments

I express my sincere gratitude to my co-authors mentioned in the publications on this project regarding studies of the thallium effects on isolated mitochondria. The research was carried out within the state assignment of FASO of Russia (theme No. AAAA-A18-118012290142-9).

References

Adiele, R.C., Stevens, D., Kamunde, C., 2012. Differential inhibition of electron transport chain enzyme complexes by cadmium and calcium in isolated rainbow trout (*Oncorhynchus mykiss*) hepatic mitochondria. Toxicol. Sci. 127 (1), 110–119. https://doi.org/10.1093/toxsci/kfs091.

Aldakkak, M., Stowe, D.F., Dash, R.K., Camara, A.K., 2013. Mitochondrial handling of excess Ca^{2+} is substrate-dependent with implications for reactive oxygen species generation. Free Radic. Biol. Med. 56, 193–203. https://doi.org/10.1016/j.freeradbiomed.2012.09.020.

Allshire, A., Bernardi, P., Saris, N.E., 1985. Manganese stimulates calcium flux through the mitochondrial uniporter. Biochim. Biophys. Acta 807 (2), 202–209. https://doi.org/10.1016/0005-2728(85)90123-9.

Aoyama, H., Yoshida, M., Yamamura, Y., 1988. Induction of lipid peroxidation in tissues of thallous malonate-treated hamster. Toxicology 53 (1), 11–18. https://doi.org/10.1016/0300-483x(88)90232-6.

Ardehali, H., O'Rourke, B., 2005. Mitochondrial K_{ATP} channels in cell survival and death. J. Mol. Cell. Cardiol. 39 (1), 7–16. https://doi.org/10.1016/j.yjmcc.2004.12.003.

Azzi, A., Azzone, G.F., 1966. Swelling and shrinkage phenomena in liver mitochondria. IV. Reversible swelling changes linked to transport of monovalent cations stimulated by valinomycin. Biochim. Biophys. Acta 113 (3), 445–456. https://doi.org/10.1016/s0926-6593(66)80003-6.

Azzone, G.F., Massair, S., Pozzan, T., 1976a. Mechanism of active shrinkage in mitochondria II Coupling between weak electrolyte fluxes. Biochim. Biophys. Acta 423 (1), 15–26. https://doi.org/10.1016/0005-2728(76)90097-9.

Azzone, G.F., Massair, S., Pozzan, T., 1976b. Mechanism of active shrinkage in mitochondria II Coupling between strong electrolyte fluxes. Biochim. Biophys. Acta 423 (1), 27–41. https://doi.org/10.1016/0005-2728(76)90098-0.

Baines, C.P., 2009. The molecular composition of the mitochondrial permeability transition pore. J. Mol. Cell. Cardiol. 46 (6), 850–857. https://doi.org/10.1016/j.yjmcc.2009.02.007.

Barrera, H., Gomez-Puyou, A., 1975. Characteristics of the movement of K^+ across the mitochondrial membrane and the inhibitory action of Tl^+. J. Biol. Chem. 250 (14), 5370–5374.

Barroso-Moguel, R., Villeda-Hernández, J., Méndez-Armenta, M., Ríos, C., Monroy-Noyola, A., 1994. Combined D-penicillamine and Prussian blue as antidotal treatment against thallotoxicosis in rats: evaluation of cerebellar lesions. Toxicology 89 (1), 15–24. https://doi.org/10.1016/0300-483x(94)90129-5.

Basso, E., Petronilli, V., Forte, M.A., Bernardi, P., 2008. Phosphate is essential for inhibition of the mitochondrial permeability transition pore by cyclosporin A and by cyclophilin D ablation. J. Biol. Chem. 283 (39), 26307–26311. https://doi.org/10.1074/jbc.C800132200.

Bates, T.E., Heales, S.J., Davies, S.E., Boakye, P., Clark, J.B., 1994. Effects of 1-methyl-4-phenylpyridinium on isolated rat brain mitochondria: evidence for a primary involvement of energy depletion. J. Neurochem. 63 (2), 640–648. https://doi.org/10.1046/j.1471-4159.1994.63020640.x.

Beatrice, M.C., Palmer, J.W., Pfeiffer, D.R., 1980. The relationship between mitochondrial membrane permeability, membrane potential, and the retention of Ca^{2+} by mitochondria. J. Biol. Chem. 255 (18), 8663–8671.

Belyaeva, E.A., Glazunov, V.V., Korotkov, S.M., 2002. Cyclosporin A-sensitive permeability transition pore is involved in Cd^{2+}-induced dysfunction of isolated rat liver mitochondria: doubts no more. Arch. Biochem. Biophys. 405 (2), 252–264. https://doi.org/10.1016/s0003-9861(02)00400-9.

Belyaeva, E.A., Glazunov, V.V., Korotkov, S.M., 2004a. Cd^{2+} -promoted mitochondrial permeability transition: a comparison with other heavy metals. Acta Biochim. Pol. 51 (2), 545–551.

Belyaeva, E.A., Glazunov, V.V., Korotkov, S.M., 2004b. Cd^{2+} versus Ca^{2+}-produced mitochondrial membrane permeabilization: a proposed direct participation of respiratory complexes I and III. Chem. Biol. Interact. 150 (3), 253–270. https://doi.org/10.1016/j.cbi.2004.09.019.

Belyaeva, E.A., Glazunov, V.V., Nikitina, E.R., Korotkov, S.M., 2001. Bivalent metal ions modulate Cd^{2+} effects on isolated rat liver mitochondria. J. Bioenerg. Biomembr. 33 (4), 303–318. https://doi.org/10.1023/a:1010607307099.

Belyaeva, E.A., Korotkov, S.M., 2003. Mechanism of primary Cd^{2+}-induced rat liver mitochondria dysfunction: discrete modes of Cd^{2+} action on calcium and thiol-dependent domains. Toxicol. Appl. Pharmacol. 192 (1), 56–68. https://doi.org/10.1016/s0041-008x(03)00255-2.

Belyaeva, E.A., Korotkov, S.M., Saris, N.E., 2011. In vitro modulation of heavy metal-induced rat liver mitochondria dysfunction: a comparison of copper and mercury with cadmium. J. Trace Elem. Med. Biol. 25 (Suppl. 1), S63–S73. https://doi.org/10.1016/j.jtemb.2010.10.007.

Bernardes, C.F., Meyer-Fernandes, J.R., Basseres, D.S., Castilho, R.F., Vercesi, A.E., 1994. Ca^{2+}-dependent permeabilization of the inner mitochondrial membrane by 4,4′-diisothiocyanatostilbene-2,2′-disulfonic acid (DIDS). Biochim. Biophys. Acta 1188 (1–2), 93–100. https://doi.org/10.1016/0005-2728(94)90026-4.

Bernardes, C.F., Meyer-Fernandes, J.R., Martins, O.B., Vercesi, A.E., 1997. Inhibition of succinic dehydrogenase and F_0F_1-ATP synthase by 4,4′-diisothiocyanatostilbene-2,2′-disulfonic acid (DIDS). Z. Naturforsch C. J. Biosci. 52 (11–12), 799–806. https://doi.org/10.1515/znc-1997-11-1212.

Bernardi, P., 1999. Mitochondrial transport of cations: channels, exchangers, and permeability transition. Physiol. Rev. 79 (4), 1127–1155.

Bernardi, P., 2013. The mitochondrial permeability transition pore: a mystery solved? Front. Physiol. 4 (95), 1–12. https://doi.org/10.1152/physrev.1999.79.4.1127.

Bernardi, P., Carraro, M., Lippe, G., 2021. The mitochondrial permeability transition: recent progress and open questions. FEBS Lett. https://doi.org/10.1111/febs.16254.

Bernardi, P., Pietrobon, D., 1982. On the nature of Pi-induced Mg^{2+}-prevented Ca^{2+} release in rat liver mitochondria. FEBS Lett. 139 (1), 9–12. https://doi.org/10.1016/0014-5793(82)80475-4.

Bernardi, P., Vassanelli, S., Veronese, P., Colonna, R., Szabó, I., Zoratti, M., 1992. Modulation of the mitochondrial permeability transition pore. Effect of protons and divalent cations. J. Biol. Chem. 267 (5), 2934–2939.

Bernardi, P., Veronese, P., Petronilli, V., 1993. Modulation of the mitochondrial cyclosporin A-sensitive permeability transition pore. I. Evidence for two separate Me^{2+} binding sites with opposing effects on the pore open probability. J. Biol. Chem. 268 (2), 1005–1010.

Beutner, G., Alavian, K.N., Jonas, E.A., Porter Jr., G.A., 2017. The mitochondrial permeability transition pore and ATP synthase. Handb. Exp. Pharmacol. 240, 21–46. https://doi.org/10.1007/164_2016_5.

Bindoli, A., Callegaro, M.T., Barzon, E., Benetti, M., Rigobello, M.P., 1997. Influence of the redox state of pyridine nucleotides on mitochondrial sulfhydryl groups and permeability transition. Arch. Biochem. Biophys. 342 (1), 22–28. https://doi.org/10.1006/abbi.1997.9986.

Bonora, M., Morganti, C., Morciano, G., Pedriali, G., Lebiedzinska-Arciszewska, M., et al., 2017. Mitochondrial permeability transition involves dissociation of F_1F_0 ATP synthase dimers and C-ring conformation. EMBO Rep. 18 (7), 1077–1089. https://doi.org/10.15252/embr.201643602.

Bragadin, M., Toninello, A., Bindoli, A., Rigobello, M.P., Canton, M., 2003. Thallium induces apoptosis in Jurkat cells. Ann. N. Y. Acad. Sci. 1010, 283–291. https://doi.org/10.1196/annals.1299.049.

Bragadin, M., Viola, E.R., 1997. Ni^{2+} as a competitive inhibitor of calcium transport in mitochondria. J. Inorg. Biochem. 66 (4), 227–229. https://doi.org/10.1016/s0162-0134(96)00207-3.

Bravo, C., Chávez, E., Rodríguez, J.S., Moreno-Sánchez, R., 1997. The mitochondrial membrane ermeability transition induced by inorganic phosphate or inorganic arsenate. A comparative study. Comp. Biochem. Physiol. B Biochem. Mol. Biol. 117 (1), 93–99. https://doi.org/10.1016/s0305-0491(96)00257-x.PMID: 9180017.

Brawer, J.R., McCarthy, G.F., Gornitsky, M., Frankel, D., Mehindate, K., Schipper, H.M., 1998. Mercuric chloride induces a stress response in cultured astrocytes characterized by mitochondrial uptake of iron. Neurotoxicol 19 (6), 767–776.

Brierley, G.P., Jurkowitz, M., Chávez, E., Jung, D.W., 1977. Energy-dependent contraction of swollen heart mitochondria. J. Biol. Chem. 252 (22), 7932–7939.

Brierley, G.P., Jurkowitz, M., Scott, K.M., Merola, A.J., 1970. Ion transport by heart mitochondria. XX. Factors affecting passive osmotic swelling of isolated mitochondria. J. Biol. Chem. 245 (20), 5404–5411.

Brierley, G.P., Knight, V.A., Settlemire, C.T., 1968. Ion transport by heart mitochondria. XII. Activation of monovalent cation uptake by sulfhydrly group reagents. J. Biol. Chem. 243 (19), 5035–5043.

Broekemeier, K.M., Dempsey, M.E., Pfeiffer, D.R., 1989. Cyclosporin A is a potent inhibitor of the inner membrane permeability transition in liver mitochondria. J. Biol. Chem. 264 (14), 7826–7830.

Broekemeier, K.M., Pfeiffer, D.R., 1995. Inhibition of the mitochondrial permeability transition by cyclosporin A during long time frame experiments: relationship between pore opening and the activity of mitochondrial phospholipases. Biochemistry 34 (50), 16440–16449. https://doi.org/10.1021/bi00050a027.

Brown, A.M., Kristal, B.S., Effron, M.S., Shestopalov, A.I., Ullucci, P.A., et al., 2000. Zn^{2+} inhibits alpha-ketoglutarate-stimulated mitochondrial respiration and the isolated alpha-ketoglutarate dehydrogenase complex. J. Biol. Chem. 275 (18), 13441–13447. https://doi.org/10.1074/jbc.275.18.13441.

Brustovetsky, N., 2020. The role of adenine nucleotide translocase in the mitochondrial permeability transition. Cells 9 (12), 2686. https://doi.org/10.3390/cells9122686.

Bunni, M.A., Baindur, S.R., Douglas, K.T., Drake, A.F., 1987. Organothallium(III) reagents for modification of biomacromolecules: irreversible labelling of phosphoglycerate kinase from rabbit muscle. Biochim. Biophys. Acta 923 (3), 421–430. https://doi.org/10.1016/0304-4165(87)90050-x.

Bustamante, J., Nutt, L., Orrenius, S., Gogvadze, V., 2005. Arsenic stimulates release of cytochrome c from isolated mitochondria via induction of mitochondrial permeability transition. Toxicol. Appl. Pharmacol. 207 (2), 110–116. https://doi.org/10.1016/j.taap.2005.01.024. Suppl. 1.

Cameron, I., McNamee, P.M., Markham, A., Morgan, R.M., Wood, M., 1986. The effects of cadmium on succinate and NADH-linked substrate oxidations in rat hepatic mitochondria. J. Appl. Toxicol. 6 (5), 325–330. https://doi.org/10.1002/jat.2550060505.

Caplan, A.I., Carafoli, E., 1965. The effect of Sr^{2+} on swelling and ATP-linked contraction of mitochondria. Biochim. Biophys. Acta 104 (2), 317–329. https://doi.org/10.1016/0304-4165(65)90338-7.

Carafoli, E., 1965. Active accumulation of Sr^{2+} by rat liver mitochondria. 3. Stimulation of respiration by Sr^{2+} and its stoichiometry. Biochim. Biophys. Acta 97 (1), 107–117. https://doi.org/10.1016/0304-4165(65)90274-6.

Carraro, M., Checchetto, V., Szabó, I., Bernardi, P., 2019. F-ATP synthase and the permeability transition pore: fewer doubts, more certainties. FEBS Lett. 593 (13), 1542–1553. https://doi.org/10.1002/1873-3468.13485.

Castilho, R.F., Kowaltowski, A.J., Vercesi, A.E., 1996. The irreversibility of inner mitochondrial membrane permeabilization by Ca^{2+} plus prooxidants is determined by the extent of membrane protein thiol cross-linking. J. Bioenerg. Biomembr. 28 (6), 523–539. https://doi.org/10.1007/BF02110442.

Cavanagh, J.B., 1991. What have we learnt from Graham Frederick Young?: reflections on the mechanism of thallium neurotoxicity. Neuropathol. Appl. Neurobiol. 7 (1), 3–9. https://doi.org/10.1111/j.1365-2990.1991.tb00687.x.

Chandler, H.A., Scott, M., 1986. A review of thallium toxicology. J. R. Nav. Med. Serv. 72 (2), 75–79.

Chappell, J.B., Greville, G.D., 1954. Effect of silver ions on mitochondrial adenosine triphosphatase. Nature 174 (4437), 930–931. https://doi.org/10.1038/174930b0.

Chávez, E., 1973. Ion regulation of ATP hydrolysis in mitochondria. FEBS Lett. 33 (1), 129–131.

Chávez, E., Holguín, J.A., 1988. Mitochondrial calcium release as induced by Hg^{2+}. J. Biol. Chem. 263 (8), 3582–3587.

Chávez, E., Holguin, J.A., Zazueta, C., Bravo, C., 1989. Induction of mitochondrial Ca^{2+} uptake by mersalyl. Int. J. Biochem. 21 (11), 1241–1244. https://doi.org/10.1016/0020-711x(89)90010-4.

Chávez, E., Jay, D., Bravo, C., 1987. The mechanism of lead-induced mitochondrial Ca^{2+} efflux. J. Bioenerg. Biomembr. 19 (3), 285–295. https://doi.org/10.1007/BF00762418.

Chávez, E., Meléndez, E., Zazueta, C., Reyes-Vivas, H., Perales, S.G., 1997. Membrane permeability transition as induced by dysfunction of the electron transport chain. Biochem. Mol. Biol. Int. 41 (5), 961–968. https://doi.org/10.1080/15216549700202021.

Chávez, E., Zazueta, C., Osornio, A., Holguín, J.A., Miranda, M.E., 1991. Protective behavior of captopril on Hg^{2+}-induced toxicity on kidney mitochondria. In vivo and in vitro experiments. J. Pharmacol. Exp. Ther. 256 (1), 385–390.

Cheng, Y., Debska-Vielhaber, G., Siemen, D., 2010. Interaction of mitochondrial potassium channels with the permeability transition pore. FEBS Lett. 584 (10), 2005–2012. https://doi.org/10.1016/j.febslet.2009.12.038.

Cheng, Y., Gu, X.Q., Bednarczyk, P., Wiedemann, F.R., Haddad, G.G., Siemen, D., 2008. Hypoxia increases activity of the BK-channel in the inner mitochondrial membrane and reduces activity of the permeability transition pore. Cell. Physiol. Biochem. 22 (1–4), 127–136. https://doi.org/10.1159/000149790.

Chinopoulos, C., Szabadkai, G., 2013. What makes you can also break you: mitochondrial permeability transition pore formation by the c subunit of the F_1F_0 ATP-synthase? Front. Oncol. 3 (25), 1–2. https://doi.org/10.3389/fonc.2013.00025.

Collison, M.W., Beidler, D., Grimm, L.M., Thomas, J.A., 1986. A comparison of protein S-thiolation (protein mixed-disulfide formation) in heart cells treated with t-butyl hydroperoxide or diamide. Biochim. Biophys. Acta 885 (1), 58–67. https://doi.org/10.1016/0167-4889(86)90038-8.

Cornelius, G., Gärtner, W., Haynes, D.H., 1974. Cation complexation by valinomycin- and nigericin-type inophores registered by the fluorescence signal of Tl^+. Biochemistry 13 (15), 3052–3057. https://doi.org/10.1021/bi00712a009.

Costa, A.D., Garlid, K.D., 2008. Intramitochondrial signaling: interactions among mito-K_{ATP}, PKCepsilon, ROS, and MPT. Am. J. Physiol. Heart Circ. Physiol. 295 (2), H874–H882. https://doi.org/10.1152/ajpheart.01189.2007.

Costa, A.D., Jakob, R., Costa, C.L., Andrukhiv, K., West, I.C., Garlid, K.D., 2006b. The mechanism by which the mitochondrial ATP-sensitive K^+ channel opening and H_2O_2 inhibit the mitochondrial permeability transition. J. Biol. Chem. 281 (30), 20801–20808. https://doi.org/10.1074/jbc.M600959200.

Costa, A.D., Quinlan, C.L., Andrukhiv, A., West, I.C., Jabůrek, M., Garlid, K.D., 2006a. The direct physiological effects of mitoK_{ATP} opening on heart mitochondria. Am. J. Physiol. Heart Circ. Physiol. 290 (1), H406–H415. https://doi.org/10.1152/ajpheart.00794.2005.

Costantini, P., Belzacq, A.S., Vieira, H.L., Larochette, N., de Pablo, M.A., et al., 2000. Oxidation of a critical thiol residue of the adenine nucleotide translocator enforces Bcl-2-independent permeability transition pore opening and apoptosis. Oncogene 19 (2), 307–314. https://doi.org/10.1038/sj.onc.1203299.

Costantini, P., Colonna, R., Bernardi, P., 1998. Induction of the mitochondrial permeability transition by N-ethylmaleimide depends on secondary oxidation of critical thiol groups. Potentiation by copper-ortho-phenanthroline without dimerization of the adenine nucleotide translocase. Biochim. Biophys. Acta 1365 (3), 385–392. https://doi.org/10.1016/s0005-2728(98)00090-5.

Crestanello, J.A., Doliba, N.M., Babsky, A.M., Doliba, N.M., Niibori, K., et al., 2000. Opening of potassium channels protects mitochondrial function from calcium overload. J. Surg. Res. 94 (2), 116–123. https://doi.org/10.1006/jsre.2000.5979.

Cvjetko, P., Cvjetko, I., Pavlica, M., 2010. Thallium toxicity in humans. Arh. Hig. Rada Toksikol. Croatia 61 (1), 111–119. https://doi.org/10.2478/10004-1254-61-2010-1976.

Dahlem, Y.A., Wolf, G., Siemen, D., Horn, T.F., 2006. Combined modulation of the mitochondrial ATP-dependent potassium channel and the permeability transition pore causes prolongation of the biphasic calcium dynamics. Cell Calcium 39 (5), 387–400. https://doi.org/10.1016/j.ceca.2006.01.001.

Danilewicz, M., Danilewicz, M., Kurnatowski, A., 1979. Histopathological changes in the kidney of experimental animals caused by lethal doses of thallium (in Polish). Med. Pr. 30 (4), 257–265.

De Stefani, D., Rizzuto, R., Pozzan, T., 2016. Enjoy the trip: calcium in mitochondria back and forth. Annu. Rev. Biochem. 85, 161–192. https://doi.org/10.1146/annurev-biochem-060614-034216.

Devin, A., Guérin, B., Rigoulet, M., 1996. Dependence of flux size and efficiency of oxidative phosphorylation on external osmolarity in isolated rat liver mitochondria: role of adenine nucleotide carrier. Biochim. Biophys. Acta 1273 (1), 13–20. https://doi.org/10.1016/0005-2728(95)00130-1.

Devin, A., Guérin, B., Rigoulet, M., 1997a. Control of oxidative phosphorylation in rat liver mitochondria: effect of ionic media. Biochim. Biophys. Acta 1319 (2–3), 293–300. https://doi.org/10.1016/s0005-2728(96)00172-7.

Devin, A., Guérin, B., Rigoulet, M., 1997b. Response of isolated rat liver mitochondria to variation of external osmolarity in KCl medium: regulation of matrix volume and oxidative phosphorylation. J. Bioenerg. Biomembr. 29 (6), 579–590. https://doi.org/10.1023/a:1022435102552.

Di Simplicio, P., Lupis, E., Rossi, R., 1996. Different mechanisms of formation of glutathione-protein mixed disulfides of diamide and tert-butyl hydroperoxide in rat blood. Biochim. Biophys. Acta 1289 (2), 252–260. https://doi.org/10.1016/0304-4165(95)00160-3.

Dineley, K.E., Richards, L.L., Votyakova, T.V., Reynolds, I.J., 2005. Zinc causes loss of membrane potential and elevates reactive oxygen species in rat brain mitochondria. Mitochondrion 5 (1), 55–65. https://doi.org/10.1016/j.mito.2004.11.001.

Diwan, J.J., 1985. Ba^{2+} uptake and the inhibition by Ba^{2+} of K^+ flux into rat liver mitochondria. J. Membr. Biol. 84 (2), 165–171. https://doi.org/10.1007/BF01872214.

Diwan, J.J., 1986. Effect of quinine on mitochondrial K^+ and Mg^{2+} flux. Biochem. Biophys. Res. Commun. 135 (3), 830–836. https://doi.org/10.1016/0006-291x(86)91003-x.

Diwan, J.J., Dazé, M., Richardson, R., Aronson, D., 1979. Kinetics of Mg^{2+} flux into rat liver mitochondria. Biochemistry 18 (12), 2590–2595. https://doi.org/10.1021/bi00579a024.

Diwan, J.J., Lehrer, P.H., 1977. Inhibition of mitochondrial potassium ion flux by thallous ions. Biochem. Soc. Trans. 5 (1), 203–205. https://doi.org/10.1042/bst0050203.

Diwan, J.J., Paliwal, R., Kaftan, E., Bawa, R., 1990. A mitochondrial protein fraction catalyzing transport of the K^+ analog $T1^+$. FEBS Lett. 273 (1–2), 215–218. https://doi.org/10.1016/0014-5793(90)81088-6.

Dong, S., Zhao, Y., Liu, H., Yang, X., Wang, K., 2009. Duality of effect of La^{3+} on mitochondrial permeability transition pore depending on the concentration. Biometals 22 (6), 917–926. https://doi.org/10.1007/s10534-009-9244-1.

Dorta, D.J., Leite, S., DeMarco, K.C., Prado, I.M., Rodrigues, T., et al., 2003. A proposed sequence of events for cadmium-induced mitochondrial impairment. J. Inorg. Biochem. 97 (3), 251–257. https://doi.org/10.1016/s0162-0134(03)00314-3.

Douglas, K.T., Bunni, M.A., Baindur, S.R., 1990. Thallium in biochemistry. Int. J. Biochem. 22 (5), 429–438. https://doi.org/10.1016/0020-711x(90)90254-z.

Drahota, Z., Gazzotti, P., Carafoli, E., Rossi, C.S., 1969. A comparison of the effects of different divalent cations on a number of mitochondrial reactions linked to ion translocation. Arch. Biochem. Biophys. 130 (1), 267–273. https://doi.org/10.1016/0003-9861(69)90033-2.

Dubyts'kyĭ, L.O., Vovkanych, L.S., 2000. [Dependence of the inhibitory effect of metal cations on rat liver mitochondrial Ca^{2+} uptake on their physico-chemical properties]. Ukr. Biochem. J. 72 (1), 42–47 (Ukrainian).

Ekwall, B., Gómez-Lechón, M.J., Hellberg, S., Bondesson, I., Castell, J.V., et al., 1990. Preliminary results from the Scandinavian multicentre evaluation of in vitro cytotoxicity (MEIC). Toxicol. Vitro 4 (4–5), 688–691. https://doi.org/10.1016/0887-2333(90)90143-h.

Ellman, G.L., 1959. Tissue sulfhydryl groups. Arch. Biochem. Biophys. 82 (1), 70–77. https://doi.org/10.1016/0003-9861(59)90090-6.

Elrod, J.W., Molkentin, J.D., 2013. Physiologic functions of cyclophilin D and the mitochondrial permeability transition pore. Circ. J. 77 (5), 1111–1122. https://doi.org/10.1253/circj.cj-13-0321.

Enriquez, J.A., 2016. Supramolecular organization of respiratory complexes. Annu. Rev. Physiol. 78. https://doi.org/10.1146/annurev-physiol-021115-105031, 533e61: 1-29.

Fagian, M.M., Pereira-da-Silva, L., Martins, I.S., Vercesi, A.E., 1990. Membrane protein thiol cross-linking associated with the permeabilization of the inner mitochondrial membrane by Ca^{2+} plus prooxidants. J. Biol. Chem. 265 (32), 19955–19960.

Fan, C., Fan, M., Orlando, B.J., Fastman, N.M., Zhang, J., et al., 2018. X-ray and cryo-EM structures of the mitochondrial calcium uniporter. Nature 559 (7715), 575–579. https://doi.org/10.1038/s41586-018-0330-9.

Flesch, P., Goldstone, S.B., 1950. Effect of thallium on sulphydryl compounds *in vitro*. J. Invest. Dermatol. 15 (5), 345–347. https://doi.org/10.1038/jid.1950.111.

Flora, S.J., Mittal, M., Mehta, A., 2008. Heavy metal induced oxidative stress & its possible reversal by chelation therapy. Indian J. Med. Res. 128 (4), 501–523.

Fontaine, E., Bernardi, P., 1999. Progress on the mitochondrial permeability transition pore: regulation by complex I and ubiquinone analogs. J. Bioenerg. Biomembr. 31 (4), 335–345. https://doi.org/10.1023/a:1005475802350.

Fontaine, E., Eriksson, O., Ichas, F., Bernardi, P., 1998. Regulation of the permeability transition pore in skeletal muscle mitochondria. Modulation by electron flow through the respiratory chain complex I. J. Biol. Chem. 273 (20), 12662–12668. https://doi.org/10.1074/jbc.273.20.12662.

Galván-Arzate, S., Pedraza-Chaverrí, J., Medina-Campos, O.N., Maldonado, P.D., Vázquez-Román, B., et al., 2005. Delayed effects of thallium in the rat brain: regional changes in lipid peroxidation and behavioral markers, but moderate alterations in antioxidants, after a single administration. Food Chem. Toxicol. 43 (7), 1037–1045. https://doi.org/10.1016/j.fct.2005.02.006.

Galván-Arzate, S., Santamaría, A., 1998. Thallium toxicity. Toxicol. Lett. 99 (1), 1–13. https://doi.org/10.1016/s0378-4274(98)00126-x.

García, N., Correa, F., Chávez, E., 2005. On the role of the respiratory complex I on membrane permeability transition. J. Bioenerg. Biomembr. 37 (1), 17–23. https://doi.org/10.1007/s10863-005-4119-9.

García, N., Martínez-Abundis, E., Pavón, N., Chávez, E., 2007. On the opening of an insensitive cyclosporine A non-specific pore by phenylarsine plus mersalyl. Cell Biochem. Biophys. 49 (2), 84–90. https://doi.org/10.1007/s12013-007-0047-0.

García, N., Zazueta, C., Martínez-Abundis, E., Pavón, N., Chávez, E., 2009. Cyclosporin A is unable to inhibit carboxyatractyloside-induced permeability transition in aged mitochondria. Comp. Biochem. Physiol. C Toxicol. Pharmacol. 149 (3), 374–381. https://doi.org/10.1016/j.cbpc.2008.09.006.

Garlid, K.D., Halestrap, A.P., 2012. The mitochondrial K_{ATP} channel-fact or fiction? J. Mol. Cell. Cardiol. 52 (3), 578–583. https://doi.org/10.1016/j.yjmcc.2011.12.011.

Garlid, K.D., Paucek, P., 2003. Mitochondrial potassium transport: the K^+ cycle. Biochim. Biophys. Acta 1606 (1–3), 23–41. https://doi.org/10.1016/s0005-2728(03)00108-7.

Giorgio, V., Bisetto, E., Soriano, M.E., Dabbeni-Sala, F., Basso, E., et al., 2009. Cyclophilin D modulates mitochondrial F_0F_1-ATP synthase by interacting with the lateral stalk of the complex. J. Biol. Chem. 284 (49), 33982–33988. https://doi.org/10.1074/jbc.M109.020115.

Giorgio, V., von Stockum, S., Antoniel, M., Fabbro, A., Fogolari, F., et al., 2013. Dimers of mitochondrial ATP synthase form the permeability transition pore. Proc. Natl. Acad. Sci. U.S.A. 110 (15), 5887–5892. https://doi.org/10.1073/pnas.1217823110.

Giustarini, D., Campoccia, G., Fanetti, G., Rossi, R., Giannerini, F., et al., 2000. Minor thiols cysteine and cysteinylglycine regulate the competition between glutathione and protein SH groups in human platelets subjected to oxidative stress. Arch. Biochem. Biophys. 380 (1), 1–10. https://doi.org/10.1006/abbi.2000.1847.

Goering, P.L., Klaassen, C.D., 1985. Mechanism of manganese-induced tolerance to cadmium lethality and hepatotoxicity. Biochem. Pharmacol. 34 (9), 1371–1379. https://doi.org/10.1016/0006-2952(85)90673-2.

Goyer, R.A., Clarkson, T.W., 2001. Toxic effects of metals. Thallium. In: Klaasen, C.D. (Ed.), Casarett and Doull's Toxicology: The Basic Science of Poisons, sixth ed. Macmillan Publishing Co., Inc., New York, pp. 855–856.

Gross, P., Runne, E., Wilson, J.W., 1948. Studies on the effect of thallium poisoning of the rat; the influence of cystine and methionine on alopecia and survival periods. J. Invest. Dermatol. 10 (3), 119–134. https://doi.org/10.1038/jid.1948.20.

Gunter, T.E., Gerstner, B., Lester, T., Wojtovich, A.P., Malecki, J., et al., 2010. An analysis of the effects of Mn^{2+} on oxidative phosphorylation in liver, brain, and heart mitochondria using state 3 oxidation rate assays. Toxicol. Appl. Pharmacol. 249 (1), 65–75. https://doi.org/10.1016/j.taap.2010.08.018.

Gunter, T.E., Sheu, S.S., 2009. Characteristics and possible functions of mitochondrial Ca^{2+} transport mechanisms. Biochim. Biophys. Acta 1787 (11), 1291–1308. https://doi.org/10.1016/j.bbabio.2008.12.011.

Gunter, T.E., Yule, D.I., Gunter, K.K., Eliseev, R.A., Salter, J.D., 2004. Calcium and mitochondria. FEBS Lett. 567 (1), 96–102. https://doi.org/10.1016/j.febslet.2004.03.071.

Gutiérrez-Aguilar, M., Baines, C.P., 2015. Structural mechanisms of cyclophilin D-dependent control of the mitochondrial permeability transition pore. Biochim. Biophys. Acta 1850 (10), 2041–2047. https://doi.org/10.1016/j.bbagen.2014.11.009.

Hakala, J.E., 1984. Thallium poisoning in a dog. Mod. Vet. Pract. 65 (10), 783–786.

Halestrap, A.P., 2009. What is the mitochondrial permeability transition pore? J. Mol. Cell. Cardiol. 46 (6), 821–831. https://doi.org/10.1016/j.yjmcc.2009.02.021.

Halestrap, A.P., 2010. A pore way to die: the role of mitochondria in reperfusion injury and cardioprotection. Biochem. Soc. Trans. 38 (4), 841–860.

Halestrap, A.P., 2014. The c ring of the F1Fo ATP synthase forms the mitochondrial permeability transition pore: a critical appraisal. Front. Oncol. 4, 234. https://doi.org/10.3389/fonc.2014.00234. eCollection 2014.

Halestrap, A.P., Brenner, C., 2003. The adenine nucleotide translocase: a central component of the mitochondrial permeability transition pore and key player in cell death. Curr. Med. Chem. 10 (16), 1507–1525. https://doi.org/10.2174/0929867033457278.

Halestrap, A.P., Davidson, A.M., 1990. Inhibition of Ca^{2+}-induced large amplitude swelling of liver and heart mitochondria by cyclosporin A is probably caused by the inhibitor binding to mitochondrial matrix peptidyl-prolyl *cis-trans* isomerase and preventing it interacting with the adenine nucleotide translocase. Biochem. J. 268, 153–160. https://doi.org/10.1042/bj2680153.

Halestrap, A.P., McStay, G.P., Clarke, S.J., 2002. The permeability transition pore complex: another view. Biochimie 84, 153–166. https://doi.org/10.1016/s0300-9084(02)01375-5.

Halestrap, A.P., Richardson, A.P., 2015. The mitochondrial permeability transition: a current perspective on its identity and role in ischaemia/reperfusion injury. J. Mol. Cell. Cardiol. 78, 129–141. https://doi.org/10.1016/j.yjmcc.2014.08.018.

Halestrap, A.P., Woodfield, K.Y., Connern, C.P., 1997. Oxidative stress, thiol reagents, and membrane potential modulate the mitochondrial permeability transition by affecting nucleotide binding to the adenine nucleotide translocase. J. Biol. Chem. 272 (6), 3346–3354. https://doi.org/10.1074/jbc.272.6.3346.

Hansen, R.E., Roth, D., Winther, J.R., 2009. Quantifying the global cellular thiol-disulfide status. Proc. Natl. Acad. Sci. U.S.A. 106 (2), 422–427. https://doi.org/10.1073/pnas.0812149106.

Hanzel, C.E., Verstraeten, S.V., 2009. Tl(I) and Tl(III) activate both mitochondrial and extrinsic pathways of apoptosis in rat pheochromocytoma (PC12) cells. Toxicol. Appl. Pharmacol. 236 (1), 59–70. https://doi.org/10.1016/j.taap.2008.12.029.

Harris, E.J., Al-Shaikhaly, M., Baum, H., 1979. Stimulation of mitochondrial calcium ion efflux by thiol-specific reagents and by thyroxine. The relationship to adenosine diphosphate retention and to mitochondrial permeability. Biochem. J. 182 (2), 455–464. https://doi.org/10.1042/bj1820455.

Harris, E.J., Zaba, B., 1977. The phosphate requirement for Ca^{2+}-uptake by heart and liver mitochondria. FEBS Lett. 79 (2), 284–290. https://doi.org/10.1016/0014-5793(77)80804-1.

Hasan, M., Ali, S.F., 1981. Effects of thallium, nickel, and cobalt administration of the lipid peroxidation in different regions of the rat brain. Toxicol. Appl. Pharmacol. 57 (1), 8–13. https://doi.org/10.1016/0041-008x(81)90019-3.

Hasan, M., Haider, S.S., 1989. Acetyl-homocysteine thiolactone protects against some neurotoxic effects of thallium. Neurotoxicol 10 (2), 257–261.

Hashimoto, M., Majima, E., Goto, S., Shinohara, Y., Terada, H., 1999. Fluctuation of the first loop facing the matrix of the mitochondrial ADP/ATP carrier deduced from intermolecular cross-linking of Cys_{56} residues by bifunctional dimaleimides. Biochemistry 38 (3), 1050−1056. https://doi.org/10.1021/bi9822978.

Hausenloy, D.J., Yellon, D.M., Mani-Babu, S., Duchen, M.R., 2004. Preconditioning protects by inhibiting the mitochondrial permeability transition. Am. J. Physiol. Heart Circ. Physiol. 287 (2), H841−H849. https://doi.org/10.1152/ajpheart.00678.2003.

Haworth, R.A., Hunter, D.R., 1979. The Ca^{2+}-induced membrane transition in mitochondria. II. Nature of the Ca^{2+} trigger site. Arch. Biochem. Biophys. 195 (2), 460−467. https://doi.org/10.1016/0003-9861(79)90372-2.

Heaton, G.M., Nicholls, D.G., 1976. The calcium conductance of the inner membrane of rat liver mitochondria and the determination of the calcium electrochemical gradient. Biochem. J. 156 (3), 635−646. https://doi.org/10.1042/bj1560635.

Heinen, A., Aldakkak, M., Stowe, D.F., Rhodes, S.S., Riess, M.L., et al., 2007b. Reverse electron flow-induced ROS production is attenuated by activation of mitochondrial Ca^{2+}-sensitive K^+ channels. Am. J. Physiol. Heart Circ. Physiol. 293 (3), H1400−H1407. https://doi.org/10.1152/ajpheart.00198.2007.

Heinen, A., Camara, A.K., Aldakkak, M., Rhodes, S.S., Riess, M.L., Stowe, D.F., 2007a. Mitochondrial Ca^{2+}-induced K^+ influx increases respiration and enhances ROS production while maintaining membrane potential. Am. J. Physiol. Cell Physiol. 292 (1), C148−C156. https://doi.org/10.1152/ajpcell.00215.2006.

Heinen, A., Winning, A., Schlack, W., Hollmann, M.W., Preckel, B., et al., 2008. The regulation of mitochondrial respiration by opening of mK_{Ca} channels is age-dependent. Eur. J. Pharmacol. 578 (2−3), 108−113. https://doi.org/10.1016/j.ejphar.2007.09.008.

Herman, M.M., Bensch, K.G., 1967. Light and electron microscopic studies of acute and chronic thallium intoxication in rats. Toxicol. Appl. Pharmacol. 10 (2), 199−222. https://doi.org/10.1016/0041-008x(67)90104-4.

Hernández-Esquivel, L., Zazueta, C., Buelna-Chontal, M., Hernández-Reséndiz, S., Pavón, N., Chávez, E., 2011. Protective behavior of tamoxifen against Hg^{2+}-induced toxicity on kidney mitochondria: in vitro and in vivo experiments. J. Steroid Biochem. Mol. Biol. 127 (3−5), 345−350. https://doi.org/10.1016/j.jsbmb.2011.07.004.

Hoffman, R.S., 2003. Thallium toxicity and the role of Prussian blue in therapy. Toxicol. Rev. 22 (1), 29−40. https://doi.org/10.2165/00139709-200322010-00004.

Holmuhamedov, E.L., Jahangir, A., Oberlin, A., Komarov, A., Colombini, M., Terzic, A., 2004. Potassium channel openers are uncoupling protonophores: implication in cardioprotection. FEBS Lett. 568 (1−3), 167−170. https://doi.org/10.1016/j.febslet.2004.05.031.

Holmuhamedov, E.L., Wang, L., Terzic, A., 1999. ATP-sensitive K^+ channel openers prevent Ca^{2+} overload in rat carDiac mitochondria. J. Physiol. 519 (2), 347−360. https://doi.org/10.1111/j.1469-7793.1999.0347m.x.

Houston, M.C., 2011. Role of mercury toxicity in hypertension, cardiovascular disease, and stroke. J. Clin. Hypertens. Greenwich. 13 (8), 621−627. https://doi.org/10.1111/j.1751-7176.2011.00489.x.

Huang, C.C., Shih, T.S., Liu, C.H., 2012. Acute thallium intoxication: an experience in Taiwan. Curr. Top. Toxicol. 8, 33−41.

Hughes, B.P., Exton, J.H., 1983. Effect of micromolar concentrations of manganese ions on calcium-ion cycling in rat liver mitochondria. Biochem. J. 212 (3), 773−782. https://doi.org/10.1042/bj2120773.

Hunter Jr., F.E., Ford, L., 1955. Inactivation of oxidative and phosphorylative systems in mitochondria by preincubation with phosphate and other ions. J. Biol. Chem. 216 (1), 357−369.

Ibrahim, D., Froberg, B., Wolf, A., Rusyniak, D.E., 2006. Heavy metal poisoning: clinical presentations and pathophysiology. Clin. Lab. Med. 26 (1), 67–97. https://doi.org/10.1016/j.cll.2006.02.003.

Ichas, F., Mazat, J.P., 1998. From calcium signaling to cell death: two conformations for the mitochondrial permeability transition pore. Switching from low- to high-conductance state. Biochim. Biophys. Acta 1366 (1–2), 33–50. https://doi.org/10.1016/s0005-2728(98)00119-4.

Innis, R., Moses, H., 1978. Clinical conferences at the Johns Hopkins Hospital. Thallium poisoning. Johns Hopkins Med. J. 142 (1), 27–31.

Inoue, T., Suzuki, Y., Yoshimaru, T., Ra, C., 2009. Ca^{2+}-dependent mast cell death induced by Ag (I) via cardiolipin oxidation and ATP depletion. J. Leukoc. Biol. 86 (1), 167–179. https://doi.org/10.1189/jlb.1108691.

Jung, D.W., Farooqui, T., Utz, E., Brierley, G.P., 1984. Effects of quinine on K^+ transport in heart mitochondria. J. Bioenerg. Biomembr. 16 (5–6), 379–390. https://doi.org/10.1007/BF00743233.

Jung, D.W., Brierley, G.P., 1981. On the relationship between the uncoupler-induced efflux of K^+ from heart mitochondria and the oxidation-reduction state of pyridine nucleotides. J. Biol. Chem. 256 (20), 10490–10496.

Jung, D.W., Brierley, G.P., 1984. The permeability of uncoupled heart mitochondria to potassium ion. J. Biol. Chem. 259 (11), 6904–6911.

Jurkowitz, M.S., Geisbuhler, T., Jung, D.W., Brierley, G.P., 1983. Ruthenium red-sensitive and -insensitive release of Ca^{2+} from uncoupled heart mitochondria. Arch. Biochem. Biophys. 223 (1), 120–128. https://doi.org/10.1016/0003-9861(83)90577-5.

Kang, S.H., Park, W.S., Kim, N., Youm, J.B., Warda, M., et al., 2007. Mitochondrial Ca^{2+}-activated K^+ channels more efficiently reduce mitochondrial Ca^{2+} overload in rat ventricular myocytes. Am. J. Physiol. Heart Circ. Physiol. 293 (1), H307–H313. https://doi.org/10.1152/ajpheart.00789.2006.

Kapoor, S.C., van Rossum, G.D., O'Neill, K.J., Mercorella, I., 1985. Uptake of inorganic lead in vitro by isolated mitochondria and tissue slices of rat renal cortex. Biochem. Pharmacol. 34 (9), 1439–1448. https://doi.org/10.1016/0006-2952(85)90682-3.

Kelley, C., 1999. Cadmium therapeutic agents. Curr. Pharm. Des. 5 (4), 229–240.

Kiliç, G.A., Kutlu, M., 2010. Effects of exogenous metallothionein against thallium-induced oxidative stress in rat liver. Food Chem. Toxicol. 48 (3), 980–987. https://doi.org/10.1016/j.fct.2010.01.013.

Klingenberg, M., 2008. The ADP and ATP transport in mitochondria and its carrier. Biochim. Biophys. Acta 1778 (10), 1978–2021. https://doi.org/10.1016/j.bbamem.2008.04.011.

Koike, H., Shinohara, Y., Terada, H., 1991. Why is inorganic phosphate necessary for uncoupling of oxidative phosphorylation by Cd^{2+} in rat liver mitochondria? Biochim. Biophys. Acta 1060 (1), 75–81. https://doi.org/10.1016/s0005-2728(05)80121-5.

Kokoszka, J.E., Waymire, K.G., Levy, S.E., Sligh, J.E., Cai, J., et al., 2004. The ADP/ATP translocator is not essential for the mitochondrial permeability transition pore. Nature 427 (6973), 461–465. https://doi.org/10.1038/nature02229.

Kone, B.C., Brenner, R.M., Gullans, S.R., 1990. Sulfhydryl-reactive heavy metals increase cell membrane K^+ and Ca^{2+} transport in renal proximal tubule. J. Membr. Biol. 113 (1), 1–12. https://doi.org/10.1007/BF01869600.

Kone, B.C., Kaleta, M., Gullans, S.R., 1988. Silver ion (Ag^+)-induced increases in cell membrane K^+ and Na^+ permeability in the renal proximal tubule: reversal by thiol reagents. J. Membr. Biol. 102 (1), 11–19. https://doi.org/10.1007/BF01875349.

Konji, V., Montag, A., Sandri, G., Nordenbrand, K., Ernster, L., 1985. Transport of Ca^{2+} and Mn^{2+} by mitochondria from rat liver, heart and brain. Biochimie 67 (12), 1241–1250.

Korge, P., Honda, H.M., Weiss, J.N., 2002. Protection of carDiac mitochondria by diazoxide and protein kinase C: implications for ischemic preconditioning. Proc. Natl. Acad. Sci. U.S.A. 99 (5), 3312–3317. https://doi.org/10.1073/pnas.052713199.

Korge, P., Yang, L., Yang, J.H., Wang, Y., Qu, Z., Weiss, J.N., 2011. Protective role of transient pore openings in calcium handling by carDiac mitochondria. J. Biol. Chem. 286 (40), 34851–34857. https://doi.org/10.1074/jbc.M111.239921.

Korotkov, S., Konovalova, S., Emelyanova, L., Brailovskaya, I., 2014b. Y^{3+}, La^{3+}, and some bivalent metals inhibited the opening of the Tl^{+}-induced permeability transition pore in Ca^{2+}-loaded rat liver mitochondria. J. Inorg. Biochem. 141 (1), 1–9. https://doi.org/10.1016/j.jinorgbio.2014.08.004.

Korotkov, S.M., 2009. Effects of Tl^{+} on ion permeability, membrane potential and respiration of isolated rat liver mitochondria. J. Bioenerg. Biomembr. 41 (3), 277–287. https://doi.org/10.1007/s10863-009-9225-7.

Korotkov, S.M., 2021. Effects of Tl^{+} on the inner membrane thiol groups, respiration, and swelling in succinate-energized rat liver mitochondria were modified by thiol reagents. Biometals 34 (5), 987–1006. https://doi.org/10.1007/s10534-021-00329-6.

Korotkov, S.M., Brailovskaya, I.V., 2001. Tl^{+} increases the permeability of the inner membranes of rat liver mitochondria for monovalent cations. Doklady Biochem. Biophys. 378 (1–6), 145–149. https://doi.org/10.1023/a:1011538906694.

Korotkov, S.M., Brailovskaya, I.V., Kormilitsyn, B.N., Furaev, V.V., 2014a. Tl^{+} showed negligible interaction with inner membrane sulfhydryl groups of rat liver mitochondria, but formed complexes with matrix proteins. J. Biochem. Mol. Toxicol. 28 (4), 149–156. https://doi.org/10.1002/jbt.21547.

Korotkov, S.M., Brailovskaya, I.V., Shumakov, A.R., Emelyanova, L.V., 2015a. Closure of mitochondrial potassium channels favors opening of the Tl^{+}-induced permeability transition pore in Ca^{2+}-loaded rat liver mitochondria. J. Bioenerg. Biomembr. 47 (3), 243–254. https://doi.org/10.1007/s10863-015-9611-2.

Korotkov, S.M., Emelyanova, L.V., Konovalova, S.A., Brailovskaya, I.V., 2015b. Tl^{+} induces the permeability transition pore in Ca^{2+}-loaded rat liver mitochondria energized by glutamate and malate. Toxicol. Vitro 29 (5), 1034–1041. https://doi.org/10.1016/j.tiv.2015.04.006.

Korotkov, S.M., Emel'yanova, L.V., Yagodina, O.V., 2008a. Inorganic phosphate stimulates the toxic effects of Tl^{+} in rat liver mitochondria. J. Biochem. Mol. Toxicol. 22 (3), 148–157. https://doi.org/10.1002/jbt.20215.

Korotkov, S.M., Glazunov, V.V., Rozengart, E.V., Suvorov, A.A., Nikitina, E.R., 1999. Effects of Cd^{2+} and two cadmium organic complexes on isolated rat liver mitochondria. J. Biochem. Mol. Toxicol. 13 (3–4), 149–157. https://doi.org/10.1002/(sici)1099-0461(1999)13:3/4<149::aid-.

Korotkov, S.M., Glazunov, V.V., Yagodina, O.V., 2007. Increase in the toxic effects of Tl^{+} on isolated rat liver mitochondria in the presence of nonactin. J. Biochem. Mol. Toxicol. 21 (2), 81–91. https://doi.org/10.1002/jbt.20163.

Korotkov, S.M., Konovalova, S.A., Brailovskaya, I.V., 2015c. Diamide accelerates opening of the Tl^{+}-induced permeability transition pore in Ca^{2+}-loaded rat liver mitochondria. Biochem. Biophys. Res. Commun. 468 (1–2), 360–364. https://doi.org/10.1016/j.bbrc.2015.10.091.

Korotkov, S.M., Konovalova, S.A., Brailovskaya, I.V., Saris, N.E., 2016a. To involvement the conformation of the adenine nucleotide translocase in opening the Tl^{+}-induced permeability transition pore in Ca^{2+}-loaded rat liver mitochondria. Toxicol. Vitro 32 (1), 320–332. https://doi.org/10.1016/j.tiv.2016.01.015.

Korotkov, S.M., Konovalova, S.A., Nesterov, V.P., Brailovskaya, I.V., 2018. Mersalyl prevents the Tl^{+}-induced permeability transition pore opening in the inner membrane of Ca^{2+}-loaded rat liver mitochondria. Biochem. Biophys. Res. Commun. 495 (2), 1716–1721. https://doi.org/10.1016/j.bbrc.2017.12.023.

Korotkov, S.M., Lapin, A.V., 2003. Thallium induces opening of the mitochondrial permeability transition pore in the inner membrane of rat liver mitochondria. Doklady Biochem. Biophys. 392, 247–252. https://doi.org/10.1023/a:1026182511897.

Korotkov, S.M., Nesterov, V.P., Belostotskaya, G.B., Brailovskaya, I.V., Novozhilov, A.V., Sobol, C.V., 2020. Influence of Tl^{+} on the movement of Ca^{2+} and Na^{+} ions across membranes of rat neonatal cardiomyocytes and rat heart mitochondria. Ukr. Biochem. J. 92 (1), 41–55.

Korotkov, S.M., Nesterov, V.P., Brailovskaya, I.V., Furaev, V.V., Novozhilov, A.V., 2013. Tl^{+} induces both cationic and transition pore permeability in the inner membrane of rat heart mitochondria. J. Bioenerg. Biomembr. 45 (6), 531–539. https://doi.org/10.1007/s10863-013-9526-8.

Korotkov, S.M., Nesterov, V.P., Ryabchikov, N.N., 2006. Study of the Mechanism of action of diazoxide on rat heart mitochondria under calcium loading. Doklady Biochem. Biophys. 408 (1–6), 133–137. https://doi.org/10.1134/S1607672906030070.

Korotkov, S.M., Nesterov, V.P., Ryabchikov, N.N., 2008b. Study of the effect of SH-groups on respiration and swelling of rat heart mitochondria. Doklady Biochem. Biophys. 421 (1), 171–175. https://doi.org/10.1134/s1607672908040030.

Korotkov, S.M., Nikitina, E.R., Glazunov, V.V., Yagodina, O.V., 2005. Effect of thallium ions on isolated mitochondria of rat liver in the presence of nonactin. Doklady Biochem. Biophys. 401, 97–103. https://doi.org/10.1007/s10628-005-0043-5.

Korotkov, S.M., Saris, N.E., 2011. Influence of Tl^{+} on mitochondrial permeability transition pore in Ca^{2+}-loaded rat liver mitochondria. J. Bioenerg. Biomembr. 43 (2), 149–162. https://doi.org/10.1007/s10863-011-9341-z.

Korotkov, S.M., Skul'skiĭ, I.A., 1996. Changes in the effect of Cd^{2+} on the respiration of isolated rat liver mitochondria after their preincubation with Ca^{2+}, Sr^{2+}, Ba^{2+}, Mn^{2+} and ruthenium red. Tsitologiia 38 (4–5), 500–509 (Russian). https://www.elibrary.ru/item.asp?id=14935301.

Korotkov, S.M., Skul'skiĭ, I.A., Glazunov, V.V., 1998. Cd^{2+} effects on respiration and swelling of rat liver mitochondria were modified by monovalent cations. J. Inorg. Biochem. 70 (1), 17–23. https://doi.org/10.1016/s0162-0134(98)00008-7.

Korotkov, S.M., Sobol, K.V., Shemarova, I.V., Furaev, V.V., Novozhilov, A.V., Nesterov, V.P., 2019. Effects of Nd^{3+} on calcium-dependent processes in isolated rat heart mitochondria and frog heart muscle. Biochemistry (Moscow). Suppl. A. Membr. Cell Biol. 13 (2), 161–167. https://doi.org/10.1134/s1990747819070018.

Korotkov, S.M., Sobol, K.V., Shemarova, I.V., Furaev, V.V., Shumakov, A.R., Nesterov, V.P., 2016b. A comparative study of the effects of Pr^{3+} and La^{3+} ions on calcium dependent processes in frog cardiac muscle and rat heart mitochondria. Biophysics 61 (5), 733–740. https://doi.org/10.1134/S0006350916050122.

Kowaltowski, A.J., Castilho, R.F., 1997. Ca^{2+} acting at the external side of the inner mitochondrial membrane can stimulate mitochondrial permeability transition induced by phenylarsine oxide. Biochim. Biophys. Acta 1322 (2–3), 221–229. https://doi.org/10.1016/s0005-2728(97)00078-9.

Kowaltowski, A.J., Vercesi, A.E., Castilho, R.F., 1997. Mitochondrial membrane protein thiol reactivity with N-ethylmaleimide or mersalyl is modified by Ca^{2+}: correlation with mitochondrial permeability transition. Biochim. Biophys. Acta 1318 (3), 395–402. https://doi.org/10.1016/s0005-2728(96)00111-9.

Krabbendam, I.E., Honrath, B., Culmsee, C., Dolga, A.M., 2018. Mitochondrial Ca^{2+}-activated K^+ channels and their role in cell life and death pathways. Cell Calcium 69, 101−111. https://doi.org/10.1016/j.ceca.2017.07.005.

Kupsch, K., Parvez, S., Siemen, D., Wolf, G., 2007. Modulation of the permeability transition pore by inhibition of the mitochondrial K_{ATP} channel in liver vs brain mitochondria. J. Membr. Biol. 215 (2−3), 69−74. https://doi.org/10.1007/s00232-007-9006-3.

Kushnareva, Y.E., Sokolove, P.M., 2000. Prooxidants open both the mitochondrial permeability transition pore and a low-conductance channel in the inner mitochondrial membrane. Arch. Biochem. Biophys. 376 (2), 377−388. https://doi.org/10.1006/abbi.2000.1730.

Lapidus, R.G., Sokolove, P.M., 1994. The mitochondrial permeability transition. Interactions of spermine, ADP, and inorganic phosphate. J. Biol. Chem. 269 (29), 18931−18936.

Laskowski, M., Augustynek, B., Kulawiak, B., Koprowski, P., Bednarczyk, P., et al., 2016. What do we not know about mitochondrial potassium channels? Biochim. Biophys. Acta 1857 (8), 1247−1257. https://doi.org/10.1016/j.bbabio.2016.03.007.

Le Quoc, K., Le Quoc, D., Gaudemer, Y., 1979. Influence of the energetic state of mitochondria on the inhibition of oxidative phosphorylation by N-ethylmaleimide. Biochim. Biophys. Acta 546 (2), 356−364. https://doi.org/10.1016/0005-2728(79)90053-7.

Leblondel, G., Allain, P., 1984. A thiol oxidation interpretation of the Cu^{2+} effects on rat liver mitochondria. J. Inorg. Biochem. 21 (3), 241−251. https://doi.org/10.1016/0162-0134(84)83007-x.

Lee, W.K., Spielmann, M., Bork, U., Thévenod, F., 2005. Cd^{2+}-induced swelling-contraction dynamics in isolated kidney cortex mitochondria: role of Ca^{2+} uniporter, K^+ cycling, and protonmotive force. Am. J. Physiol. Cell Physiol. 289 (3), C656−C664. https://doi.org/10.1152/ajpcell.00049.2005.

Lehninger, A.L., Carafoli, E., 1971. The interaction of La^{3+} with mitochondria in relation to respiration-coupled Ca^{2+} transport. Arch. Biochem. Biophys. 143 (2), 506−515. https://doi.org/10.1016/0003-9861(71)90235-9.

Lê-Quôc, D., Lê-Quôc, K., 1989. Relationships between the NAD(P) redox state, fatty acid oxidation, and inner membrane permeability in rat liver mitochondria. Arch. Biochem. Biophys. 273 (2), 466−478. https://doi.org/10.1016/0003-9861(89)90506-7.

Lê-Quôc, K., Lê-Quôc, D., 1985. Crucial role of sulfhydryl groups in the mitochondrial inner membrane structure. J. Biol. Chem. 260 (12), 7422−7428.

Leung, A.W., Varanyuwatana, P., Halestrap, A.P., 2008. The mitochondrial phosphate carrier interacts with cyclophilin D and may play a key role in the permeability transition. J. Biol. Chem. 283 (39), 26312−26323. https://doi.org/10.1074/jbc.M805235200.

Leverve, X.M., Fontaine, E., 2001. Role of substrates in the regulation of mitochondrial function in situ. IUBMB Life 52 (3−5), 221−229. https://doi.org/10.1080/15216540152846037.

Li, B., Chauvin, C., De Paulis, D., De Oliveira, F., Gharib, A., et al., 2012. Inhibition of complex I regulates the mitochondrial permeability transition through a phosphate-sensitive inhibitory site masked by cyclophilin D. Biochim. Biophys. Acta 1817 (9), 1628−1634. https://doi.org/10.1016/j.bbabio.2012.05.011.

Li, B., Gao, T.-M., 2016. Functional role of mitochondrial and nuclear BK channels. Int. Rev. Neurobiol. 128, 163−191. https://doi.org/10.1016/bs.irn.2016.03.018.

Li, Y., Johnson, N., Capano, M., Edwards, M., Crompton, M., 2004. Cyclophilin-D promotes the mitochondrial permeability transition but has opposite effects on apoptosis and necrosis. Biochem. J. 383 (Pt 1), 101−109. https://doi.org/10.1042/BJ20040669.

Ligeti, E., Bodnar, J., Karoly, E., Lindner, E., 1981. Ni^{2+}, a new inhibitor of mitochondrial calcium transport. Biochim. Biophys. Acta 656 (2), 177−182. https://doi.org/10.1016/0005-2787(81)90084-8.

Ligeti, E., Fonyo, A., 1984. Reactivity of the sulphydryl groups of the mitochondrial phosphate carrier. FEBS J. 139 (2), 279−285. https://doi.org/10.1111/j.1432-1033.1984.tb08005.x.

Lim, K.H., Javadov, S.A., Das, M., Clarke, S.J., Suleiman, M.S., Halestrap, A.P., 2002. The effects of ischaemic preconditioning, diazoxide and 5-hydroxydecanoate on rat heart mitochondrial volume and respiration. J. Physiol. 545, 961−974. https://doi.org/10.1113/jphysiol.2002.031484. Part 3.

Lock, J.T., Sinkins, W.G., Schilling, W.P., 2012. Protein S-glutathionylation enhances Ca^{2+}-induced Ca^{2+} release via the IP3 receptor in cultured aortic endothelial cells. J. Physiol. 590 (15), 3431−3447. https://doi.org/10.1113/jphysiol.2012.232645.

Lukács, G.L., Fonyó, A., 1985. Ba^{2+} ions inhibit the release of Ca^{2+} ions from rat liver mitochondria. Biochim. Biophys. Acta 809 (2), 160−166. https://doi.org/10.1016/0005-2728(85)90058-1.

Lund, B.O., Miller, D.M., Woods, J.S., 1993. Studies on Hg(II)-induced H_2O_2 formation and oxidative stress in vivo and in vitro in rat kidney mitochondria. Biochem. Pharmacol. 45 (10), 2017−2024. https://doi.org/10.1016/0006-2952(93)90012-l.

Ma, L., Liu, J.Y., Dong, J.X., Xiao, Q., Zhao, J., Jiang, F.L., 2017. Toxicity of Pb^{2+} on rat liver mitochondria induced by oxidative stress and mitochondrial permeability transition. Toxicol. Res. 6 (6), 822−830. https://doi.org/10.1039/c7tx00204a.

Majima, E., Koike, H., Hong, Y.M., Shinohara, Y., Terada, H., 1993. Characterization of cysteine residues of mitochondrial ADP/ATP carrier with the SH-reagents eosin 5-maleimide and N-ethylmaleimide. J. Biol. Chem. 268 (29), 22181−22187.

McStay, G.P., Clarke, S.J., Halestrap, A.P., 2002. Role of critical thiol groups on the matrix surface of the adenine nucleotide translocase in the mechanism of the mitochondrial permeability transition pore. Biochem. J. 367 (Part 2), 541−548. https://doi.org/10.1042/BJ20011672.

Melnick, R.L., Monti, L.G., Motzkin, S.M., 1976. Uncoupling of mitochondrial oxidative phosphorylation by thallium. Biochem. Biophys. Res. Commun. 69 (1), 68−73. https://doi.org/10.1016/s0006-291x(76)80273-2.

Méndez-Armenta, M., Nava-Ruiz, C., Fernández-Valverde, F., Sánchez-García, A., Rios, C., 2011. Histochemical changes in muscle of rats exposed subchronically to low doses of heavy metals. Environ. Toxicol. Pharmacol. 32 (1), 107−112. https://doi.org/10.1016/j.etap.2011.03.001.

Miccadei, S., Floridi, A., 1993. Sites of inhibition of mitochondrial electron transport by cadmium. Chem. Biol. Interact. 89 (2−3), 159−167. https://doi.org/10.1016/0009-2797(93)90006-k.

Mio, Y., Uezono, S., Kitahata, H., 2014. Anesthetic cardioprotection in relation to mitochondria: basic science. Curr. Pharm. Des. 20 (36), 5673−5680. https://doi.org/10.2174/1381612820666140204110101.

Mitchell, P., Moyle, J., 1969. Translocation of some anions cations and acids in rat liver mitochondria. FEBS J. 9 (2), 149−155. https://doi.org/10.1111/j.1432-1033.1969.tb00588.x.

Miyahara, M., Utsumi, K., 1975. Oxidative phosphorylation controlled by potassium in rat liver mitochondria. Cell Struct. Funct. 1 (1), 51−59. https://doi.org/10.1247/csf.1.51.

Montes, S., Pérez-Barrón, G., Rubio-Osornio, M., Ríos, C., Diaz-Ruíz, A., et al., 2011. Additive effect of DL-penicillamine plus Prussian blue for the antidotal treatment of thallotoxicosis in rats. Environ. Toxicol. Pharmacol. 32 (3), 349–355. https://doi.org/10.1016/j.etap.2011.07.002.

Moore, C.L., 1971. Specific inhibition of mitochondrial Ca^{2+} transport by ruthenium red. Biochem. Biophys. Res. Commun. 42 (2), 298–305. https://doi.org/10.1016/0006-291x(71)90102-1.

Morciano, G., Naumova, N., Koprowski, P., Valente, S., Sardão, V.A., et al., 2021. The mitochondrial permeability transition pore: an evolving concept critical for cell life and death. Biol. Rev. Camb. Philos. Soc. 96 (6), 2489–2521. https://doi.org/10.1111/brv.12764.

Mulkey, J.P., Oehme, F.W., 1993. A review of thallium toxicity. Vet. Hum. Toxicol. 35 (5), 445–453.

Nakashima, R.A., Garlid, K.D., 1982. Quinine inhibition of Na^+ and K^+ transport provides evidence for two cation/H^+ exchangers in rat liver mitochondria. J. Biol. Chem. 257 (16), 9252–9254.

Nalyvajko, N.V., Vovkanych, L.S., Dubitsky, L.O., 2006. Inhibitory analysis of the monovalent metals' cations interaction with the system of Na^+-dependent Ca^{2+} efflux from liver mitochondria. Ukr. Biochem. J. 78 (5), 44–50.

Nantes, I.L., Rodrigues, T., Caires, A.C., Cunha, R.L., Pessoto, F.S., et al., 2011. Specific effects of reactive thiol drugs on mitochondrial bioenergetics. J. Bioenerg. Biomembr. 43 (1), 11–18. https://doi.org/10.1007/s10863-011-9328-9.

Nicholls, D.G., Akerman, A.E., 1982. Mitochondrial calcium transport. Biochim. Biophys. Acta 683 (1), 57–88. https://doi.org/10.1016/0304-4173(82)90013-1.

Nikitina, E.R., Glazunov, V.V., 2003. Involvement of K^+-ATP-dependent channel in transport of monovalent thallium (Tl^+) across the inner membrane of rat liver mitochondria. Doklady Biochem. Biophys. 392 (1–6), 244–246. https://doi.org/10.1023/a:1026130527827.

Nishida, H., Matsumoto, A., Tomono, N., Hanakai, T., Harada, S., Nakaya, H., 2010. Biochemistry and physiology of mitochondrial ion channels involved in cardioprotection. FEBS Lett. 584 (10), 2161–2166. https://doi.org/10.1016/j.febslet.2009.12.033.

Novgorodov, S.A., Gudz, T.I., Brierley, G.P., Pfeiffer, D.R., 1994. Magnesium ion modulates the sensitivity of the mitochondrial permeability transition pore to cyclosporin A and ADP. Arch. Biochem. Biophys. 311 (2), 219–228. https://doi.org/10.1006/abbi.1994.1230.

Novgorodov, S.A., Gudz, T.I., Milgrom, Y.M., Brierley, G.P., 1992. The permeability transition in heart mitochondria is regulated synergistically by ADP and cyclosporin A. J. Biol. Chem. 267 (23), 16274–16282.

Nuñez, I.P., Fantinelli, J., Arbeláez, L.F., Mosca, S.M., 2011. Mitochondrial K_{ATP} channels participate in the limitation of infarct size by cariporide. Naunyn-Schmiedeberg's Arch. Pharmacol. 383 (6), 563–571. https://doi.org/10.1007/s00210-011-0632-z.

Oehme, F.M., 1978. Mechanisms of Heavy Metal Inorganic Toxicities. Toxicity of Heavy Metals in the Environment. Part 1. Marcel Dekker, New York, pp. 69–85.

Ogata, M., Mori, T., Izushi, F., Etoh, K., Sakai, R., et al., 1983. Classification of potentially toxic chemicals based on their effects on mitochondrial respiration. Physiol. Chem. Phys. Med. NMR 15 (3), 229–232.

Osorio-Rico, L., Santamaria, A., Galván-Arzate, S., 2017. Thallium toxicity: general issues, neurological symptoms, and neurotoxic mechanisms. Adv. Neurobiol. 18, 345–353. https://doi.org/10.1007/978-3-319-60189-2_17.

Osorio-Rico, L., Villeda-Hernández, J., Santamaría, A., Königsberg, M., Galván-Arzate, S., 2015. The n-methyl-d-aspartate receptor antagonist MK-801 prevents thallium-induced behavioral and biochemical alterations in the rat brain. Int. J. Toxicol. 34 (6), 505–513. https://doi.org/10.1177/1091581815603936.

Pachauri, V., Saxena, G., Mehta, A., Mishra, D., Flora, S.J., 2009. Combinational chelation therapy abrogates lead-induced neurodegeneration in rats. Toxicol. Appl. Pharmacol. 240 (2), 255–264. https://doi.org/10.1016/j.taap.2009.07.001.

Paillard, M., Csordás, G., Huang, K.T., Várnai, P., Joseph, S.K., Hajnóczky, G., 2018. MICU1 interacts with the D-Ring of the MCU pore to control its Ca^{2+} flux and sensitivity to Ru360. Mol. Cell. 72 (4), 778–785. https://doi.org/10.1016/j.molcel.2018.09.008.

Panov, A., Filippova, S., Lyakhovich, V., 1980. Adenine nucleotide translocase as a site of regulation by ADP of the rat liver mitochondria permeability to H^+ and K^+ ions. Arch. Biochem. Biophys. 199 (2), 420–426. https://doi.org/10.1016/0003-9861(80)90298-2.

Panov, A., Schonfeld, P., Dikalov, S., Hemendinger, R., Bonkovsky, H.L., Brooks, B.R., 2009. The neuromediator glutamate, through specific substrate interactions, enhances mitochondrial ATP production and reactive oxygen species generation in nonsynaptic brain mitochondria. J. Biol. Chem. 284 (21), 14448–14456. https://doi.org/10.1074/jbc.M900985200.

Panov, A.V., 2015. Practical Mitochondriology. Researchgate (Russian). https://www.researchgate.net/publication/282178984_PRAKTICESKAA_MITOHONDRIOLOGIA.

Parr, D.R., Harris, E.J., 1976. The effect of lead on the calcium-handling capacity of rat heart mitochondria. Biochem. J. 158 (2), 289–294. https://doi.org/10.1042/bj1580289.

Pasdois, P., Beauvoit, B., Costa, A.D., Vinassa, B., Tariosse, L., et al., 2007. Sarcoplasmic ATP-sensitive potassium channel blocker HMR1098 protects the ischemic heart: implication of calcium, complex I, reactive oxygen species and mitochondrial ATP-sensitive potassium channel. J. Mol. Cell. Cardiol. 42 (3), 631–642. https://doi.org/10.1016/j.yjmcc.2006.12.014.

Peng, T.I., Jou, M.J., 2010. Oxidative stress caused by mitochondrial calcium overload. Ann. N. Y. Acad. Sci. 1201, 183–188. https://doi.org/10.1111/j.1749-6632.2010.05634.x.

Perrin, D.D., 1979. Stability Constants of Metal-Ion Complexes. Part B. Organic Ligands. Pergamon, Oxford, pp. 1–1263.

Petronilli, V., Costantini, P., Scorrano, L., Colonna, R., Passamonti, S., Bernardi, P., 1994. The voltage sensor of the mitochondrial permeability transition pore is tuned by the oxidation-reduction state of vicinal thiols. Increase of the gating potential by oxidants and its reversal by reducing agents. J. Biol. Chem. 269 (24), 16638–16642. https://doi.org/10.1074/jbc.271.12.6746.

Petronilli, V., Sileikyte, J., Zulian, A., Dabbeni-Sala, F., Jori, G., et al., 2009. Switch from inhibition to activation of the mitochondrial permeability transition during hematoporphyrin-mediated photooxidative stress. Unmasking pore-regulating external thiols. Biochim. Biophys. Acta 1787 (7), 897–904. https://doi.org/10.1016/j.bbabio.2009.03.014.

Portenhauser, R., Schäfer, G., Trolp, R., 1971. Inhibition of mitochondrial metabolism by the diabetogenic thiaDiazine diazoxide. II. Interaction with energy conservation and ion transport. Biochem. Pharmacol. 20 (10), 2623–2632. https://doi.org/10.1016/0006-2952(71)90171-7.

Pourahmad, J., Eskandari, M.R., Daraei, B., 2010. A comparison of hepatocyte cytotoxic mechanisms for thallium (I) and thallium (III). Environ. Toxicol. 25 (5), 456–467. https://doi.org/10.1002/tox.20590.

Prick, J.J.G., 1979. Thallium poisoning. In: Vinken, P.J., Bruyn, G.W. (Eds.), Handbook of clinical neurology, Intoxication of the Nervous System, vol. 36. North-Holland, New York, pp. 239–278.

Puga Molina, L.C., Verstraeten, S.V., 2008. Thallium(III)-mediated changes in membrane physical properties and lipid oxidation affect cardiolipin-cytochrome c interactions. Biochim. Biophys. Acta 1778 (10), 2157–2164. https://doi.org/10.1016/j.bbamem.2008.04.013.

Queliconi, B.B., Wojtovich, A.P., Nadtochiy, S.M., Kowaltowski, A.J., Brookes, P.S., 2011. Redox regulation of the mitochondrial K_{ATP} channel in cardioprotection. Biochim. Biophys. Acta 1813 (7), 1309–1315. https://doi.org/10.1016/j.bbamcr.2010.11.005.

Ramsden, D.B., 2007. Thallium. In: Waring, R.H., Steventon, G.B., Mitchel, S.C. (Eds.), Molecules of Death (2nd). Imperial College Press, London, pp. 415–422.

Rasheed, B.K., Diwan, J.J., Sanadi, D.R., 1984. Activation of potassium ion transport in mitochondria by cadmium ion. FEBS J. 144 (3), 643–647. https://doi.org/10.1111/j.1432-1033.1984.tb08513.x.

Reed, K.C., Bygrave, F.L., 1974a. Accumulation of lanthanum by rat liver mitochondria. Biochem. J. 138 (2), 239–252. https://doi.org/10.1042/bj1380239.

Reed, K.C., Bygrave, F.L., 1974b. The inhibition of mitochondrial calcium transport by lanthanides and ruthenium red. Biochem. J. 140 (2), 143–155. https://doi.org/10.1042/bj1400143.

Ricchelli, F., Sileikytė, J., Bernardi, P., 2011. Shedding light on the mitochondrial permeability transition. Biochim. Biophys. Acta 1807 (5), 482–490. https://doi.org/10.1016/j.heliyon.2021.e07087.

Riess, M.L., Camara, A.K., Heinen, A., Eells, J.T., Henry, M.M., Stowe, D.F., 2008. K_{ATP} channel openers have opposite effects on mitochondrial respiration under different energetic conditions. J. Cardiovasc. Pharmacol. 51 (5), 483–491. https://doi.org/10.1097/FJC.0b013e31816bf4a4.

Rigobello, M.P., Turcato, F., Bindoli, A., 1995. Inhibition of rat liver mitochondrial permeability transition by respiratory substrates. Arch. Biochem. Biophys. 319 (1), 225–230. https://doi.org/10.1006/abbi.1995.1286.

Rigoni, F., Mathien-Shire, Y., Deana, R., 1980. Effect of ruthenium red on calcium efflux from rat liver mitochondria. FEBS Lett. 120 (2), 255–258. https://doi.org/10.1016/0014-5793(80)80310-3.

Rigoni, F., Panato, L., Deana, R., 1984. Calcium efflux parallel to total phosphate retention in rat liver mitochondria. Int. J. Biochem. 16 (11), 1121–1125. https://doi.org/10.1016/0020-711x(84)90004-1.

Rikans, L.E., Yamano, T., 2000. Mechanisms of cadmium-mediated acute hepatotoxicity. J. Biochem. Mol. Toxicol. 14 (2), 110–117. https://doi.org/10.1002/(sici)1099-0461(2000)14:2<110::aid-jbt7>3.0.co;2-j.

Riley, M.V., Lehninger, A.L., 1964. Changes in sulfhydryl groups of rat liver mitochondria during swelling and contraction. J. Biol. Chem. 239 (6), 2083–2089.

Riley Jr., W.W., Pfeiffer, D.R., 1985. Relationships between Ca^{2+} release, Ca^{2+} cycling, and Ca^{2+}-mediated permeability changes in mitochondria. J. Biol. Chem. 260 (23), 12416–12425.

Rousou, A.J., Ericsson, M., Federman, M., Levitsky, S., McCully, J.D., 2004. Opening of mitochondrial K_{ATP} channels enhances cardioprotection through the modulation of mitochondrial matrix volume, calcium accumulation, and respiration. Am. J. Physiol. Heart Circ. Physiol. 287 (5), H1967–H1976. https://doi.org/10.1152/ajpheart.00338.2004.

Rugolo, M., Siliprandi, D., Siliprandi, N., Toninello, A., 1981. Parallel efflux of Ca^{2+} and Pi in energized rat liver mitochondria. Biochem. J. 200 (3), 481–486. https://doi.org/10.1042/bj2000481.

Rusyniak, D.E., Arroyo, A., Acciani, J., Froberg, B., Kao, L., Furbee, B., 2010. Heavy metal poisoning: management of intoxication and antidotes. Experientia (Suppl) 100, 365–396. https://doi.org/10.1007/978-3-7643-8338-1_11.

Salvi, M., Battaglia, V., Brunati, A.M., La Rocca, N., Tibaldi, E., et al., 2007. Catalase takes part in rat liver mitochondria oxidative stress defense. J. Biol. Chem. 282 (33), 24407–24415. https://doi.org/10.1074/jbc.M701589200.

Samartsev, V.N., Chezganova, S.A., Polishchuk, L.S., Paydyganov, A.P., Vidyakina, O.V., Zeldi, I.P., 2003. Temperature dependence of rat liver mitochondrial respiration with uncoupling of oxidative phosphorylation by fatty acids. Influence of inorganic phosphate. Biochemistry (Moscow) 68 (6), 618–626. https://doi.org/10.1023/a:1024657507814.

Sanadi, D.R., Hughes, J.B., Joshi, S., 1981. Activation of potassium-dependent H^+ efflux from mitochondria by cadmium and phenylarsine oxide. J. Bioenerg. Biomembr. 13 (5–6), 425–431. https://doi.org/10.1007/BF00743214.

Saris, N.E., Bernardi, P., 1983. Inhibition by Sr^{2+} of specific mitochondrial Ca^{2+}-efflux pathways. Biochim. Biophys. Acta 725 (1), 19–24. https://doi.org/10.1016/0005-2728(83)90219-0.

Saris, N.E., Skul'skiĭ, I.A., Savina, M.V., Glasunov, V.V., 1981. Mechanism of mitochondrial transport of thallous ions. J. Bioenerg. Biomembr. 13 (1–2), 51–59. https://doi.org/10.1007/BF00744746.

Saris, N.E., van den Bosch, H., 1988. Interaction of Sr^{2+} with Ca^{2+}-induced Ca^{2+} release in mitochondria. J. Bioenerg. Biomembr. 20 (6), 749–757. https://doi.org/10.1007/BF00762551.

Saris, N.E.L., Järvisalo, J., 1977. Transport and binding of cadmium in kidney mitochondria. In: Brown, S.S. (Ed.), Clinical Chemistry and Toxicology of Metals. Elsevier, Amsterdam, pp. 109–112.

Schäfer, G., Wegener, C., Portenhauser, R., Bojanovski, D., 1969. Diazoxide, an inhibitor of succinate oxidation. Biochem. Pharmacol. 18 (10), 2678–2681.

Scott, K.M., Hwang, K.M., Jurkowitz, M., Brierley, G.P., 1971. Ion transport by heart mitochondria. 23. The effects of lead on mitochondrial reactions. Arch. Biochem. Biophys. 147 (2), 557–567. https://doi.org/10.1016/0003-9861(71)90413-9.

Shalbuyeva, N., Brustovetsky, T., Bolshakov, A., Brustovetsky, N., 2006. Calcium-dependent spontaneously reversible remodeling of brain mitochondria. J. Biol. Chem. 281 (49), 37547–37558. https://doi.org/10.1074/jbc.M607263200.

Shalbuyeva, N., Brustovetsky, T., Brustovetsky, N., 2007. Lithium desensitizes brain mitochondria to calcium, antagonizes permeability transition, and diminishes cytochrome C release. J. Biol. Chem. 282 (25), 18057–18068. https://doi.org/10.1074/jbc.M702134200.

Shemarova, I.V., Korotkov, S.M., Demina, I.N., Nesterov, V.P., 2010. Action of oxidative reactions in mitochondria on contractivity of heart muscle. Effects of Ni^{2+} J. Evol. Biochem. Physiol. 46 (2), 167–172. https://doi.org/10.1134/S0022093010020055.

Shemarova, I.V., Korotkov, S.M., Nesterov, V.P., 2013. Action of La^{3+} on the systems providing contractility of vertebrate myocardium. J. Evol. Biochem. Physiol. 49 (4), 423–428. https://doi.org/10.1134/S0022093013040063.

Shilo, S., Aronis, A., Komarnitsky, R., Tirosh, O., 2003. Selenite sensitizes mitochondrial permeability transition pore opening in vitro and in vivo: a possible mechanism for chemo-protection. Biochem. J. 370 (Part 1), 283–290. https://doi.org/10.1042/BJ20021022.

Singh, B.K., Tripathi, M., Pandey, P.K., Kakkar, P., 2011. Alteration in mitochondrial thiol enhances calcium ion dependent membrane permeability transition and dysfunction in vitro: a cross-talk between mtThiol, Ca^{2+}, and ROS. Mol. Cell. Biochem. 357 (1–2), 373–385. https://doi.org/10.1007/s11010-011-0908-0.
Skul'skiĭ, I.A., 1977. Transport of monovalent thallium ions across mitochondrial membranes. Doklady Akad. Nauk SSSR. 232 (4), 945–948 (Russian). PMID: 852379.
Skul'skiĭ, I.A., Allakhverdov, B.L., Burovina, I.V., Lapin, A.V., 1980a. X-ray local microanalysis in cytology. Sh. Effect of thallium on the ionic composition and transport properties of frog skin epithelial cells. Cytology 22 (4), 409–414 (Russian).
Skul'skiĭ, I.A., Glazunov, V.V., 1981. Non-dependence of the membrane potential of mitochondria on their energetic state. Doklady Akad. Nauk SSSR. 258 (6), 1496–1498 (Russian). PMID: 6894897.
Skul'skiĭ, I.A., Glazunov, V.V., 1982. Effect of valinomycin and nonactin on monovalent cation transport in mitochondria. Tsitologiia 24 (2), 183–187 (Russian). PMID: 6896103.
Skul'skiĭ, I.A., Glazunov, V.V., Savina, M.V., 1979a. [Study of Tl^+ as a probe to assess the membrane potential of mitochondria]. In: Proceedings of Fourth All-USSR Biochemical Congress. Science, Moscow, pp. 124–125 (Russian).
Skul'skiĭ, I.A., Glazunov, V.V., Savina, M.V., 1979b. Proof of the possibility of univalent cation active transport in rat liver mitochondria. Doklady Akad. Nauk SSSR. 246 (2), 504–508 (Russian). PMID: 582434.
Skul'skiĭ, I.A., Glazunov, V.V., Savina, M.V., 1980b. [Transport of monovalent cations in mitochondria]. Mol. Biol. (Ukraine) 25, 47–55 (Russian).
Skul'skiĭ, I.A., Ivanova, T.I., Savina, M.V., 1984. Action of the salts of univalent thallium on the liver mitochondria of the rat. Zh. Evolut. Biokh. Fiziol. 20 (4), 353–355 (Russian). PMID: 6485653.
Skul'skiĭ, I.A., Korotkov, S.M., Glazunov, V.V., Ivanova, T.I., Savina, M.V., 1988. Effect of cadmium ions on the respiration, cation transport and morphology of the liver mitochondria in the rat and the lamprey. Tsitologiia 30 (8), 956–962 (Russian). PMID: 2849819.
Skulachev, V.P., 2001. Oxygen and the phenomena of programmed death. Biomed. J. 2 (2), 116–126 (Russian). http://www.medline.ru/public/art/tom2/art20.phtml.
Skulskii, I.A., Saris, N.E., Savina, M.V., Glasunov, V.V., 1981. Uptake of thallous ions by mitochondria is stimulated by nonactin but not by respiration alone. FEBS J. 120 (2), 263–266. https://doi.org/10.1111/j.1432-1033.1981.tb05698.x.
Skulskii, I.A., Savina, M.V., Glasunov, V.V., Saris, N.E., 1978. Electrophoretic transport of Tl^+ in mitochondria. J. Membr. Biol. 44 (2), 187–194. https://doi.org/10.1007/BF01976038.
Spencer, P.S., Peterson, E.R., Madrid, R., Raine, C.S., 1973. Effects of thallium salts on neuronal mitochondria in organotypic cord-ganglia-muscle combination cultures. J. Cell Biol. 58 (1), 79–95. https://doi.org/10.1083/jcb.58.1.79.
Stacey, N.H., Klaassen, C.D., 1981. Interaction of metal ions with cadmium-induced cellular toxicity. J. Toxicol. Environ. Health A. 7 (1), 149–158. https://doi.org/10.1080/15287398109529966.
Stohs, S.J., Bagchi, D., Hassoun, E., Bagchi, M., 2001. Oxidative mechanisms in the toxicity of chromium and cadmium ions. J. Environ. Pathol. Toxicol. Oncol. 20 (2), 77–88.
Taber, P., 1964. Chronic thallium poisoning: rapid diagnosis and treatment. J. Pediatr. 65 (3), 461–463. https://doi.org/10.1016/s0022-3476(64)80414-5.
Tang, H.W., Yan, H.L., Hu, X.H., Liang, Y.X., Shen, X.Y., 1996. Lead cytotoxicity in primary cultured rat astrocytes and Schwann cells. J. Appl. Toxicol. 16 (3), 187–196. https://doi.org/10.1002/(SICI)1099-1263(199605)16:3<187::AID-JAT329>3.0.CO;2-Y.

Tedeschi, H., 1981. The transport of cations in mitochondria. Biochim. Biophys. Acta 639 (3–4), 157–196. https://doi.org/10.1016/0304-4173(81)90009-4.

Testai, L., Martelli, A., Marino, A., D'antongiovanni, V., Ciregia, F., et al., 2013. The activation of mitochondrial BK potassium channels contributes to the protective effects of naringenin against myocardial ischemia/reperfusion injury. Biochem. Pharmacol. 85 (11), 1634–1643. https://doi.org/10.1016/j.bcp.2013.03.018.

Toth, P.P., Sell, J.E., Holland, J.F., Suelter, C.H., 1991. The effect of inorganic phosphate on chick (*Gallus domesticus*) heart mitochondrial volume and cation content. Comp. Biochem. Physiol. B 100 (3), 547–554. https://doi.org/10.1016/0305-0491(91)90218-3.

Truhaut, R., 1960. The toxicology of thallium. J. Occup. Med. 2, 334–336. https://doi.org/10.1097/00043764-196007000-00008.

Truter, M.R., 1975. Recognition of metal cations by biological systems. Philos. Trans. R. Soc. Lond. B Biol. Sci. 272 (915), 29–41. https://doi.org/10.1098/rstb.1975.0068.

Tsai, M.F., Jiang, D., Zhao, L., Clapham, D., Miller, C., 2014. Functional reconstitution of the mitochondrial Ca^{2+}/H^{+} antiporter Letm1. J. Gen. Physiol. 143 (1), 67–73. https://doi.org/10.1085/jgp.201311096.

Vainio, H., Mela, L., Chance, B., 1970. Energy dependent bivalent cation translocation in rat liver mitochondria. FEBS J. 12 (2), 387–391. https://doi.org/10.1111/j.1432-1033.1970.tb00863.x.

Vasington, F.D., Gazzotti, P., Tiozzo, R., Carafoli, E., 1972. The effect of ruthenium red on Ca^{2+} transport and respiration in rat liver mitochondria. Biochim. Biophys. Acta 256 (1), 43–54. https://doi.org/10.1016/0005-2728(72)90161-2.

Verstraeten, S.V., 2006. Relationship between thallium(I)-mediated plasma membrane fluidification and cell oxidants production in Jurkat T cells. Toxicology 222 (1–2), 95–102. https://doi.org/10.1016/j.tox.2006.01.028.

Viarengo, A., Nicotera, P., 1991. Possible role of Ca^{2+} in heavy metal cytotoxicity. Comp. Biochem. Physiol. C 100 (1–2), 81–84. https://doi.org/10.1016/0742-8413(91)90127-f.

Waldmeier, P.C., Feldtrauer, J.J., Qian, T., Lemasters, J.J., 2002. Inhibition of the mitochondrial permeability transition by the nonimmunosuppressive cyclosporine derivative NIM811. Mol. Pharmacol. 62 (1), 22–29. https://doi.org/10.1124/mol.62.1.22.

Wang, L., Cherednichenko, G., Hernandez, L., Halow, J., Camacho, S.A., et al., 2001. Preconditioning limits mitochondrial Ca^{2+} during ischemia in rat hearts: role of K_{ATP} channels. Am. J. Physiol. Heart Circ. Physiol. 280 (5), H2321–H2328. https://doi.org/10.1152/ajpheart.2001.280.5.H2321.

Wehrle, J.P., Pedersen, P.L., 1981. Phosphate transport in rat liver mitochondria: location of sulfhydryl groups essential for transport activities. J. Bioenerg. Biomembr. 13 (5–6), 285–294. https://doi.org/10.1007/BF00743206.

Weinberg, J.M., Harding, P.G., Humes, H.D., 1982. Mitochondrial bioenergetics during the initiation of mercuric chloride-induced renal injury. I. Direct effects of in vitro mercuric chloride on renal cortical mitochondrial function. J. Biol. Chem. 257 (1), 60–67.

Wilson, D.E., Asimakis, G.K., 1987. Phosphate-induced efflux of adenine nucleotides from rat-heart mitochondria: evaluation of the roles of the phosphate/hydroxyl exchanger and the dicarboxylate carrier. Biochim. Biophys. Acta 893 (3), 470–479. https://doi.org/10.1016/0005-2728(87)90098-3.

Wojtovich, A.P., Nadtochiy, S.M., Urciuoli, W.R., Smith, C.O., Grunnet, M., et al., 2013. A non-cardiomyocyte autonomous mechanism of cardioprotection involving the SLO1 BK channel. PeerJ 1, 1–23. https://doi.org/10.7717/peerj.48 e48.

Wojtovich, A.P., Williams, D.M., Karcz, M.K., Lopes, C.M., Gray, D.A., et al., 2010. A novel mitochondrial K_{ATP} channel assay. Circ. Res. 106 (7), 1190–1196. https://doi.org/10.1161/CIRCRESAHA.109.215400.

Woods, J.S., Fowler, B.A., 1986. Alteration of hepatocellular structure and function by thallium chloride: ultrastructural, morphometric, and biochemical studies. Toxicol. Appl. Pharmacol. 83 (2), 218–229. https://doi.org/10.1016/0041-008x(86)90299-1.

Wudarczyk, J., Debska, G., Lenartowicz, E., 1999. Zinc as an inducer of the membrane permeability transition in rat liver mitochondria. Arch. Biochem. Biophys. 363 (1), 1–8. https://doi.org/10.1006/abbi.1998.1058.

Zazueta, C., Sánchez, C., García, N., Correa, F., 2000. Possible involvement of the adenine nucleotide translocase in the activation of the permeability transition pore induced by cadmium. Int. J. Biochem. Cell Biol. 32 (10), 1093–1101. https://doi.org/10.1016/s1357-2725(00)00041-8.

Zhao, J., Zhou, Z.Q., Jin, J.C., Yuan, L., He, H., et al., 2014. Mitochondrial dysfunction induced by different concentrations of gadolinium ion. Chemosphere 100, 194–199. https://doi.org/10.1016/j.chemosphere.2013.11.031.

Zierold, K., 2000. Heavy metal cytotoxicity studied by electron probe X-ray microanalysis of cultured rat hepatocytes. Toxicol. Vitro 14 (6), 557–563. https://doi.org/10.1016/s0887-2333(00)00049-7.

Zoccarato, F., Rugolo, M., Siliprandi, D., Siliprandi, N., 1981. Correlated effluxes of adenine nucleotides, Mg^{2+} and Ca^{2+} induced in rat-liver mitochondria by external Ca^{2+} and phosphate. FEBS J. 114 (2), 195–199. https://doi.org/10.1111/j.1432-1033.1981.tb05136.x.

Zook, B.C., Holzworth, J., Thornton, G.W., 1968. Thallium poisoning in cats. J. Am. Vet. Med. Assoc. 153 (3), 285–299.

Zoratti, M., Szabò, I., 1995. The mitochondrial permeability transition. Biochim. Biophys. Acta 1241 (2), 139–176. https://doi.org/10.1016/0304-4157(95)00003-a.

Zorov, D.B., Juhaszova, M., Yaniv, Y., Nuss, H.B., Wang, S., Sollott, S.J., 2009. Regulation and pharmacology of the mitochondrial permeability transition pore. Cardiovasc. Res. 83 (2), 213–225. https://doi.org/10.1093/cvr/cvp151.

Index

Note: 'Page numbers followed by "f" indicate figures and "t" indicate tables."

A

Acute alopecia, 15
Adenine nucleotide translocase (ANT), 144–145, 206t, 207f, 219t
  absorbance change, succinate-energized rat liver mitochondria, 201–205, 203f
  Ca^{2+}-induced swelling, 201–205
  cysteines and, 221, 222f
  fluorescence change, 214f
  mitochondrial swelling, 215
  N-ethylmaleimide, 201–205, 205f
  oxygen consumption rates, 206, 208f, 209t, 210f, 211t, 212f
  thiol reagents, on absorbance change, 200–201, 201f
  Zn^{2+}-induced mitochondrial swelling, 217
Albuminuria, 17–18
Alopecia, 31–32
Antioxidants, 54–55, 87
Aquatic protozoa, 66–67
Arianta arbustorum, 8–9
Ataxia, 20
Atomic absorption spectroscopy, 7–8

B

Bacillus megaterium, 66–67
Bioaccumulation, 9–10
Boucher doses, 10–11

C

Caenorhabditis elegans, 8–9
Capillaroscopy, 30–31
Ceriodaphnia dubia, 32–33
Chlorella vulgaris, 67
Chromosomal aberrations, 56–57
Coproporphyrins, 89–90
Cyclophilin D (CyP-D), 145
Cytoplasm, 84–85
Cytotoxic poisons, 3–4, 18

D

Danio rerio, 56–57
Delta-aminolevulinic acid, 30–31
Desferrioxamine, 25–26

E

Eisenia fetida, 8–9
Electroretinography, 30–31
Energized mitochondria, 116
  contraction, 116, 134–136
  quinine inhibition, 118–120
  respiration and swelling, 111
Escherichia coli, 83–84
Eye diseases, 17

F

Furosemide, 25–26

G

Genotoxic effects, 56–59
Glutathione, 54–55

H

Hematuria, 17–18
Hepatoprotective effects, 54–55
Hyalella azteca, 9
Hypotension, 20

I

Ion permeability/potential, 112–120
  ADP and nonactin, 118f
  Cd^{2+}-induced swelling, 116
  energized mitochondria, 116–117

Ion permeability/potential (*Continued*)
inner membrane passive ion permeability, 117–118
quinine, on swelling, 117f
safranin O fluorescence, 119f
succinate injection, 118–120
Tl^+
absorbance, 112–113, 115t
passive swelling, 112–113, 114f
swelling, 112–113, 113f
Isolated mitochondria
heavy metal effects, 108–112
thallium effects, 109–112
succinate injection, 110–111
univalent thallium, 111

K
Kidney damage, 53–54

L
Leukocytes, 57–58

M
Malondialdehyde, 54–55
Mammalian cells
high electron density granules, 84–85
intracellular mechanisms, 89–90
metallothioneins, 87
mitochondria, 84–85
thallium quinolinate, 85–86
Maximum permissible concentration (MPC), 4–5
Metallothioneins, 87
Micrococcus lysodeikticus, 66–67
Mitochondria. *See also specific topics*
granules, 84–85
mammalian cells, 84–85
thallium uncoupled oxidative phosphorylation, 84–85
Mitochondrial permeability transition pore (MPTP), 144–164, 166f, 168f, 173f, 174t–175t, 178t
absorbance, 150t, 152t, 167t, 169t, 172t
adenine nucleotide translocase (ANT), 145, 206t, 207f, 219t
absorbance change, succinate-energized rat liver mitochondria, 201–205, 203f
Ca^{2+}-induced swelling, 201–205
cysteines and, 221, 222f
fluorescence change, 214f
mitochondrial swelling, 215
N-ethylmaleimide, 201–205, 205f
oxygen consumption rates, 206, 208f, 209t, 210f, 211t, 212f
thiol reagents, on absorbance change, 200–201, 201f
Zn^{2+}-induced mitochondrial swelling, 217
Ca^{2+}, sites and, 179–188
absorbance, 181, 182f
bivalent and trivalent metal ions, 182f
inhibition, 188f
La^{3+}, 185f
manganese and nickel salts, 181–183
rate of potential decrease, 161t
and rotenone, 155f
ruthenium red, 182f
safranin O fluorescence, 157, 160f, 183f
cyclophilin D (CyP-D), 145
heavy metals, 156
oxygen consumption rates, 176f
phosphate symporter (PiC), 145, 221–234
absorbance, 223, 228t
biochemical mechanisms, 235f
Ca^{2+} and, 223, 230f
mersalyl (MSL), 223, 225f, 227f, 230f, 231t
safranin O fluorescence change, 232f
potassium channels and, 188–199
absorbance, 196, 197t
diazoxide, 190–192, 191f
fluorescence change, 192–193, 193f
5-hydroxydecanoate and $Ca2^+$, 190–192, 191f
inner membrane potential and voltage-dependent calcium uptake, 189–190, 189f
modulators and, 190–192, 190f
safranin O fluorescence, 195f
safranin O fluorescence, 177f
thallium nitrate, 159t
Tl^+ and $Ca2^+$ influence, 147f–148f, 158f, 160f
voltage-dependent anion channel (VDAC), 145
Mitochondrial proteins, 138–144

oxidative phosphorylation, 140
protein denaturation, 143
Tl^+, 139f
Mitochondrial respiration, 129t–130t
ADP and quinine, 125f
DNP, energized rat liver mitochondria, 135f
inorganic phosphate and nonactin, 133f
oxygen consumption rates, 136–137
succinate-energized rat liver mitochondria, 134f
thallium nitrate, 121–126, 126t–127t
Tl^+, 121–126, 123f–125f
uncoupled respiration, 134–136
Mitotic activity, 58–59
Monovalent thallium ions, 79–80
Muscle tissues, 62–64
Mutagenic effects, 56–59
Myocardial infarction, 64

N

Nervous system, 59–62
optic nerve damages, 59
pharmacological agents, 60–61
polyneuritis, 59
potassium channels, 60
Rb^+ transport, 62
Nitrobacter agilis, 66–67
Nonenergized mitochondria, 109, 118f
high-amplitude swelling, 110
preswollen, 110–111

O

Opioids, 8
Oxidative phosphorylation, 134–136

P

Phosphate symporter (PiC), 145, 221–234
absorbance, 223, 228t
biochemical mechanisms, 235f
Ca^{2+} and, 223, 230f
mersalyl (MSL), 223, 225f, 227f, 230f, 231t
safranin O fluorescence change, 232f
Pimephales promelas, 32–33, 58–59
Planorbarius corneus, 62–63
Plasmapheresis, 26
Polyneuritis, 59
Polyneuropathy, 16
Proteinuria, 53–54
Protoporphyrin-9, 89–90
Prussian blue, 27–30
Pseudokirchneriella subcapitata, 32–33
Purkinje fibers, 62–63

R

Rana esculenta, 62–63
Rana pipiens, 60–63
Rat heart mitochondria (RHM), 111
Rat liver mitochondria (RLM), 108
Reactive oxygen species (ROS), 108–109
Rodenticide, 8–9
Rotenone, 154–155

S

Saccharomyces cerevisiae, 66–67
Serum nitric oxide, 54–55
Shewanella putrefaciens, 5
Sodium polystyrene sulfonate, 25
Spermatogenesis, 58–59
Sperm motility, in animals, 58–59
Staphylococcus aureus, 66–67
Streptococcus faecalis, 66–67
Synechococcus leopoliensis, 66–67

T

Teratogenic effects, 56–59
Thallium
accumulation
animals, 8–9
humans, 7–8
plants, 9–10
in tumor tissues, 66
analysis, in biological samples, 32–33
compounds
chemical and toxic properties of, 1–4
comparative toxicity, humans, 10–11
distribution of, 4–5
genotoxic effects, 56–59
mutagenic effects, 56–59
release routes, 5–7

Thallium (*Continued*)
teratogenic effects, 56–59
toxicological characteristics, 13–15
human body route of, 11–13
intoxication, 84
poisoning. *See* Thallium poisoning
Thallium acetate, 59–60
in cats, 20
isolated mitochondria, 108
Prussian blue, 92
Thallium carbonate, 57–58
Thallium nitrate, 121–126, 126t
inner membrane potential, 112–138
oxygen consumption rates, 159t
swelling, 109–110, 112–113, 136–137, 161–162
Thallium poisoning
animal body, symptoms and characteristics of, 19–24
classification of, 15–18
acute alopecia, 15
albuminuria, 17–18
cytotoxic poisons, 18
eye diseases, 17
hematuria, 17–18
professional and nonprofessional, 15
weight loss, 16
diagnostics, 30–32
liver and muscles, 53–54
prevention of, 24–26
furosemide, 25–26
sodium polystyrene sulfonate, 25
thioacetamide, 26
Prussian blue, 27–30
Thallium quinolinate, 85–86
Thallium sulfate, 8–9, 55
intravenous administration, 64–65
rats, 21
Thallium transport, cell membranes
channel blockers, 81–82
erythrocytes, 82–83
HEK293 cells, 81
Na^+/K^+ - ATPase, 80
thallium isotopes, 82
thallium salts, 82–83
voltage-gated potassium channels, 80–81
Thallotoxicosis, 27–30
Thioacetamide, 26
Thiol groups, in mitochondrial proteins, 138–144
Three-part model, 54–55
Tl^+
mitochondrial permeability transition pore (MPTP), 144–179
adenine nucleotide translocase, 199–221
calcium sites, 179–188
phosphate symporter, 221–234
potassium channels, 188–199
Tl^{3+} interaction and, 91–93
Tl^{3+}, 91–93
^{201}Tl isotopes, 57, 82
^{204}Tl isotopes, 55, 57, 82
Toxicology
animals, 54–55
aquatic protozoa, 66–67
blood and kidneys, 64–65
frog skin, 65–66
hemodynamics, 64–65
human body, 53–54
muscle tissues, 62–64
nervous system, 59–62
unicellular organisms, 66–67
Tridentate chelator deferasirox, 29–30
Tumor tissues, 66

U
Ulva lactuca, 67
Ulva rigida, 67
Unicellular organisms, 66–67

V
Voltage-dependent anion channel (VDAC), 145

X
Xenopus laevis, 80

Z
Zebrafish, 9–10, 56–57

FSC
www.fsc.org
MIX
Papier | Fördert
gute Waldnutzung
FSC® C083411

Kolibri 360 GmbH
Ferdinand-Jühlke-Straße 7
99095 Erfurt, Deutschland
produktsicherheit@kolibri360.de